UNDERSTANDING CONSTRUCTION DRAWINGS

Third Edition

MARK W. HUTH

and

WALTER WELLS

Delmar
Thomson Learning™

Africa • Australia • Canada • Denmark • Japan • Mexico • New Zealand • Philippines
Puerto Rico • Singapore • Spain • United Kingdom • United States

NOTICE TO THE READER

Publisher does not warrant or guarantee any of the products described herein or perform any independent analysis in connection with any of the product information contained herein. Publisher does not assume, and expressly disclaims, any obligations to obtain and include information other than that provided to it by the manufacturer.

The reader is expressly warned to consider and adopt all safety precautions that might be indicated by the activities herein and to avoid all potential hazards. By following the instructions contained herein, the reader willingly assumes all risks in connection with such instructions.

The Publisher makes no representation or warranties of any kind, including but not limited to, the warranties of fitness for particular purpose or merchantability, nor are any such representations implied with respect to the material set forth herein, and the publisher takes no responsibility with respect to such material. The publisher shall not be liable for any special, consequential, or exemplary damages resulting, in whole or part, from the readers' use of, or reliance upon, this material.

Delmar Staff:
Business Unit Director: Alar Elken
Executive Editor: Sandy Clark
Acquisitions Editor: Tom Schin
Editorial Assistant: Fionnuala McAvey
Executive Marketing Manager: Maura Theriault
Channel Manager: Mona Caron
Marketing Coordinator: Paula Collins
Executive Production Manager: Mary Ellen Black
Production Manager: Larry Main
Art Director: Rachel Baker
Technology Project Manager: Tom Smith

Online Services

Delmar Online
To access a wide variety of Delmar products and services on the World Wide Web, point your browser to:
   http://www.delmar.com
   or email: info@delmar.com

For more information, contact Delmar, 3 Columbia Circle, PO Box 15015, Albany, NY 12212-0515; or find us on the World Wide Web at http://www.delmar.com

Asia
Thomson Learning
60 Albert Street, #15-01
Albert Complex
Singapore 189969

Australia/New Zealand
Nelson/Thomson Learning
102 Dodds Street
South Melbourne, Victoria 3205
Australia

Canada
Nelson/Thomson Learning
1120 Birchmount Road
Scarborough, Ontario
Canada M1K 5G4

International Headquarters
Thomson Learning
International Division
290 Harbor Drive, 2nd Floor
Stamford, CT 0692-7477
USA

Japan
Thompson Learning
Palaceside Building 5F
1-1-1 Hitotsubashi, Chiyoda-ku
Tokyo 100 0003
Japan

Latin America
Thomson Learning
Seneca, 53
Colonia Polanco
11560 Mexico D. F. Mexico

Spain
Thomson Learning
Calle Magallanes, 25
28015-Madrid
Espana

UK/Europe/Middle East
Thomson Learning
Berkshire House
168-173 High Holborn
London
WC1V 7AA United Kingdom

Thomas Nelson & Sons Ltd.
Nelson House
Mayfield Road
Walton-on-Thames
KT 12 5PL United Kingdom

Library of Congress Cataloging-in-Publication Data
Huth, Mark W.
   Understanding construction drawings / Mark Huth and Walter Wells.—3rd ed.
      p. cm.
   ISBN 0-7668-1580-3 (alk. paper)
   1. Structural drawing.   2. Dwellings—Drawings.   I. Wells, Walter G. II. Title.

T355.H87 1999
692'.1—dc21   99-047047

Contents

Preface

Understanding Construction Drawings is carefully designed to help you learn to read the drawings that are used to communicate information about buildings. This textbook includes drawings for buildings that were designed for construction in several parts of North America. The diversity of building classifications and geographic locations ensures that you are ready to work on construction jobs anywhere in the industry. In this third edition new content has been added to make the book even more useful to anyone studying the electrical trade. Even if you do not work in these trades, you may be involved with electrical, air conditioning, heating, or plumbing drawings. Everyone who works in building construction should be able to read and understand the drawings of the major trades.

The book is divided into four major parts and several units within each part. The parts relate to the prints in the separate drawing package. The drawing package contains prints for four buildings: a simple two-family duplex that is very easy to understand, a more complex single-family home, one unit of a townhouse that uses different materials and methods than the first two buildings, and an addition to a school. Each of the parts of the book is based on one of these buildings. The individual units are made up of four elements: Unit Objectives, the main body of the unit, a "Check Your Progress" checklist, and Assignment questions. The Unit Objectives appear at the beginning of the unit, so that you will know what to look for as you study the unit. The body is the presentation of content with many illustrations and references to the prints for the building being studied in that part. The Check Your Progress list gives you a quick check on how well you understood the main points in the unit. Each unit has 10 to 20 questions that require you to understand the content of the unit and usually require you to apply that understanding to actually reading the drawings for the building being studied. There are more than 600 questions in the textbook.

At the back of the text you will find several helpful aids for studying construction drawings. The Math Reviews are an innovative feature that have helped many construction students through a difficult area. These are concise reviews of the basic math you are likely to encounter throughout the building construction field. As math is required in this textbook reference is made to the appropriate Math Review. All of the math that is needed to complete the end-of-unit assignments in this book is covered in these Math Reviews. The glossary defines all of the new technical terms introduced throughout the textbook. Each of these terms is defined where it is first used, but if you need to refresh your memory, turn to the glossary. There is also a complete list of construction abbreviations used on the prints, along with their meanings. Another appendix explains the most commonly used symbols for materials and small equipment.

The textbook is supported by an Instructor's Guide that gives answers to all of the Assignment questions in the textbook and explains how the answers were found or calculated. The Instructor's Guide also contains more than 500 additional questions that can be used for tests, supplemental assignments, and review. The answer to each of these questions is given, along with an explanation of the answer.

I am grateful to all who contributed to this textbook. The students who studied from and the instructors who taught from the first two editions and provided a wealth of valuable feedback played an instrumental role in shaping this edition. Several companies provided expertise and contributed illustrations. They are named in the captions to the illustrations they provided. I would especially like to thank the architects and engineers who supplied the drawings for the drawing packet:

Robert Kurzon, A.I.A. for the Duplex and Lake House

Clark Forrest Butts, Architect at Berkus-Group Architects for Hidden Valley

Carl Griffith, A.I.A. at Cataldo, Waters, and Griffith Architects, P.C. and HA2F Consultants in Engineering for the School addition

I would also like to thank the instructors who reviewed the manuscript for the third edition, providing guidance in making it the best print reading textbook it could be.

Phil Marks
New England Institute of Technology
Warwick, RI

Stan Flippen
IEC Dallas Chapter
Irving, TX

Cliff Redinger
Rocky Mountain Chapter of IEC
Denver, CO

Robert Jones
IEC Houston Chapter
Houston, TX

Part I

DRAWINGS— THE LANGUAGE OF INDUSTRY

Part I helps you develop a foundation upon which to build skills and knowledge in reading the drawings used in the construction industry. The topics of the various units in this section are the basic concepts upon which all construction drawings are read and interpreted. The details of construction will be explored in Parts II, III, and IV.

Many of the assignment questions in this part refer to the drawings of the Duplex included in the drawing packet that accompanies this textbook. The Duplex was designed as income property for a small investor. It was built on a corner lot in a small city in upstate New York. The Duplex is an easy-to-understand building. Its one-story, rectangular design requires only a minimum of views; you can quickly become familiar with the Duplex drawings.

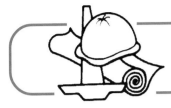

UNIT 1 The Design-Construction Sequence and the Design Professions

OBJECTIVES

After completing this unit, you will be able to perform the following tasks:

- Name the professions included in the design and planning of a house or light commercial building.

- List the major functions of each of these professions in the design and planning process.

- Identify the profession or agency that should be contacted for specific information about a building under construction.

The construction industry employs about 15 percent of the working people in the United States and Canada. More than 60 percent of these workers are involved in new construction. The rest are involved in repairing, remodeling, and maintenance. As the needs of our society change, the demand for different kinds of construction increases. Homeowners and businesses demand more energy-efficient buildings. The shift toward automation and the use of computers in business and industry mean that more offices are needed. Our national centers of commerce and industry are shifting. These are only a few of the reasons that new housing starts are considered important indicators of our economic health.

There are four main classifications of construction: light, heavy, industrial, and civil. *Light construction* includes single-family homes, small apartment buildings, condominiums, and small commercial buildings, Figure 1-1. *Heavy construction* includes high-rise office and apartment buildings, hotels, large stores and shopping centers, and other large buildings. *Industrial construction* includes structures other than buildings, such as refineries and paper mills, that are built for industry. *Civil construction* is more closely linked with the land and refers to highways, bridges, airports, and dams, for example. This book refers specifically to light construction. The principles apply to drawings of all types of construction, however.

Figure 1-1 Light contruction
Courtesy of Wyerhaeuser Co.

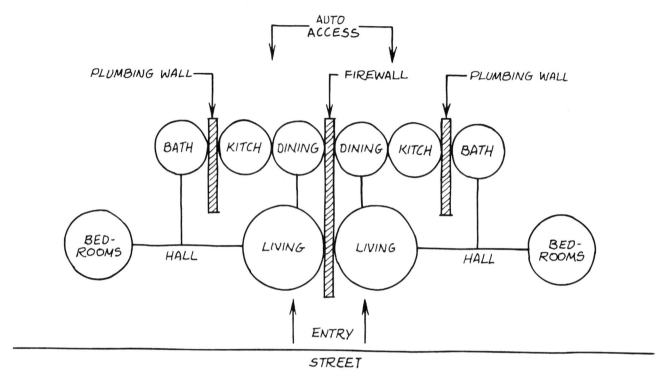

Figure 1-2   Balloon sketch of Duplex

THE DESIGN PROCESS

The design process starts with the owner. The owner has definite ideas about what is needed, but may not be expert at describing that need or desire in terms the builder can understand. The owner contacts an architect to help describe the building.

The architect serves as the owner's agent throughout the design and construction process. Architects combine their knowledge of construction—of both the mechanics and the business—with artistic or aesthetic knowledge and ability. They design buildings for appearance and use.

The architect helps the owner determine how much space is needed, how many rooms are needed for now and in the future, what type of building best suits the owner's life-style or business needs, and what the costs will be. As the owner's needs take shape, the architect makes rough sketches to describe the planned building. At first these may be balloon diagrams, Figure 1-2, to show traffic flow and the number of rooms. Eventually, the design of the building begins to take shape, Figure 1-3.

Before all of the details of the design can be finalized, other construction professionals become involved. Most communities have building codes which specify requirements to insure that buildings are safe from fire hazards, earthquakes, termites, surface water, and other concerns of the community. There are

several organizations that publish model building codes, Figure 1-4. These are called *model codes* because they are simply models to be followed by building departments. The codes have no authority until they are adopted by a government agency. A community may adopt the total model code or may choose specific parts of the code. This then becomes the *local building code.*

The local building code is administered by a building department of the local government. Building inspectors, working for the building department, review the architect's plans before construction begins and inspect the construction throughout its progress to insure that the code is followed.

Most communities also have zoning laws. A *zoning law* divides the community into zones where only certain types of buildings are permitted. Zoning laws prevent such problems as factories and shopping centers being built in the same neighborhood as homes.

Building departments usually require that very specific procedures are followed for each construction project. A building permit is required before construction begins. The building permit notifies the building department about planned construction. Then, the building department can make sure that the building complies with all the local zoning laws and building codes. When the building department approves the completed construction, it issues a *certificate of occupancy.* This certificate is not issued

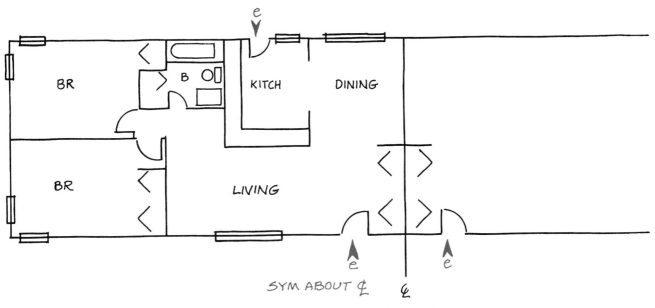

Figure 1-3 **Straight line sketch of Duplex**

until the building department is satisfied that the construction has been completed according to the local code. The owner is not permitted to move into the new building until the certificate of occupancy has been issued.

If the building is more complex than a home or simple frame building, engineers may be hired to help design the structural, mechanical, electrical, or other aspects of the building. Consulting engineers specialize in certain aspects of construction and are employed by architects to provide specific services. Finally, architects and their consultants prepare construction drawings that show all aspects of the building. These drawings tell the contractor specifically what to build.

Many homes are built from stock plans available from catalogs of house designs, building materials dealers, or magazines, Figure 1-5. However, many states

require a registered architect to approve the design and supervise the construction.

STARTING CONSTRUCTION

After the architect and the owner decide on a final design, the owner obtains financing. The most common way of financing a home is through a mortgage. A *mortgage* is a guarantee that the loan will be paid in installments. If the loan is not paid, the lender has the right to sell the building in order to recover the money owed. In return for the use of the lenders money, the borrower pays interest—a percentage of the outstanding balance of the loan.

When financing has been arranged (sometimes before it is finalized), a contractor is hired. Usually a general contractor is hired with overall responsibility for completing the project. The general contractor in turn

Figure 1-4 **Building codes can be ordered from catalogs.**

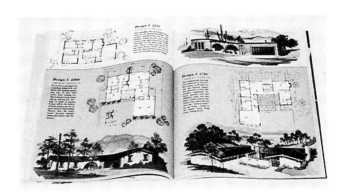

Figure 1-5 **Stock plans can be ordered from catalogs.**

hires subcontractors to complete certain parts of the project. All stages of construction may be subcontracted. The parts of home construction most often subcontracted are plumbing and heating, electrical, drywall, painting and decorating, and landscaping. The relationships of all of the members of the design and construction team are shown in Figure 1-6. The utility installer should carefully investigate ALL of the drawings, especially the architectural drawings, in order to determine the installation locations of their equipment.

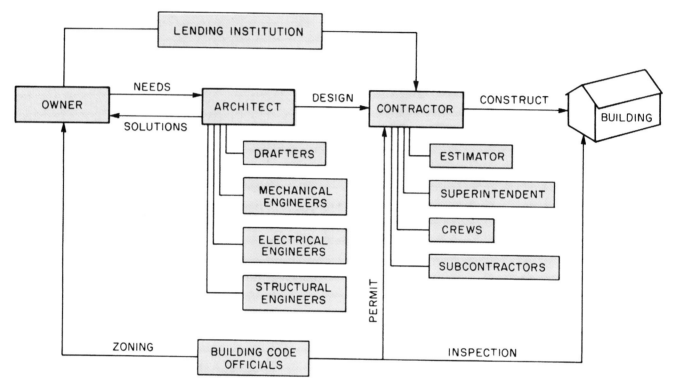

Figure 1-6   Design and construction team

✓ CHECK YOUR PROGRESS

Can you perform these tasks?

☐ List construction design professions.

☐ Describe what work is done by each of these professions.

☐ Name the profession responsible for each major part of the design-construction progress.

 ASSIGNMENT

1. Who acts as the owner's agent while the building is being constructed?

2. Who designs the structural aspects of a commercial building?

3. Who would normally hire an electrical engineer for the design of a store?

4. Who is generally responsible for obtaining financing for a small building?

5. To whom would the general contractor go if there were a problem with the foundation design for a home?

6. If local building codes require specific features for earthquake protection, who is responsible for seeing that they are included in a home design?

7. Whom would the owner inform about last-minute changes in the interior trim when the building is under construction?

8. What regulations specify what parts of the community are to be reserved for single-family homes only?

9. Who issues the building permit?

10. What regulations are intended to insure that all new construction is safe?

UNIT 2   Views

After completing this unit, you will be able to perform the following tasks:

- Recognize oblique, isometric, and orthographic drawings.
- Draw simple isometric sketches.
- Identify plan views, elevations, and sections.

ISOMETRIC DRAWINGS

A useful type of pictorial drawing for construction purposes is the *isometric drawing*. In an isometric drawing, vertical lines are drawn vertically and horizontal lines are drawn at an angle of 30 degrees from horizontal, Figure 2-1. All lines on one of these isometric axes are drawn in proportion to their actual length. Isometric drawings tend to look out of proportion because we are used to seeing the object appear smaller as it gets farther away.

Isometric drawings are often used to show construction details, Figure 2-2. The ability to draw simple isometric sketches is a useful skill for communicating on the job site. Try sketching a brick in isometric as shown in Figure 2-3.

Step 1.   Sketch a Y with the top lines about 30° from horizontal.
Step 2.   Sketch the bottom edges parallel to the top edges.
Step 3.   Mark off the width on the left top and bottom edges. This will be about twice the height.

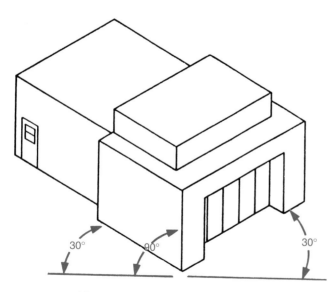

Figure 2-1   Isometric of building.

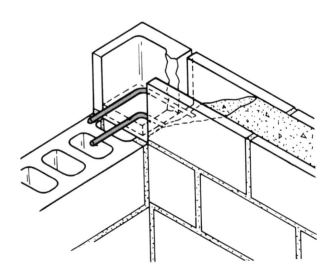

Figure 2-2   Isometric construction detail
Courtesy of Stearns Manufacturing Company, Inc.

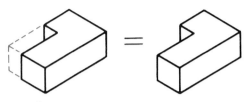

(a) GABLE ROOF BUILDING

(a) STEP 1

(b) STEP 2

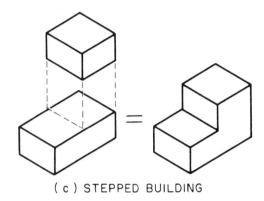

(b) ELL SHAPED BUILDING

(c) STEP 3

(d) STEP 4

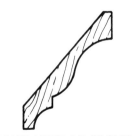

(c) STEPPED BUILDING

Figure 2-4   Variations on the isometric brick

(e) STEP 5

Figure 2-3   Sketching an isometric brick

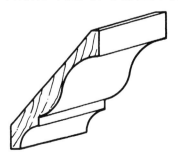

(a)   FRONT VIEW OF CROWN MOLDING

Step 4.   Mark off the length on the right top and bottom edges. The length will be about twice the width.

Step 5.   Sketch the two remaining vertical lines and the back edges.

Other isometric shapes can be sketched by adding to or subtracting from this basic isometric brick, Figure 2-4. Angled surfaces are sketched by locating their edges; then, connecting them.

(b)   OBLIQUE VIEW OF CROWN MOLDING

Figure 2-5   Oblique drawing

OBLIQUE DRAWINGS

When an irregular shape is to be shown in a pictorial drawing, an *oblique drawing* may be best. In oblique drawings, the most irregular surface is drawn in proportion as though it were flat against the drawing surface. Parallel lines are added to show the depth of the drawing, Figure 2-5.

ORTHOGRAPHIC PROJECTION

To show all information accurately and to keep all lines and angles in proportion, most construction drawings are drawn by *orthographic projection*. Orthographic projection is most often explained by imagining the object to be drawn inside a glass box. The corners and the lines representing the edges of the object are then projected onto the sides of the box, Figure 2-6. If the box is unfolded, the images projected onto its sides will be on a single plane, as on a sheet of paper, Figure 2-7. In other words, in orthographic projection each view of an object shows only one side (or top or bottom) of the object.

All surfaces that are parallel to the plane of projection (the surface of the box) are shown in proportion to their actual size and shape. However, surfaces that are not parallel to the plane of projection are not shown in proportion. For example, both of the roofs in the top views of Figure 2-8 appear to be the same size and shape, but they are quite different. To find the actual shape of the roof you must look at the end view.

In construction drawings, the views are called plans and elevations. A *plan view* shows the layout of

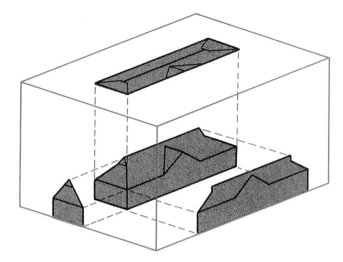

Figure 2-6   Duplex inside a glass box: method of orthographic projection of roof, front side.

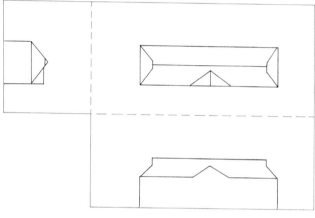

Figure 2-7

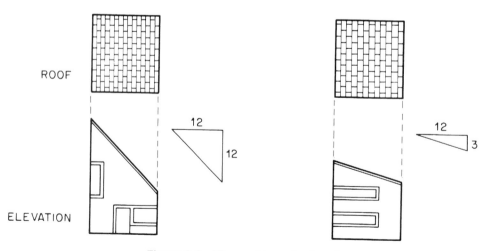

Figure 2-8   Views of two shed roofs

the object as viewed from above, Figure 2-9. A set of drawings for a building usually includes plan views of the site (lot), the floor layout, and the foundation. *Elevations* are drawing that show height. For example, a drawing that shows what would be seen standing in front of a house is a building elevation, Figure 2-10. Elevations are also used to show cabinets and interior features.

Because not all features of construction can be seen in plan views and elevations from the outside of a building, many construction drawings are section views. A section view, usually referred to simply as a *section,* shows what would be exposed if a cut were made through the object, Figure 2-11. Actually, a floor plan is a type of section view, Figure 2-12.

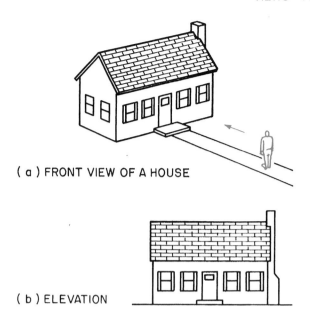

(a) FRONT VIEW OF A HOUSE

(b) ELEVATION

Figure 2-10   Building elevation

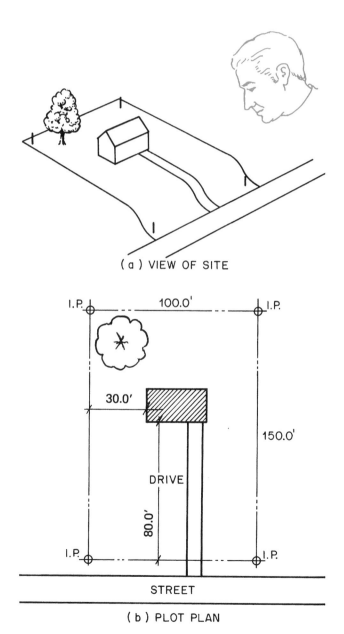

(a) VIEW OF SITE

(b) PLOT PLAN

Figure 2-9   Plan view

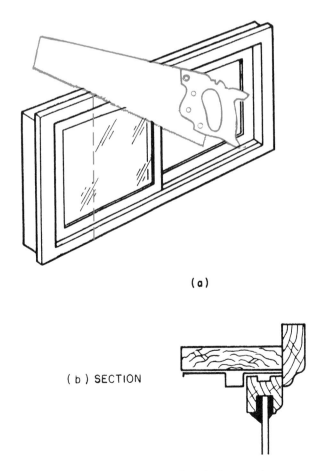

(a)

(b) SECTION

Figure 2-11   Section of a window sash

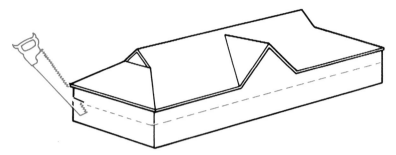

(a) An imaginary cut is made at a level that passes through all windows and doors.

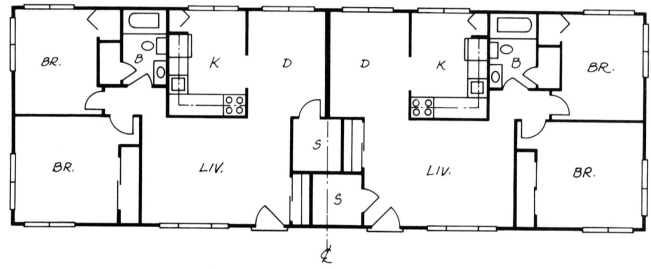

(b) The floor plan shows what is left when the top is removed.

Figure 2-12   A floor plan is actually a section view of the building.

✓ CHECK YOUR PROGRESS

Can you perform these tasks?

☐ Identify oblique drawings.

☐ Identify isometric drawings.

☐ Identify orthographic drawings.

☐ Identify plan views

☐ Identify elevation views.

☐ Identify section views

ASSIGNMENT

1.   Identify each of the drawings in Figure 2-13 as oblique, isometric, or orthographic.

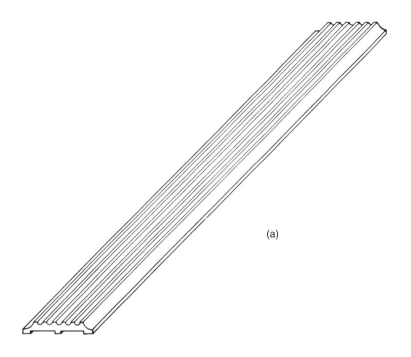

(a)

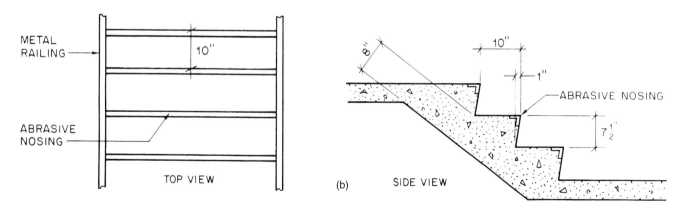

METAL
RAILING

10"

ABRASIVE
NOSING

TOP VIEW

8"

10"

1"

ABRASIVE NOSING

$7\frac{1}{2}$"

(b)   SIDE VIEW

Figure 2-13

2.   Identify each of the drawings in Figure 2-14 as elevation, plan, or section.

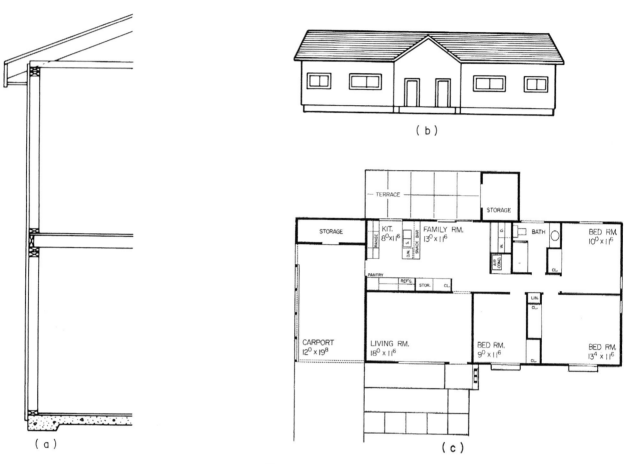

Figure 2-14

3.   In the view of the house shown in Figure 2-15, which lines are true length?

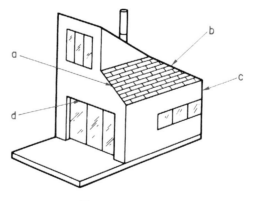

Figure 2-15

4.   What type of pictorial drawings are easiest to draw on the job site?

5.   What type of drawings are used for working drawings?

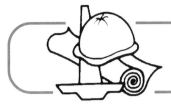

UNIT 3 Scales

OBJECTIVES

After completing this unit, you will be able to perform the following tasks:

• Identify the scale used on a construction drawing.

• Read an architect's scale.

SCALE DRAWINGS

Because construction projects are too large to be drawn full size on a sheet of paper, everything must be drawn proportionately smaller than it really is. For example, floor plans for a house are frequently drawn $1/48$ of the actual size. This is called *drawing to scale*. At a scale of $1/4" = 1'\text{-}0"$, $1/4$ inch on the drawing represents 1 foot on the actual building. When it is necessary to fit a large object on a drawing, a small scale is used.

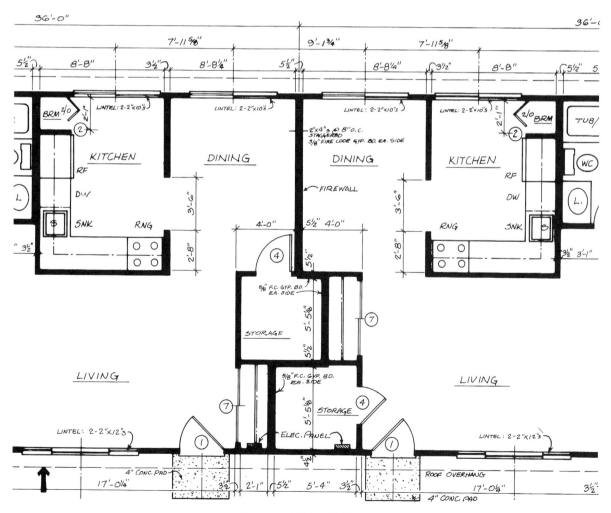

Courtesy of Robert C. Kurzon

Figure 3-1 Portion of a floor plan with a firewall

Smaller objects and drawings that must show more detail are drawn to a larger scale. The floor plan in Figure 3-1 was drawn to a scale of ¼" = 1'-0". The detail drawing in Figure 3-2 was drawn to a scale of 3" = 1'-0" to show the construction of one of the walls on the floor plan.

The scale to which a drawing is made is noted on the drawing. The scale is usually indicated alongside or beneath the title of the view. On some drawings, the scale is shown by including a drawing that looks something like a ruler. This graphic scale has graduations representing feet and inches drawn to the

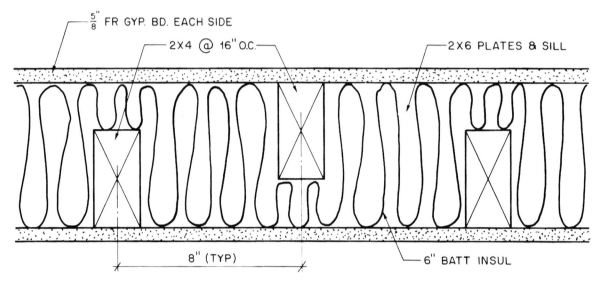

Figure 3-2   Detail (plan at firewall)

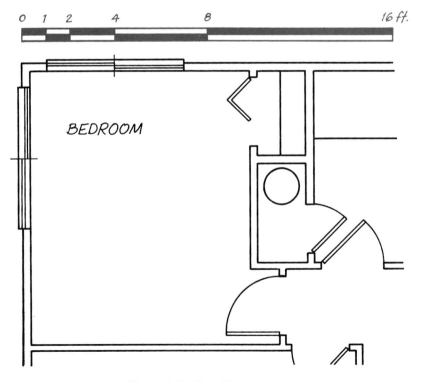

Figure 3-3   Graphic scale

scale of the view, Figure 3-3. If the drawing is enlarged or reduced, the graphic scale is also enlarged or reduced. The graduations on the scale indicator can be marked on the edge of a sheet of paper, then stepped off on the drawing, Figure 3-4. They may also be transferred to the drawing with dividers, Figure 3-5.

READING AN ARCHITECT'S SCALE

All necessary dimensions should be shown on the drawings. The instrument used to make drawings to scale is called an *architect's scale*, Figure 3-6. Measuring a drawing with an architect's or engineer's scale is a poor practice. At small scales it is especially difficult for the drafter to be precise and any tolerance introduced during drafting can be made amplified by trying to use a scale to measure a drawing. The following discussion of how to read an architect's scale is presented to ensure an understanding of the scales used on drawings. The triangular scale includes eleven scales frequently used on drawings.

Full Scale					
$3/32"$	=	1' - 0"	$3/16"$	=	1' - 0"
$1/8"$	=	1' - 0"	$1/4"$	=	1' - 0"
$3/8"$	=	1' - 0"	$3/4"$	=	1' - 0"
$1/2"$	=	1' - 0"	1"	=	1' - 0"
$1^1/2"$	=	1' - 0"	3"	=	1' - 0"

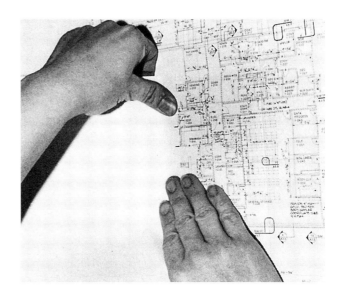

Figure 3-4   Marking the graduations on the edge of a piece of paper

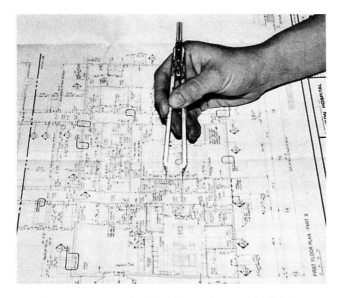

Figure 3-5   Transferring dimensions with dividers

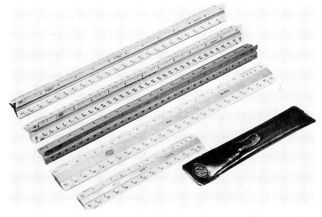

Courtesy of Teledyne Post

Figure 3-6   Architect's scales

Two scales are combined on each face, except for the full-size scale which is fully divided into sixteenths. The combined scales work together because one is twice as large as the other, and their zero points and extra divided units are on opposite ends of the scale.

The fraction, or number, near the zero at each end of the scale indicates the unit length in inches that is used on the drawing to represent one foot of the actual building. The extra unit near the zero end of the scale is subdivided into twelfths of a foot (inches) as well as fractions of inches on the larger scales.

To read the architect's scale, turn it to the ¼-inch scale. The scale is divided on the left from the zero toward the ¼ mark so that each line represents one inch. Counting the marks from the zero toward the ¼ mark, there are twelve lines marked on the scale. Each one of these lines is one inch on the ¼" = 1'-0" scale.

The fraction ⅛ is on the opposite end of the same scale, Figure 3-7. This is the ⅛-inch scale and is read in the opposite direction. Notice that the divided unit is only half as large as the one on the ¼-inch end of the scale. Counting the lines from zero toward the ⅛ mark, there are only six lines. This means that each line represents two inches at the ⅛-inch scale.

Now look at the 1½-inch scale, Figure 3-8. The divided unit is broken into twelfths of a foot (inches) and also fractional parts of an inch. Reading from the zero toward the number 1½, notice the figures 3, 6, and 9. These figures represent the measurements of 3 inches, 6 inches, and 9 inches at the 1½" = 1'-0" scale. From the zero to the first long mark that represents one inch (which is the same length as the mark shown at 3) and 4 lines. This means that each line on the scale is equal to ¼ of an inch. Reading the zero to the 3, read each line as follows: ¼, ½, ¾, 1, 1¼, 1½, 1¾, 2, 2¼, 2½, 2¾, and 3 inches.

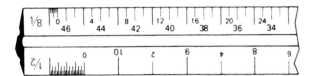 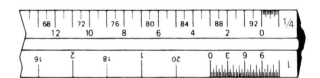

Figure 3-7   Architect's triangular scale

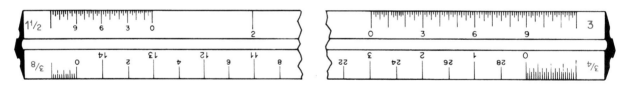

Figure 3-8   Architect's triangular scale showing 1 ½" and 3" scales

✓ CHECK YOUR PROGRESS

Can you perform these tasks?

☐ Locate the scale notations on drawings.

☐ Use an architect's scale to measure objects drawn to scale.

ASSIGNMENT

1–10.   What are the dimensions indicated on the scale in Figure 3-9?

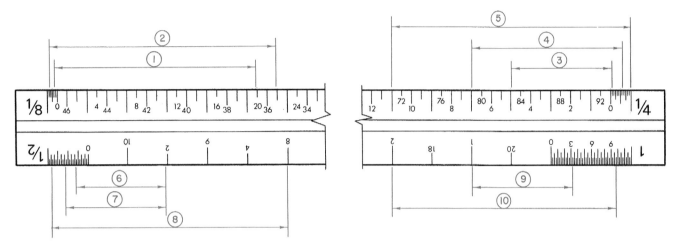

Figure 3-9

11.   What scales are used for the following views of the Duplex? (Refer to the Duplex drawings in the packet.)

a. Floor plan
b. Site plan
c. Front elevation
d. Typical wall section

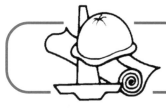

UNIT 4 Alphabet of Lines

After completing this unit, you will be able to identify and understand the meaning of the listed lines:

• Object lines

• Dashed lines (hidden and phantom)

• Extension lines and dimension lines

• Centerlines

• Leaders

• Cutting-plane lines

The fact that drawings are used in construction for the communication of information was discussed earlier. The drawings, then, serve as a language for the construction industry. The basis for any language is its alphabet. The English language uses an alphabet made up of twenty-six letters. Construction drawings use an *alphabet of lines,* Figure 4-1.

The weight or thickness of lines is sometimes varied to show their relative importance. For example, in Figure 4-2 notice that the basic outline of the building is heavier than the windows and doors. This difference in line weight sometimes helps distinguish the basic shape of an object from surface details.

OBJECT LINES

Object lines are used to show the shape of an object. All visible edges are represented by object lines. All of the lines in Figure 4-2 are object lines. Drawings usually include many solid lines that are not object lines, however. Some of these other solid lines are discussed here. Others are discussed later.

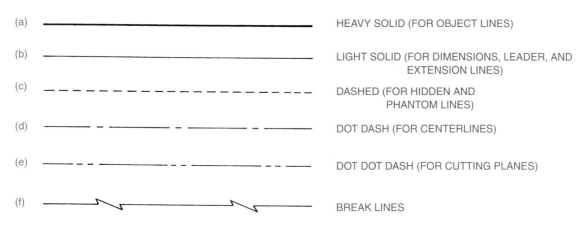

(a) HEAVY SOLID (FOR OBJECT LINES)

(b) LIGHT SOLID (FOR DIMENSIONS, LEADER, AND EXTENSION LINES)

(c) DASHED (FOR HIDDEN AND PHANTOM LINES)

(d) DOT DASH (FOR CENTERLINES)

(e) DOT DOT DASH (FOR CUTTING PLANES)

(f) BREAK LINES

Figure 4-1 Alphabet of lines

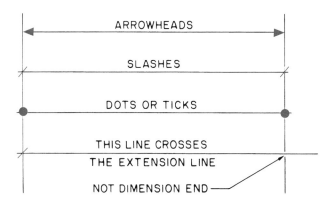

Figure 4-8   Dimension line ends

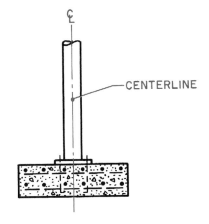

Figure 4-9   This centerline indicates that the column is symmetrical, or the same, on both sides of the centerline.

CENTERLINES

Centerlines are made up of long and short dashes. They are used to show the centers of round or cylindrical objects. Centerlines are also used to indicate that an object is *symmetrical,* or the same on both sides of the center, Figure 4-9. To show the center of a round object, two centerlines are used so that the short dashes cross in the center, Figure 4-10.

To lay out an *arc* or part of a circle, the radius must be known. The *radius* of an arc is the distance from the center to the edge of the arc. On construction drawings, the center of an arc is shown by crossing centerlines. The radius is dimensioned on a thin line from the center to the edge of the arc, Figure 4-11.

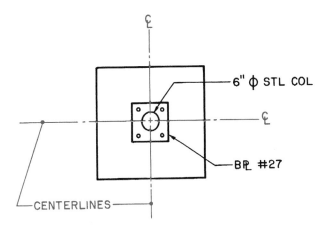

Figure 4-10   When centerlines show the center of a round object, the short dashes of two centerlines cross.

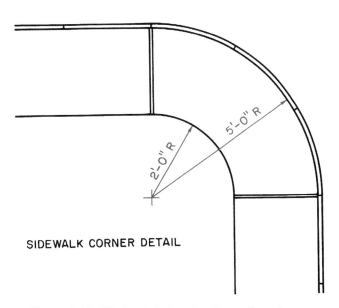

SIDEWALK CORNER DETAIL

Figure 4-11   Method of showing the radius of an arc

Rather than clutter the drawing with unnecessary lines, only the short, crossing dashes of the centerlines are shown. If the centerlines are needed to dimension the location of the center, only the needed centerlines are extended.

LEADERS

Some construction details are too small to allow enough room for clear dimensioning by the methods described earlier. To overcome this problem, the dimension is shown in a clear area of the drawing. A thin line called a *leader* shows where the dimension belongs, Figure 4-12.

CUTTING-PLANE LINES

It was established earlier that section views are needed to show interior detail. In order to show where the imaginary cut was made, a *cutting-plane line* is drawn on the view through which the cut was made, Figure 4-13. A cutting-plane line is usually a heavy line

with long dashes and pairs of short dashes. Some drafters, however, use a solid, heavy line. In either case cutting-plane lines always have some identification at their ends. Cutting-plane-line identification symbols are discussed in the next unit.

Some section views may not be referenced by a cutting-plane line on any other view. These are *typical sections* that would be the same if drawn from an imaginary cut in any part of the building, Figure 4-14.

Some section views may not be referenced by a cutting-plane line on any other view. These are *typical sections* that would be the same if drawn from an imaginary cut in any part of the building, Figure 4-14.

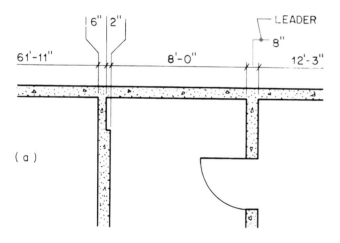
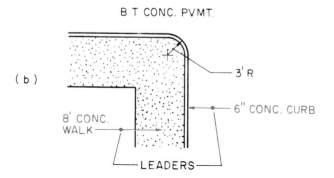

Figure 4-12   Leaders used for dimensioning

Figure 4-13   A cutting-plane line indicates where the imaginary cut is made and how it is viewed.

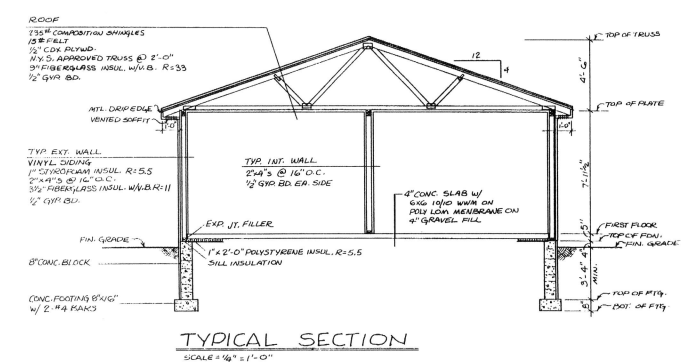

ROOF
235# COMPOSITION SHINGLES
15# FELT
1/2" CDX PLYWD.
N.Y.S. APPROVED TRUSS @ 2'-0"
9" FIBERGLASS INSUL. w/v.B. R=33
1/2" GYP. BD.

MTL. DRIP EDGE
VENTED SOFFIT
1'-0"

TYP. EXT. WALL
VINYL SIDING
1" STYROFOAM INSUL. R=5.5
2"x4"S @ 16" O.C.
3½" FIBERGLASS INSUL. W/V.B. R=11
1/2" GYP. BD.

EXP. JT. FILLER

FIN. GRADE

8" CONC. BLOCK

CONC. FOOTING 8"x16"
w/ 2-#4 BARS

TOP OF TRUSS
4'-6"
12
4
TOP OF PLATE

TYP. INT. WALL
2"x4"S @ 16" O.C.
1/2" GYP. BD. EA. SIDE

4" CONC. SLAB w/
6x6 10/10 WWM ON
POLY LOW MENBRANE ON
4" GRAVEL FILL

1'-0"
7'-11½"
5"
FIRST FLOOR
TOP OF FDN.
FIN. GRADE

1"x 2'-0" POLYSTYRENE INSUL. R=5.5
SILL INSULATION

3'-4" MIN.
TOP OF FTG.
BOT. OF FTG.

TYPICAL SECTION
SCALE = 1/4" = 1'-0"

Figure 4-14   Building section

✓ CHECK YOUR PROGRESS

Can you perform these tasks?

- ☐ Identify and explain the use of object lines.
- ☐ Identify and explain the use of hidden lines.
- ☐ Identify and explain the use of phantom lines.
- ☐ Identify and explain the use of dimension and extension lines.
- ☐ Identify and explain the use of centerlines.
- ☐ Identify and explain the use of leaders.
- ☐ Identify and explain the use of cutting-plane lines.

ASSIGNMENT

Refer to the drawings of the Two-Unit Apartment in the packet. For each of the lines numbered A5.1 through A5.10, identify the kind of line and briefly describe its purpose on these drawings. The broad arrows with A5 numbers are for use in this assignment.

Example: A5.E, object line, shows end of building.

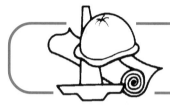

UNIT 5 Use of Symbols

OBJECTIVES

After completing this unit, you will be able to identify and understand the meaning of the listed symbols:

- Door and window symbols
- Materials symbols
- Electrical and mechanical symbols
- Reference marks for coordinating drawings
- Abbreviations

An alphabet of lines allows for clear communication through drawings; the use of standard symbols makes for even better communication. Many features of construction cannot be drawn exactly as they appear on the building. Therefore, standard symbols are used to show various materials, plumbing fixtures and fittings, electrical devices, windows, doors, and other common objects. Notes are added to drawings to give additional explanations.

It is not important to memorize all of the symbols and abbreviations used in construction before you learn to read drawings. You should, however, memorize a few of the most common symbols and abbreviations so that you may learn the principles involved in their use. Additional symbols and abbreviations can be looked up as they are needed. The illustrations shown here represent only a few of the more common symbols and abbreviations. A more complete reference is given in the Appendix.

DOOR AND WINDOW SYMBOLS

Door and window symbols show the type of door or window used and the direction the door or window opens. There are three basic ways for household doors to open—swing, slide, or fold, Figure 5-1. Within each of these basic types there are variations that can be readily understood from their symbols. The direction a swing-type door opens is shown by an arc representing the path of the door.

There are seven basic types of windows. They are named according to how they open, Figure 5-2. The symbols for hinged windows—awning, casement, and hopper—indicate the direction they open. In elevation, the symbols include dashed lines that come to a point at the hinged side, as viewed from the exterior.

The sizes of windows and doors are usually shown on a special window schedule or door schedule, but they might also be indicated by notes on the plans near their symbols. Door and window schedules are explained later. The notations of size show width first and height second. Manufacturers catalogs

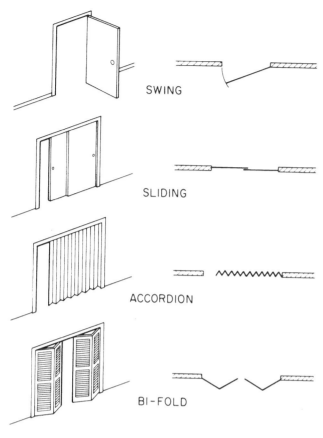

SWING

SLIDING

ACCORDION

BI-FOLD

Figure 5-1   Types of doors and their plan symbols

usually list several sets of dimensions for every window model, Figure 5-3. The glass size indicates the area that will actually allow light to pass. The rough opening size is important for the carpenter who will frame the wall into which the window will be installed. The masonry opening is important to masons. The notations on plans and schedules usually indicate nominal dimensions. A *nominal dimension* is an approximate size and may not represent any of the actual dimensions of the unit. Nominal dimensions are usually rounded off to whole inches or feet and inches and are used only as a convenient way to refer to the window or door size. The actual dimensions should be obtained from the manufacturer before construction begins.

MATERIALS SYMBOLS

The drawing of an object shows its shape and location. The outline of the drawing may be filled in with a material symbol to show what the object is made of, Figure 5-4. Many materials are represented by one

symbol in elevations and another symbol in sections. Examples of such symbols are concrete block and brick. Other materials look pretty much the same when viewed from any direction, so their symbols are drawn the same in sections and elevations.

When a large area is made up of one material, it is common to only draw the symbol in a part of the area, Figure 5-5. Some drafters simplify this even further by using a note to indicate what material is used and omitting the symbol altogether.

ELECTRICAL AND MECHANICAL SYMBOLS

The electrical and mechanical systems in a building include wiring, electrical devices, piping, pipe fittings, plumbing fixtures, registers, and heating and air-conditioning ducts. It is not practical to draw these items as they would actually appear, so standard symbols have been devised to indicate them.

The electrical system in a house includes wiring as well as devices such as switches, receptacles, light fixtures, and appliances. Wiring is indicated by lines that show how devices are connected. These lines are not shown in their actual position. They simply indicate which switches control which lights, for example. Outlets (receptacles) and switches are usually shown in their approximate positions. Major fixtures and appliances are shown in their actual positions. A few of the most common electrical symbols are shown in Figure 5-6.

Mechanical systems—plumbing and HVAC (heating, ventilating, and air conditioning)—are not usually shown in much detail on drawings for single-family homes. However, some of the most important features may be shown. Piping is shown by lines; different types of lines represent different kinds of piping. Symbols for pipe fittings are the same basic shape as the fittings they represent. A short line, or *hash mark*, represents the joint between the pipe and the fitting. Plumbing fixtures are drawn pretty much as the actual fixture appears. A few plumbing symbols are shown in Figure 5-7.

REFERENCE MARKS

A set of drawings for a complex building may include several sheets of section and detail drawings. These sections and details do not have much meaning without some way of knowing what part of the building they are meant to show. Callouts, called *reference marks,* on plans and elevations indicate where details

PLAN   ELEVATION   PICTORIAL

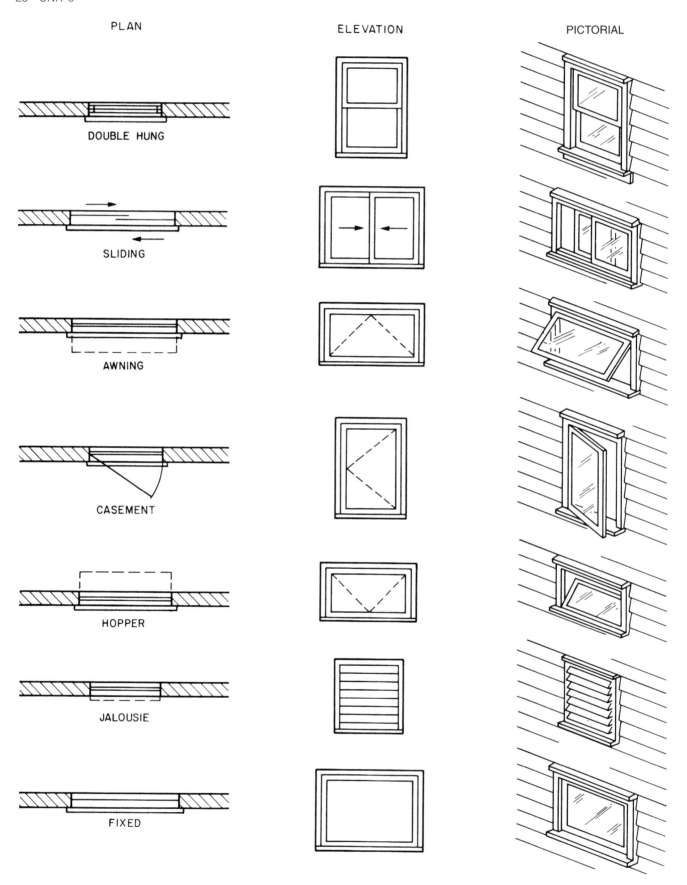

DOUBLE HUNG

SLIDING

AWNING

CASEMENT

HOPPER

JALOUSIE

FIXED

Figure 5-2   Window symbols

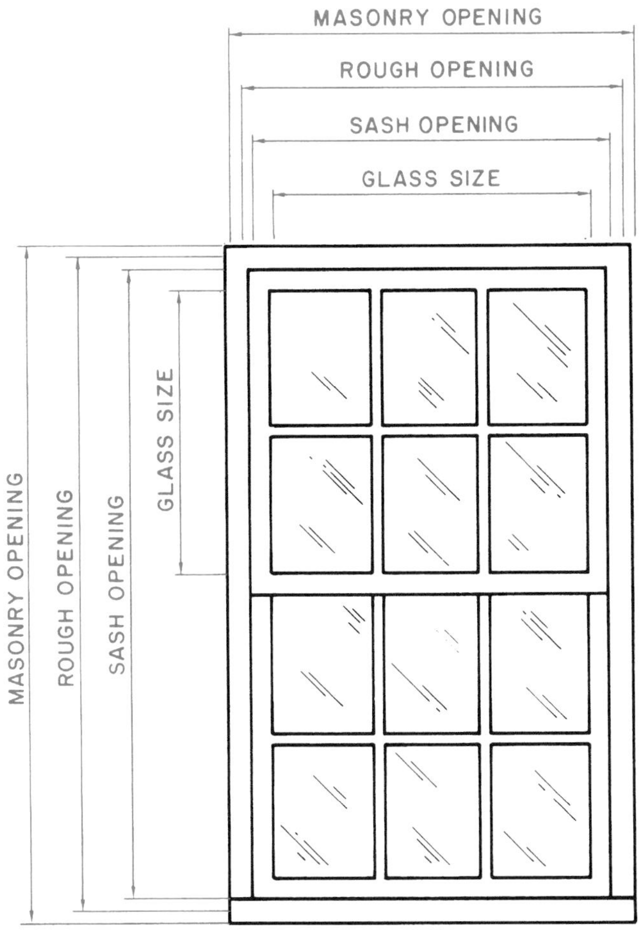

MASONRY OPENING

ROUGH OPENING

SASH OPENING

GLASS SIZE

MASONRY OPENING

ROUGH OPENING

SASH OPENING

GLASS SIZE

GLASS SIZE

Figure 5-3   Windows and doors can be measured in several ways

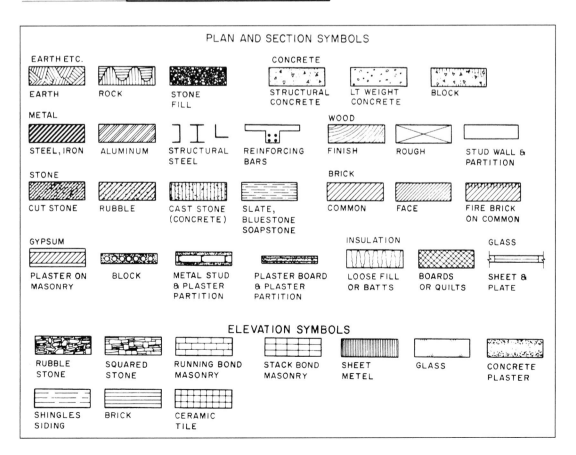

PLAN AND SECTION SYMBOLS

EARTH ETC.
EARTH   ROCK   STONE FILL

CONCRETE
STRUCTURAL CONCRETE   LT WEIGHT CONCRETE   BLOCK

METAL
STEEL, IRON   ALUMINUM   STRUCTURAL STEEL   REINFORCING BARS

WOOD
FINISH   ROUGH   STUD WALL & PARTITION

STONE
CUT STONE   RUBBLE   CAST STONE (CONCRETE)   SLATE, BLUESTONE SOAPSTONE

BRICK
COMMON   FACE   FIRE BRICK ON COMMON

GYPSUM
PLASTER ON MASONRY   BLOCK   METAL STUD & PLASTER PARTITION   PLASTER BOARD & PLASTER PARTITION

INSULATION
LOOSE FILL OR BATTS   BOARDS OR QUILTS

GLASS
SHEET & PLATE

ELEVATION SYMBOLS

RUBBLE STONE   SQUARED STONE   RUNNING BOND MASONRY   STACK BOND MASONRY   SHEET METEL   GLASS   CONCRETE PLASTER

SHINGLES SIDING   BRICK   CERAMIC TILE

Figure 5-4 Material symbols

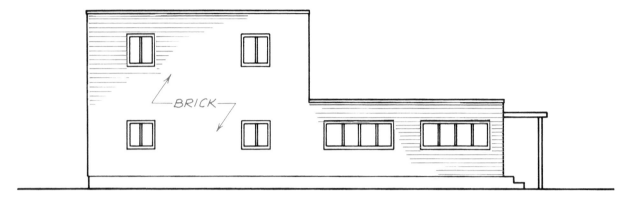

Figure 5-5   Only part of the area is covered by the brick symbol, although the entire building will be brick.

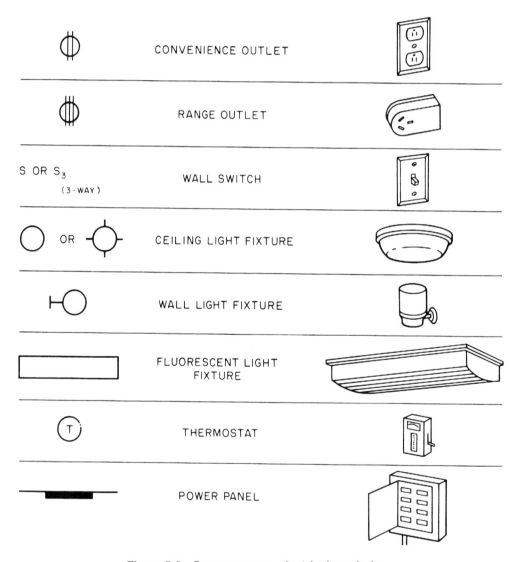

Figure 5-6   Some common electrical symbols

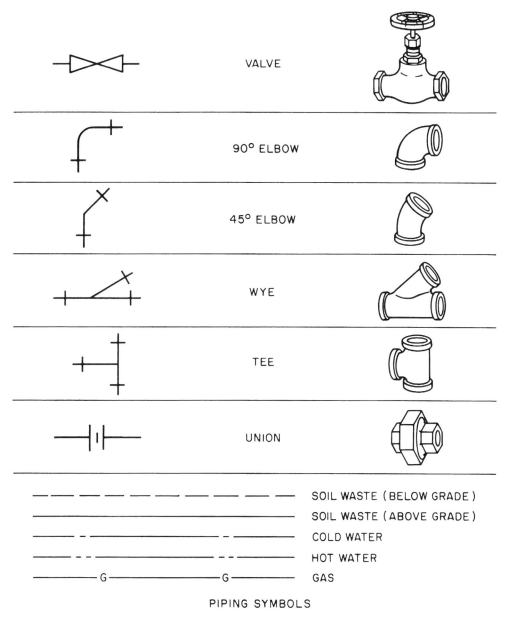

VALVE

90° ELBOW

45° ELBOW

WYE

TEE

UNION

SOIL WASTE (BELOW GRADE)

SOIL WASTE (ABOVE GRADE)

COLD WATER

HOT WATER

G ———— G ———— GAS

PIPING SYMBOLS

Figure 5-7   Some common plumbing symbols

or sections of important features have been drawn. To be able to use these reference marks for coordinating drawings, you must first understand the numbering system used on the drawings. The simplest numbering system for drawings consists of numbering the drawing sheets and naming each of the views. For example, Sheet 1 might include a site plan and foundation plan; Sheet 2, floor plans; and Sheet 3, elevations.

On large, complex sets of drawings the sheets are numbered according to the kind of drawings shown. Architectural drawing sheets are numbered A-1, A-2, and so on for all the sheets. Electrical drawings are numbered E-1, E-2, and E-3. A view number identifies each

separate drawing or view on the sheet. Figure 5-8 shows drawing 5 on sheet A-4.

Because most of the drawings for a house are architectural, and the drawing set is fairly small, letters indicating the type of drawing are not usually included. Instead, the views are numbered and a second number shows on which sheet it appears. For example, the fourth drawing on the third sheet would be 4/3, 4.3, or 4-3.

Numbering each view and the sheet on which it appears makes it easy to reference a section or detail to another drawing. The identification of a section view is given with the cutting-plane line showing where it is

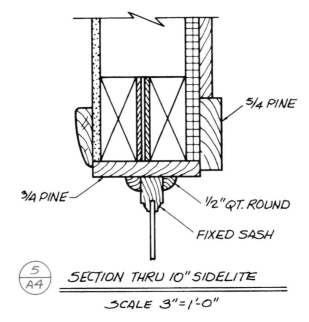

5/4 PINE

3/4 PINE

1/2" QT. ROUND

FIXED SASH

$\frac{5}{A4}$ SECTION THRU 10" SIDELITE

SCALE 3"=1'-0"

Figure 5-8 This is drawing 5 on sheet A-4.

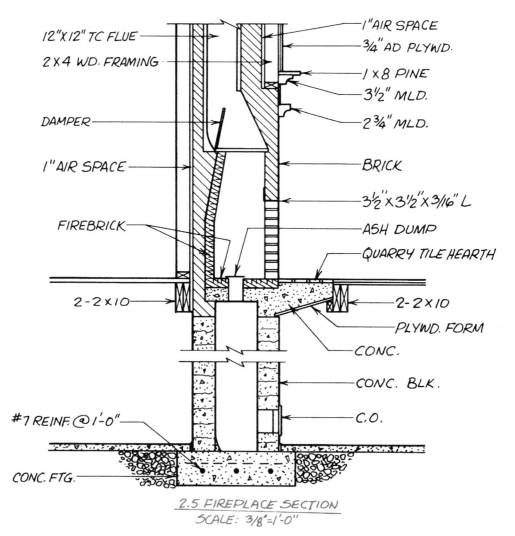

12"×12" TC FLUE

2×4 WD. FRAMING

DAMPER

1" AIR SPACE

FIREBRICK

2-2×10

#7 REINF. @ 1'-0"

CONC. FTG.

1" AIR SPACE

3/4" AD PLYWD.

1×8 PINE

$3\frac{1}{2}$" MLD.

$2\frac{3}{4}$" MLD.

BRICK

$3\frac{1}{2}$"×$3\frac{1}{2}$"×3/16" L

ASH DUMP

QUARRY TILE HEARTH

2-2×10

PLYWD. FORM

CONC.

CONC. BLK.

C.O.

2.5 FIREPLACE SECTION
SCALE: 3/8"=1'-0"

Figure 5-9 This section view is drawing 2 on sheet 5.

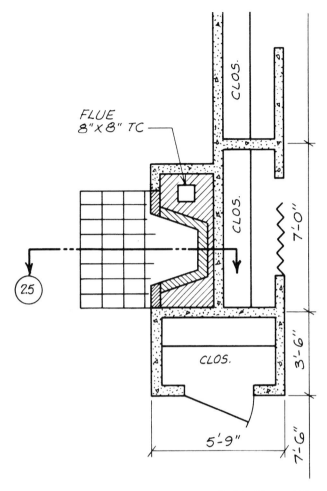

FLUE
8"x8" TC

CLOS.

CLOS.

7'-0"

2.5

CLOS.

3'-6"

5'-9"

7'-6"

Figure 5-10   Plan for fireplace detailed in Figure 5-9.

taken from. For example, the section view shown in Figure 5-9 shows the fireplace at the cutting-plane line in Figure 5-10. Notice that the cutting-plane line in Figure 5-10 indicates that the section is viewed from the top of the page toward the bottom, with the fireplace opening on the right. That is how the section view in Figure 5-9 is drawn. This numbering system is also used for details that cannot be located by a cutting-plane line. The detail drawing of the cornice (edge of the roof) in Figure 5-11 is drawing 4 on sheet A-4, Figure 5-12.

ABBREVIATIONS

Drawings for construction include many notes and labels of parts. These notes and labels are usually abbreviated as much as possible to avoid crowding the drawing. The abbreviations used on drawings are usually a shortened form of the word and are easily understood. For example, BLDG stands for building. The abbreviations used throughout this textbook and on the related drawings are defined in the Appendix.

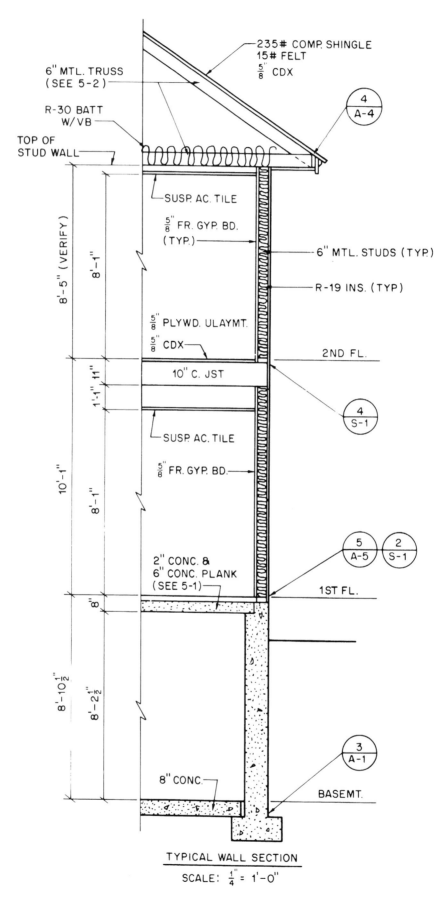

235# COMP. SHINGLE
15# FELT
$\frac{5}{8}''$ CDX

6" MTL. TRUSS
(SEE 5-2)

R-30 BATT
W/VB

TOP OF
STUD WALL

$\frac{4}{A-4}$

SUSP. AC. TILE

$\frac{5}{8}''$ FR. GYP. BD.
(TYP.)

6" MTL. STUDS (TYP.)

R-19 INS. (TYP)

8'-5" (VERIFY)

8'-1"

$\frac{5}{8}''$ PLYWD. ULAYMT.
$\frac{5}{8}''$ CDX

2ND FL.

1'-1" 11"

10" C. JST

$\frac{4}{S-1}$

SUSP. AC. TILE

10'-1"

8'-1"

$\frac{5}{8}''$ FR. GYP. BD.

$\frac{5}{A-5}$ $\frac{2}{S-1}$

2" CONC. &
6" CONC. PLANK
(SEE 5-1)

1ST FL.

8"

8'-10$\frac{1}{2}$"

8'-2$\frac{1}{2}$"

$\frac{3}{A-1}$

8" CONC.

BASEMT.

TYPICAL WALL SECTION
SCALE: $\frac{1}{4}'' = 1'-0''$

Figure 5-11   The detail of this cornice is shown in drawing 4 on sheet A-4, Figure 5-12.

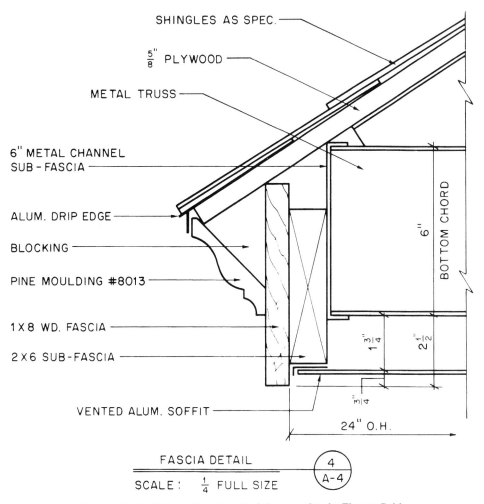

SHINGLES AS SPEC.

$\frac{5}{8}$" PLYWOOD

METAL TRUSS

6" METAL CHANNEL
SUB-FASCIA

ALUM. DRIP EDGE

BLOCKING

PINE MOULDING #8013

1 X 8 WD. FASCIA

2 X 6 SUB-FASCIA

VENTED ALUM. SOFFIT

BOTTOM CHORD

6"

$1\frac{3}{4}$"

$2\frac{1}{2}$"

$\frac{3}{4}$"

24" O.H.

FASCIA DETAIL

SCALE : $\frac{1}{4}$ FULL SIZE

4
A-4

Figure 5-12   This is the detail of the cornice in Figure 5-11.

✓ CHECK YOUR PROGRESS

Can you perform these tasks?

☐ Identify window types by their symbols.

☐ Identify materials by their symbols.

☐ Identify the most common electrical equipment by its symbols.

☐ Identify the most common plumbing equipment by its symbols.

☐ Reference details by their symbols.

☐ Define several common abbreviations used on construction drawings.

ASSIGNMENT

1. What is represented by each of these symbols?

   a.

   b.

   c.

   d.

   e.

   f.

   g. ———G———

   h. S_3

   i. (WH)

   j.

2. What is meant by each of these abbreviations?

   a. GYP. BD.

   b. FOUND.

   c. FIN. FL.

   d. O.C.

   e. REINF.

   f. EXT.

   g. COL.

   h. DIA.

   i. ELEV.

   j. CONC.

3. Where in a set of drawings would you find a detail numbered 6.4?

4. Where in a set of drawings would you find a detail numbered $\dfrac{5}{M-3}$?

UNIT 6   Plan Views

OBJECTIVES

After completing this unit, you will be able to explain the general kinds of information shown on the listed plans:

- Site plans
- Foundation plans
- Floor plans

You learned earlier in Unit 2 that plans are drawings that show an object as viewed from above. Many of the detail and section drawings in a set show parts of the building from above. Some of the plan views that show an entire building are discussed here. This brief explanation will help you feel more comfortable with plans, although it does not cover plans in depth. You will use plans frequently throughout your study of the remainder of this textbook. Each of the remaining units helps you understand plan views more thoroughly.

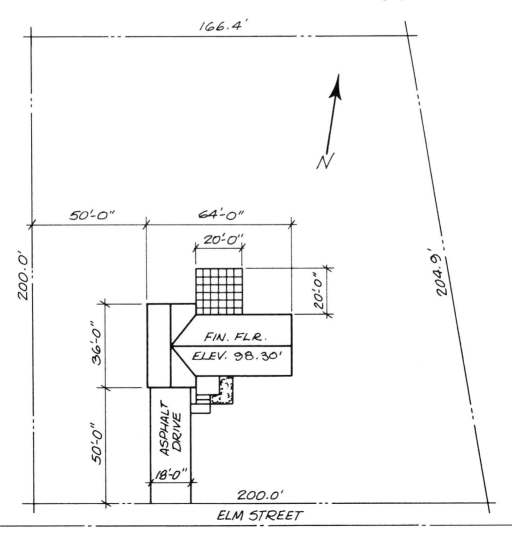

Figure 6-1   Minimum information shown on a site plan

SITE PLANS

A site plan gives information about the site on which the building is to be constructed. The boundaries of the site (property lines) are shown. The property line is usually a heavy line with one or two short dashes between longer line segments. The lengths of the boundaries are noted next to the line symbol. Property descriptions are often the result of a survey by a surveyor or civil engineer. These professionals usually work with decimal parts of feet, rather than feet and inches. Therefore, site dimensions are usually stated in tenths or hundredths of feet, Figure 6-1.

A symbol or arrow of some type indicates what compass direction the site faces. Unless this north arrow includes a correction for the difference between true north and magnetic north, it may be only an approximation. However, it is sufficient to show the general direction the site faces.

The site plan also indicates where the building is positioned on the site. As a minimum, the dimensions to the front and one side of the site are given. The overall dimensions of the building are also included. Anyone reading the site plan will have this basic information without referring to the other drawings. If the finished site is to include walks, drives, or patios, these are also described by their overall dimensions.

FOUNDATION PLANS

A foundation plan is like a floor plan, but of the foundation instead of the living spaces. It shows the foundation walls and any other structural work to be done below the living spaces.

There are two types of foundations that are commonly used in homes and other small buildings. One type has a concrete base, called the *footing,* supporting foundation walls, Figure 6-2. The other is the slab-on-grade type. A *slab-on-grade* foundation consists of a concrete slab placed directly on the soil with little or no other support. Slabs on grade are usually thickened at their edges and wherever they must support a heavy load, Figure 6-3.

When the footing-and-wall type foundation is used, girders are used to provide intermediate support to the structure above, Figure 6-4. The girder is shown on the foundation plan by phantom lines and a note describing it.

The foundation plan includes all of the dimensions necessary to lay out the footings and foundation walls. The footings follow the walls and may be shown on the plan. If they are shown, it is usually by means of hidden lines to show their outline only. In addition to the layout of the foundation walls, dimensions are given for opening windows, doors, and ventilators. Notes on the plan indicate areas that are not to be excavated, concrete-slab floors, and other important information about the foundation, Figure 6-5.

FLOOR PLANS

A floor plan is similar to a foundation plan. It is a section view taken at a height that shows the placement of walls, windows, doors, cabinets, and other important features. A separate floor plan is included for each floor of the building. The floor plans provide more information about the building than any of the other drawings.

Building Layout

The floor plans show the locations of all of the walls, doors, and windows. Therefore, the floor plans show how the building is divided into rooms, and how to get from one room to another. Before attempting to read any of the specific information on the floor plans, it is wise to familiarize yourself with the general layout of the building.

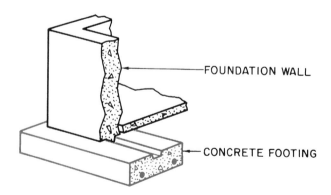

Figure 6-2   Footing and foundation wall

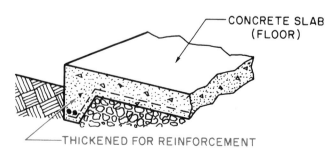

Figure 6-3   Slab-on-grade foundation

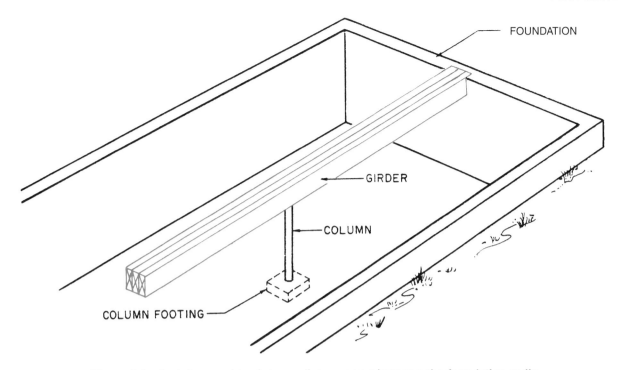

Figure 6-4   A girder provides intermediate support between the foundation walls.

To quickly familiarize yourself with a floor plan, imagine that you are walking through the house. For example, imagine yourself standing in the front door of the left side of the Duplex—plans for which are included in the drawing packed with this text. You are looking across the living room. There is a closet on your right and a large window on your left. Straight ahead is the dining room with doors into a storage room and the kitchen. Looking in the kitchen doorway (notice there is no door in this doorway), there are cabinets, a sink, and refrigerator on the opposite wall. More cabinets and a range are located on the left. Now, walk out of the kitchen and into the bedroom area. There are three doors; one leads into a large front bedroom with a long closet, another opens into a smaller bedroom, and the third opens into the bathroom. The bathroom includes a linen closet with bifold doors.

Dimensions

Dimensions are given for the sizes and locations of all walls, partitions, doors, windows, and other important features. On frame construction, exterior walls are usually dimensioned to the outside face of the wall framing. If the walls are to be covered with stucco or masonry veneer, this material is outside the dimen-sioned face of the wall frame. Interior partitions may be dimensioned to their centerlines or to the face of the studs. (*Studs* are the vertical members in a wall frame.) Windows and doors may be dimensioned about their centerlines, Figure 6-6, or to the edges of the openings.

Solid masonry construction is dimensioned entirely to the face of the masonry, Figure 6-7. Masonry openings for doors and windows are dimensioned to the edge of the openings.

Other Features of Floor Plans

The floor plan includes as much information as possible without making it cluttered and hard to read. Doors and windows are shown by their symbols as explained in Unit 5. Cabinets are shown in their proper positions. The cabinets are explained further by cabinet elevations and details, which are discussed in Unit 8. If the building includes stairs, these are shown on the floor plan. Important overhead construction is also indicated on the floor plans. If the ceiling is framed with joists, their size, direction, and spacing is shown on the floor plan. Architectural features such as exposed beams, arches in doorways, or unusual roof lines may be shown by phantom lines.

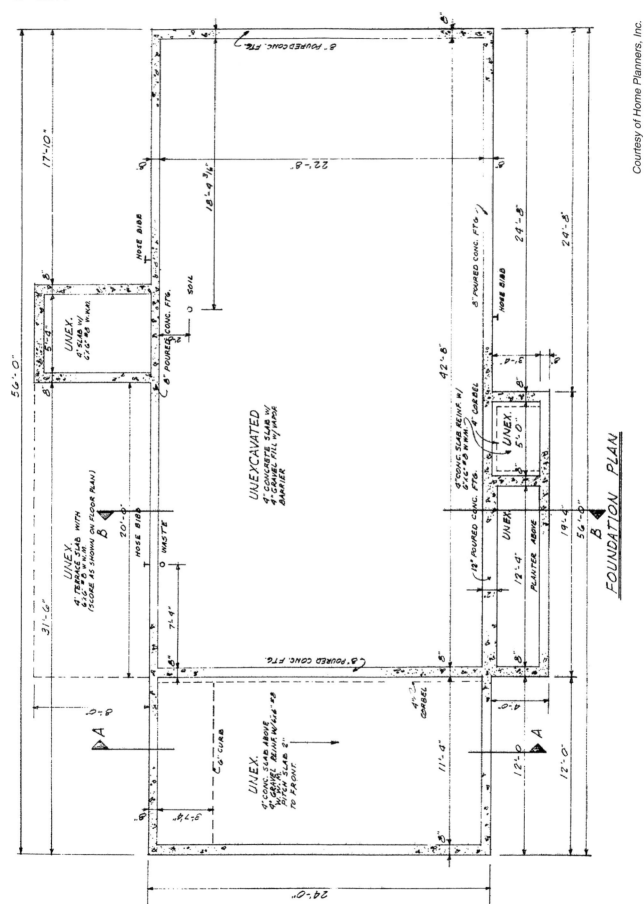

Figure 6-5 Foundation plan

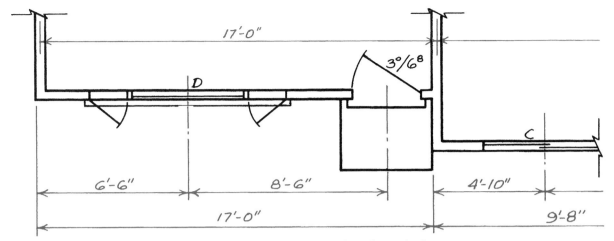

Figure 6-6 Frame construction dimensioning

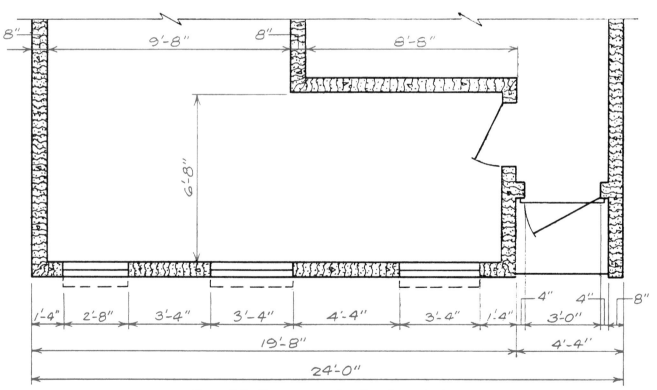

Figure 6-7 Masonry construction dimensioning

✓ CHECK YOUR PROGRESS

Can you perform these tasks?

- ☐ Describe property boundaries from a simple site plan.
- ☐ Tell which direction a site faces from the site plan
- ☐ Describe the position of a building within the site using the dimensions on a site plan.
- ☐ List the overall dimensions of a foundation from the information on a foundation plan.
- ☐ Identify girders on a foundation plan.
- ☐ Describe the locations of windows shown on plans.
- ☐ Describe the arrangement of rooms shown on a simple, one-story floor plan.
- ☐ Describe other major features shown on plans.

ASSIGNMENT

Refer to the drawings for the Two-Unit Apartment (which are included in the accompanying packet) to complete this assignment.

1. In what direction does the Apartment face?
2. What is the length and width of the Apartment site?
3. How far is the front of the Apartment from the front property line?
4. What is the overall length and width of the Apartment?
5. What are the inside dimensions of the front bedroom?
6. What is the thickness of the partitions between the two bedrooms?
7. What is the thickness of the interior wall between the two dining rooms?
8. With two exceptions, the units in the Apartment are exactly reversed. What are the two exceptions?
9. What is the distance from the west end of the Apartment to the centerline of the west, front entrance?
10. What is indicated by the small rectangle on the floor plan outside each main entrance?
11. What is the distance from the ends of the Apartment to the centerlines of the $6^0 \times 6^8$ sliding glass doors?
12. What is indicated by the dashed line just outside the front and back walls on the floor plan of the Apartment?

UNIT 7   Elevations

OBJECTIVES

After completing this unit, you will be able to perform the following tasks:

- Orient building elevations to building plans.

- Explain the kinds of information shown on elevations.

Drawings that show the height of objects are called *elevations.* However, when builders and architects refer to building elevations, they mean the exterior elevation drawings of the building, Figure 7-1. A set of working drawings usually includes an elevation of each of the four sides of the building. If the building is very complex, there may be more than four elevations. If the building is simple, there may be only two elevations—the front and one side.

ORIENTING ELEVATIONS

It is important to determine the relationship of one drawing to another. This is called *orienting* the drawings. For example, if you know which elevation is the front, you must be able to picture how it relates to the front of the floor plan.

Elevations are usually named according to compass directions, Figure 7-2. The side of the house that faces north is the north elevation, and the side that faces south is the south elevation, for example. When the elevations are named according to compass direction, they can be oriented to the floor plan, foundation plan, and site plan by the north arrow on those plans. It might help to label the edge of the plans according to the north arrow, Figure 7-3.

It is not always possible to label elevations according to compass direction, however. When drawings are prepared to be sold through a catalog or when they are for use on several sites, the compass directions cannot be included. In this case, the elevations are named according to their position as you face the building, Figure 7-4. To orient these elevations to the plans, find the front on the plans. The front is usually at the bottom of the sheet, but it can be checked by the location of the main entrance.

INFORMATION ON BUILDING ELEVATIONS

Building elevations are normally quite simple. Although the elevations do not include a lot of detailed dimensions and notes, they show the finished appearance

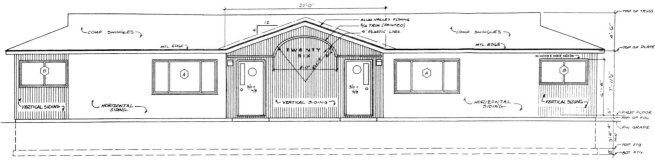

FRONT ELEVATION

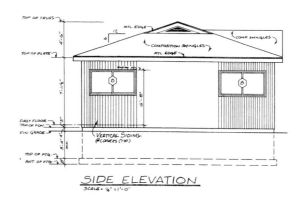

SIDE ELEVATION

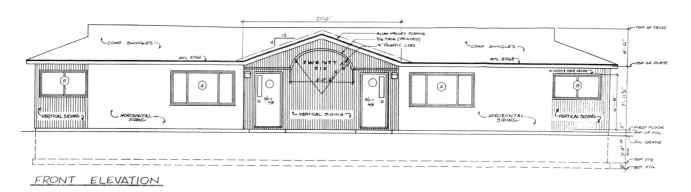

FRONT ELEVATION

Figure 7-1 Building elevations

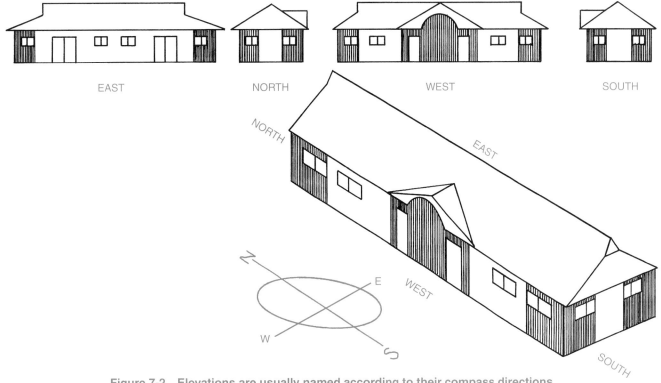

EAST   NORTH   WEST   SOUTH

Figure 7-2   Elevations are usually named according to their compass directions.

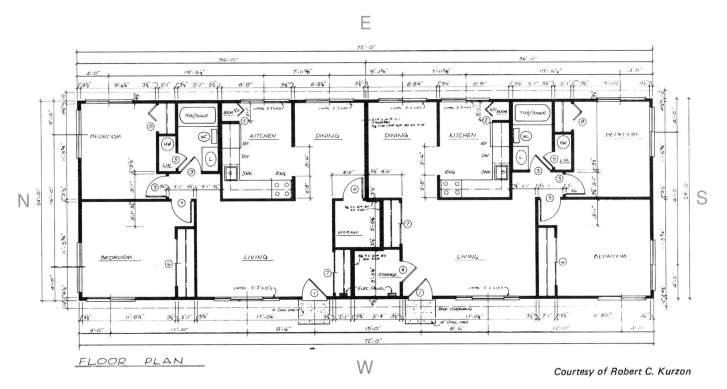

FLOOR PLAN

Figure 7-3   Plan labeled to help orientation to north arrow.

Courtesy of Robert C. Kurzon

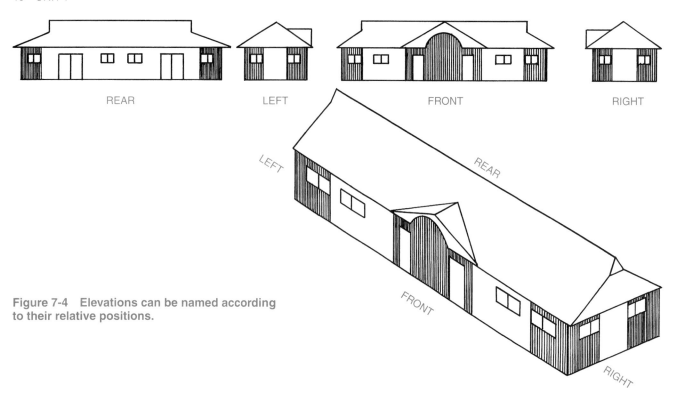

REAR   LEFT   FRONT   RIGHT

Figure 7-4   Elevations can be named according to their relative positions.

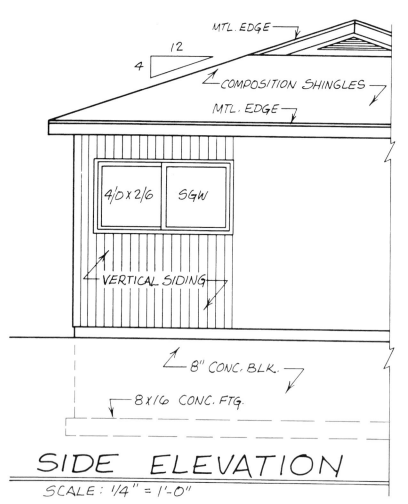

MTL. EDGE

12
4

COMPOSITION SHINGLES

MTL. EDGE

4/0 x 2/6   SGW

VERTICAL SIDING

8" CONC. BLK.

8 X 16 CONC. FTG.

SIDE ELEVATION

SCALE: 1/4" = 1'-0"

Figure 7-5   Underground portion of building is shown with dashed lines.

Courtesy of Robert C. Kurzon

of the building better than other views. Therefore, elevations are a great aid in understanding the rest of the drawing set.

The elevations show most of the building, as it will actually appear, with solid lines. However, the underground portion of the foundation is shown as hidden lines, Figure 7-5. The footing is shown as a rectangle of dashed lines at the bottom of the foundation walls.

The surface of the ground is shown by a heavy solid line, called a *grade line*. The grade line usually includes one or more notes to indicate the elevation above sea level or another reference point, Figure 7-6. *Elevation* used in this sense is altitude, or height—not a type of drawing. All references to the height of the ground or the level of key parts of the building are in terms of elevation. Methods for measuring site elevations are discussed later in Unit 9.

Some important dimensions are included on the building elevations. Most of them are given in a string at the end of one or more elevations, Figure 7-7. The dimensions most often included are listed:

- Thickness of footing

- Height of foundation walls

- Top of foundation to finished first floor

- Finished floor to ceiling or top of plate (The *plate* is the uppermost framing member in the wall.)

- Finished floor to bottom of window headers (The *headers* are the framing across the top of a window opening.)

- Roof overhang at eaves

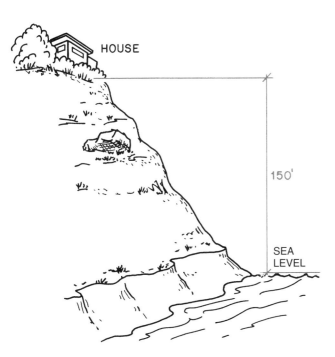

Figure 7-6 The elevation of this site is 150'.

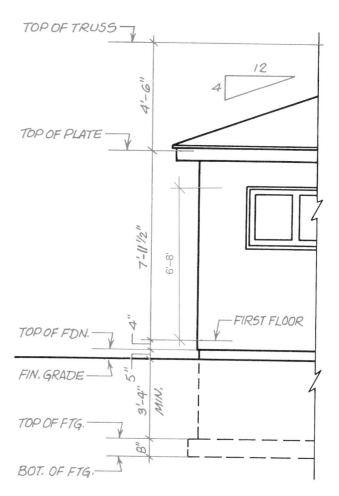

Figure 7-7 Dimensions on an elevation

✓ CHECK YOUR PROGRESS

Can you perform these tasks?

☐ Explain which side of a plan view is represented by a building elevation.

☐ Identify footings and foundations on building elevations.

☐ Find the height of a foundation wall on the building elevations.

☐ Find the dimension from the floor to the top of a wall on building elevations.

☐ Find the amount of roof overhang shown on building elevations.

☐ Describe the appearance of a building from information given on the building elevations.

ASSIGNMENT

Refer to the drawings of the Two-Unit Apartment in the packet to complete this assignment.

1. Which elevation is the north elevation?

2. In what compass direction does the left end of the Apartment face?

3. What is the dimension from the surface of the floor to the top of the wall framing?

4. What is the thickness of the floor?

5. How far does the foundation project above the ground?

6. How far below the surface of the ground does the foundation wall extend?

7. What is the total height of the foundation walls?

8. What is the minimum depth of the bottom of the footings?

UNIT 8 Sections and Details

OBJECTIVES

After completing this unit, you will be able to perform the following tasks:

- Find and explain information shown on section views.

- Find and explain information shown on large-scale details.

- Orient sections and details to the other plans and elevations.

It is not possible to show all of the details of construction on foundation plans, floor plans, and building elevations. Those drawings are meant to show the relationships of the major building elements to one another. To show how individual pieces fit together, it is necessary to use larger-scale drawings and section views. These drawings are usually grouped together in the drawing set. They are referred to as *sections and details,* Figure 8-1.

SECTIONS

Nearly all sets of drawings include, at least, a typical wall section. The typical section may be a section view of one wall, or it may be a full section of the building. Full sections are named by the direction in which the imaginary cut is made. Figure 8-2 shows a transverse section. A *transverse section* is taken from an imaginary cut across the width of the building. Transverse sections are sometimes called *cross sections.* A full section taken from lengthwise cut through the building is called a *longitudinal section,* Figure 8-3.

Full sections and wall sections normally have only a few dimensions, but have many notes with leaders to identify the parts of the wall. The following is a list of the kinds of information that are included on typical wall sections with most sets of drawings:

- Footing size and material (This may be specified by building codes.)

- Foundation wall thickness, height, and material

- Insulation, waterproofing, and interior finish for foundation walls

- Fill and waterproofing under concrete floors

- Concrete floor thickness, material, and reinforcement

- Sizes of floor framing materials

- Sizes of wall framing materials

- Wall covering (sheathing, siding, stucco, masonry, and interior wall finish) and insulation

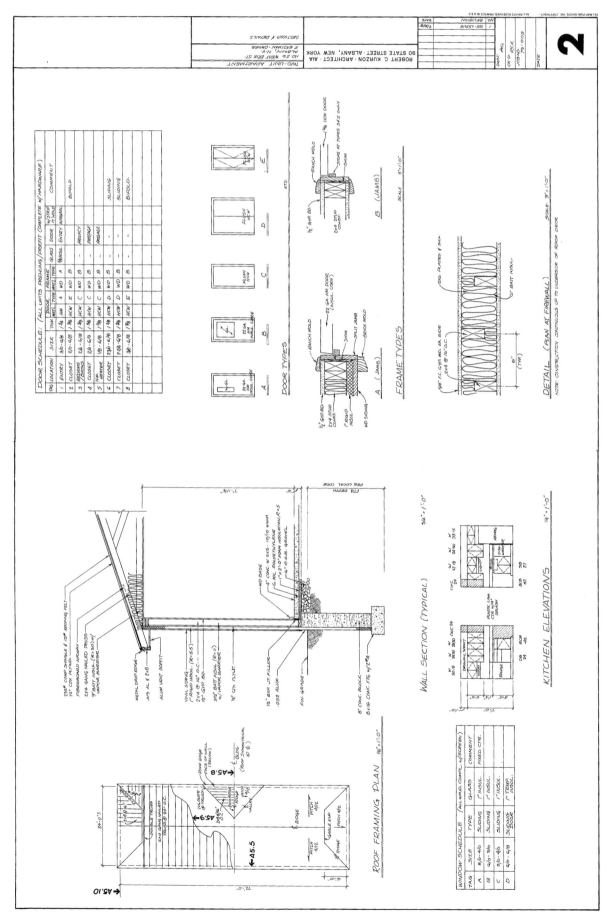

Figure 8-1 Typical sheet of sections and details for a small building

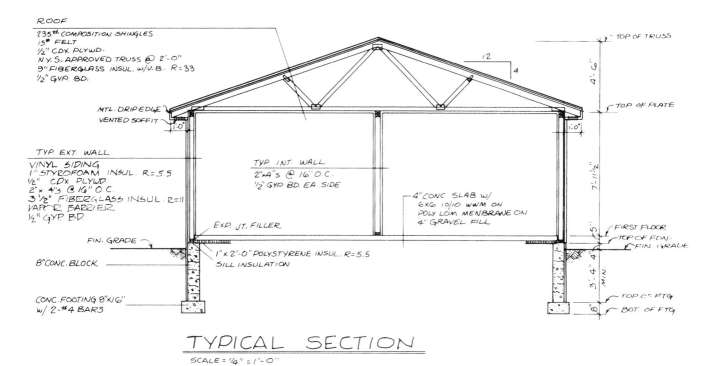

ROOF
235# COMPOSITION SHINGLES
15# FELT
½" CDX PLYWD.
N.Y.S. APPROVED TRUSS @ 2'-0"
9" FIBERGLASS INSUL. w/V.B. R=33
½" GYP. BD.

MTL. DRIP EDGE
VENTED SOFFIT

TYP. EXT. WALL
VINYL SIDING
1" STYROFOAM INSUL. R=5.5
½" CDX PLYWD.
2"x 4"s @ 16" O.C.
3 ½" FIBERGLASS INSUL. R=11
VAP. R BARRIER
½" GYP. BD.

FIN. GRADE

8" CONC. BLOCK

CONC. FOOTING 8"X16"
w/ 2-#4 BARS

TYP. INT. WALL
2"x4"s @ 16" O.C.
½" GYP. BD. EA. SIDE

4" CONC. SLAB w/
6x6 10/10 WWM ON
POLY LOM MENBRANE ON
4" GRAVEL FILL

EXP. JT. FILLER

1" x 2'-0" POLYSTYRENE INSUL. R=5.5
SILL INSULATION

12
4

TOP OF TRUSS
4'-0"
TOP OF PLATE

7'-11½"

FIRST FLOOR
5" TOP OF FDN.
FIN. GRADE

3'-4" MIN.

TOP OF FTG
8" BOT. OF FTG

TYPICAL   SECTION
SCALE = ¼" = 1'-0"

Figure 8-2   Transverse section

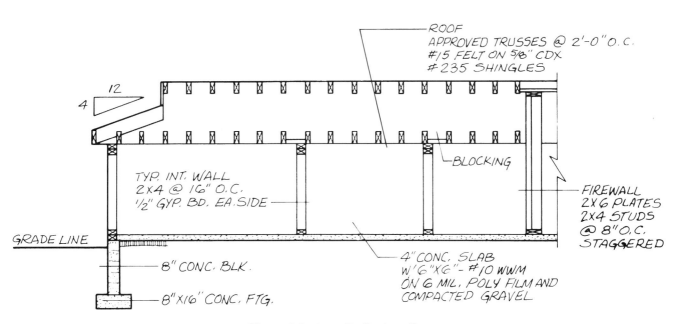

ROOF
APPROVED TRUSSES @ 2'-0" O.C.
#15 FELT ON ⅝" CDX
#235 SHINGLES

12
4

BLOCKING

TYP. INT. WALL
2X4 @ 16" O.C.
½" GYP. BD. EA. SIDE

FIREWALL
2X6 PLATES
2X4 STUDS
@ 8" O.C.
STAGGERED

GRADE LINE

8" CONC. BLK.

8"X16" CONC. FTG.

4" CONC. SLAB
W/ 6"X6"- #10 WWM
ON 6 MIL. POLY FILM AND
COMPACTED GRAVEL

Figure 8-3   Longitudinal section

- Cornice construction—materials and sizes (The *cornice* is the construction at the roof eaves.)

- Ceiling construction and insulation

Other section drawings are included as necessary to explain special features of construction. Wherever wall construction varies from the typical wall section, another wall section should be included. Section views are used to show any special construction that cannot be shown on normal plans and elevations. Figure 8-4 is an example of a special section in elevation. This section view is said to be *in elevation* because it shows the height of the ridge construction. Figure 8-5 is *in plan* because it shows the interior of the fireplace as viewed from above.

OTHER LARGE-SCALE DETAILS

Sometimes necessary information can be conveyed without showing the interior construction. A large scale may be all that is needed to show the necessary

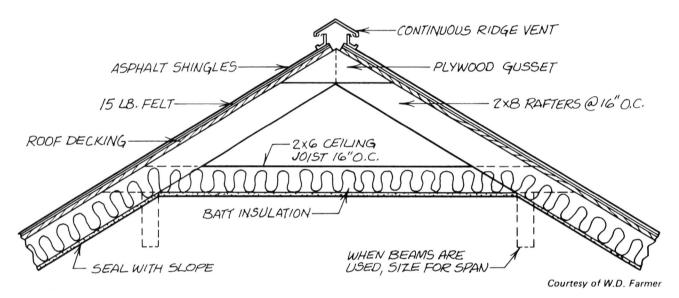

Courtesy of W.D. Farmer

Figure 8-4   Special section of ventilated ridge

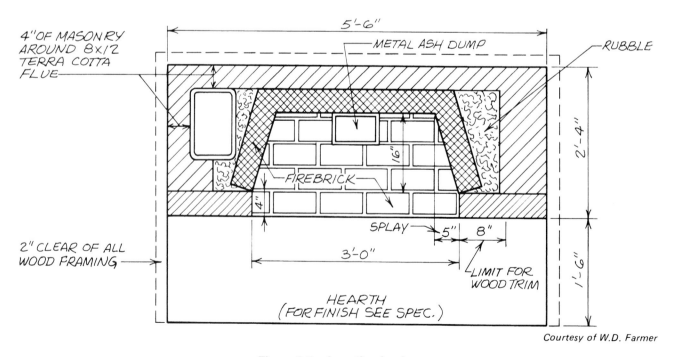

Courtesy of W.D. Farmer

Figure 8-5   A section in plan

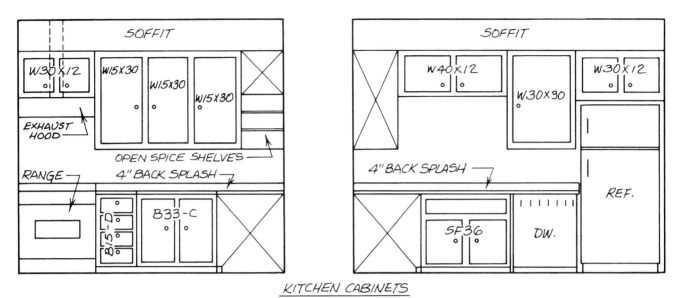

Figure 8-6 Cabinet elevations

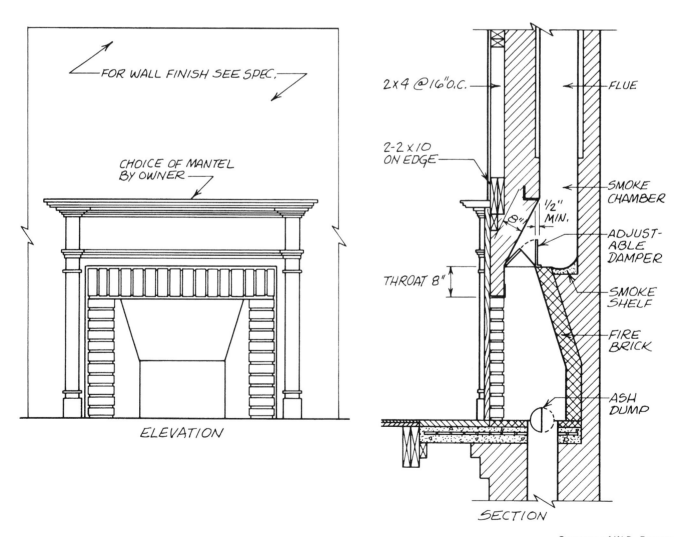

Courtesy of W.D. Farmer

Figure 8-7 Fireplace details

details. The most common examples of this are on cabinet installation drawings, Figure 8-6. Cabinet elevations show how the cabinets are located, without showing the interior construction.

Many details are best shown by combining elevations and sections or by using isometric drawings. Figure 8-7 shows an example of an elevation and a section used together to explain the construction of a fireplace. Figure 8-8 shows an isometric detail drawing that includes sections to show interior construction. Another method of showing detail is an exploded view. Figure 8-9 shows an exploded view of an electrical receptacle.

ORIENTING SECTIONS AND DETAILS

As explained earlier, some sections and details are labeled as typical. These drawings describe the construction that is used throughout most of the building.

Details and sections that refer to only one place in the building are identified by a reference mark. As was pointed out earlier in Unit 4, sections are usually referenced by a cutting-plane line. This line shows where the section was taken from. Arrows on the ends of the cutting-plane line indicate what direction the imaginary cut is viewed from. A reference mark near the arrow indicates where the detail drawing is shown. These reference marks used for orienting details may vary from one set of drawings to another. It is important, although not usually difficult, to study the drawings and learn how the architect references details. Usually a system of sheet numbers and view numbers is used. One such numbering system was explained earlier.

Some basic principles of details and sections have been discussed here. You will gain more practice later in reading details and sections.

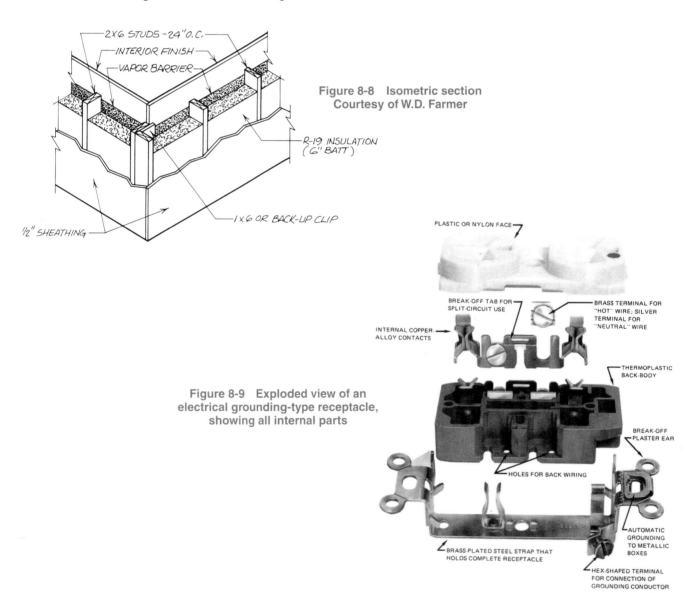

Figure 8-8 Isometric section
Courtesy of W.D. Farmer

Figure 8-9 Exploded view of an
electrical grounding-type receptacle,
showing all internal parts

✓ CHECK YOUR PROGRESS ─────────────────────────────────

Can you perform these tasks?

☐ Find the part of a building from which a section view was drawn.

☐ Explain the notes commonly included on typical wall sections.

☐ Explain whether a section view is in elevation or plan.

☐ Find the part of a building from which a large-scale detail was drawn.

 ASSIGNMENT ──────────────────────────────────

Refer to the drawings of the Two-Unit Apartment in the packet to complete the assignment.

1. What is used to show the detail of a complex design, installation, or product?

2. What kind of section drawing is the Typical Wall Section on Sheet 2?

3. What kind and size material is to be used for the foundation walls?

4. What is used between the concrete-slab floor and the exterior wall framing?

5. What kind and size of insulation is used around the foundation? Is this used on the inside or outside of the foundation?

6. What kind and size of material is to be used on the inside of the frame walls?

7. Sheet 2 includes a firewall detail. Where in the Apartment is this firewall?

8. What is the distance between the centerlines of the studs in the firewall?

9. What is the total thickness of the firewall? (Remember that a 2×6 is actually $5^{1}/_{2}"$ wide.)

10. Were the cabinet elevations drawn of the kitchen on the east side or the west side of the Apartment?

11. How would the kitchen elevations be different if they were drawn from the other kitchen?

12. What is the distance from the kitchen countertop to the bottom of the wall cabinets?

13. How far does the roof overhand project beyond the exterior walls?

14. Where are the electrical panels located?

PART 1   Test

A. Which occupation or individual listed in Column II performs the task listed in Column I?

I		II	
1.	Obtains a building permit	a.	Architect
2.	Issues the building permit	b.	Building inspector
3.	Acts as owner's representative	c.	Owner
4.	Issues certificate of occupancy	d.	Mechanical engineer
5.	Lays out rooms for traffic control	e.	Municipal building department
6.	Designs plumbing in large buildings	f.	General contractor
7.	Hires and supervises carpenters		
8.	Checks to see that codes are observed		

B. Identify each of the dimensions indicated on the illustrated scale.

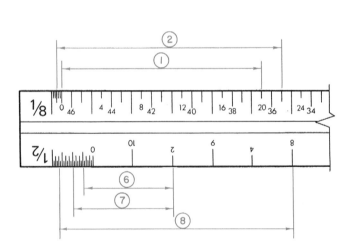
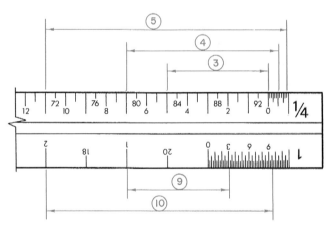

C. Which of the lines shown in Column II is most likely to be used for each purpose in Column I?

I		II	
1.	Outline of a window	a.	————————————
2.	Alternate position of a fold-down countertop	b.	————————————
3.	Centerline of a round post	c.	— — — — — —
4.	Extension line to show extent of a dimension	d.	——— — ———
5.	Buried footing	e.	——— — — ———
6.	Point at which an imaginary cut is made for a section view	f.	— — — — — — —

D. Which of the symbols shown in Column II is used for each of the objects or materials in Column I?

I

1. Awning window in elevation
2. Bifold door in plan
3. Earth
4. Rough wood
5. Batt insulation
6. Concrete
7. Ceiling light fixture
8. Finish wood
9. Shutoff valve (plumbing)
10. Hopper window in elevation

II

a.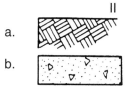

b.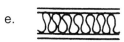

c.

d.

e.

f.

g.

h.

i.

j.

E. Select the one best answer for each question.

1. What kind of regulation controls the type of buildings allowed in each part of a community?
    a. building code
    b. zoning law
    c. specification
    d. certificate of occupancy

2. Which of the listed kinds of information can be clearly shown on construction drawings?
    a. size of parts
    b. location of parts
    c. shape of parts
    d. all of these

3. What type of drawing is the 2 × 4 shown in Illustration 3?
    a. isometric
    b. oblique
    c. perspective
    d. none of these

4. What type of drawing is the 2 × 4 shown in Illustration 4?
    a. isometric
    b. oblique
    c. perspective
    d. none of these

5. What type of drawing is the 2 × 4 shown in Illustration 5?
    a. isometric
    b. oblique
    c. perspective
    d. none of these

6. What kind of drawing is shown in Illustration 6?
    a. elevation
    b. detail
    c. rendering
    d. plan

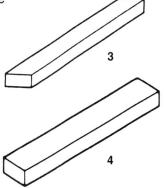

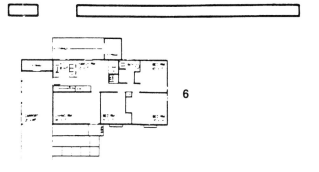

7.   What kind of drawing is shown in Illustration 7?
   a.   elevation
   b.   detail
   c.   rendering
   d.   plan

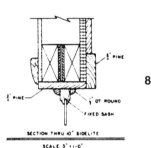

8.   What kind of drawing is shown in Illustration 8?
   a.   elevation
   b.   detail
   c.   rendering
   d.   plan

9.   If an object 12 feet long is drawn at a scale of $1/4$" = 1'-0", how long is the drawing?
   a.   48 inches   c.   3 feet
   b.   3 inches   d.   none of these

10.   If an object 1'-6" long is drawn at a scale of $1^{1}/_{2}$" = 1'-0", how long is the drawing?
   a. $2^{1}/_{4}$ inches   c.   27 inches
   b.   2 inches   d.   none of these

11.   Where in a set of drawings would you find detail number 9.6?
   a.   sheet 6   c.   ninth sheet in the mechanical section
   b.   sheet 9   d.   none of the above

12.   In the drawing key $\dfrac{4}{A-4}$, what does the letter A stand for?
   a.   architect's initial   c.   architectural
   b.   first edition of the drawings   d.   first detail on the sheet

13.   On which drawing would you expect to find the height of the foundation wall?
   a.   site plan   c.   floor plan
   b.   building elevation   d.   foundation plan

14.   On which drawing would you expect to find the set-back of the building?
   a.   site plan   c.   floor plan
   b.   building elevation   d.   foundation plan

15.   On which drawing would you expect to find the height of the window heads?
   a.   site plan   c.   floor plan
   b.   building elevation   d.   window detail

F.   Refer to the Two-Unit Apartment drawings to answer these questions.

1.   How far is the building from the west boundary?
2.   What is the dimension from the finished floor to the top of the wall plate?
3.   What is the overall length of the building at window height?
4.   What is the overall length of the building at the eaves?
5.   What is the north-to-south dimension inside the front bedrooms?
6.   What is the slope of the roof?
7.   What type of windows are used?
8.   How thick is the concrete footing?
9.   What material is the foundation wall?
10.   What is under the floor at its outer edges?

Part II

READING DRAWINGS FOR TRADE INFORMATION

In Part II, you will examine all of the information necessary to build a moderately complex single-family home. The sequence of the units in Part II follows the sequence of actual construction. In some cases, all of the information necessary for a particular phase of construction can be found on one sheet of drawings. Other phases require cross-referencing among several drawings. The relationships among the various drawings are discussed as the need to cross-reference them arises.

The assignments in this part refer to the Lake House drawings provided in the packet. The Lake House was designed as a vacation home on a lake in Virginia. The design is moderately complex, involving several floor levels and some interesting construction techniques.

UNIT 9   Clearing and Rough Grading the Site

OBJECTIVES

After completing this unit, you will be able to perform the following tasks:

- Identify work to be included in clearing a building site according to site plans.

- Interpret grading indications on a site plan.

- Interpolate unspecified site elevations.

PROPERTY BOUNDARY LINES

The boundary lines of the building site are shown on the site plan. The direction of a property line is usually expressed as a bearing angle. The *bearing* of a line is the angle between the line and North or South. Bearing angles are measured from North or South depending on which keeps the bearing under 90 degrees, Figure 9-1. Angles are expressed in degrees (°), minutes ('), and seconds ("). There are 360 degrees in a complete circle, 60 minutes in a degree, and 60 seconds in a minute.

The point of beginning (P.O.B.) may or may not be shown on the site plan. If the point of beginning is not shown on the plan, start at a convenient corner.

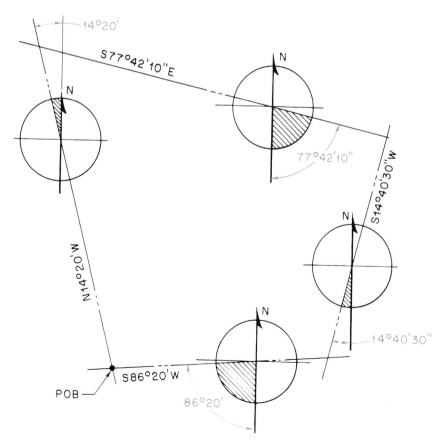

Figure 9-1   Bearing angles are always less than 90°.

61

Corners are usually marked with an iron pin (I.P.) or some permanent feature. The approximate direction of the boundaries can be found with a hand-held compass. This approximation should be accurate enough to aid in finding the marker (iron pin, manhole cover, concrete marker, or similar item) at the next corner. Proceed around the perimeter in this manner to find all corners. All construction activity should be kept within the property boundaries unless permission is first obtained from neighboring landowners.

CLEARING THE SITE

The first step in actual construction is to prepare the site. This means clearing any brush or trees that are not to be part of the finished landscape. The architect's choice of trees to remain is based on consideration of many factors. Trees and other natural features can be an important part of architecture—not only for their natural beauty, but for energy conservation. For example, deciduous trees, which lose their leaves in the winter, can be used to effectively control the solar energy striking a house. In the winter, the sun shines through the deciduous trees on the south side of a house, Figure 9-2. In the summer, the trees shade the south side of the house, Figure 9-3. The Lake House offers a good example of the importance of the selection of trees to remain on a site. This house gets a large part of its heat from its passive-solar features. The passive-solar features are described more fully later.

Trees that are to be saved are shown on the plot plan by a symbol and a note indicating their butt diameter and species, Figure 9-4. Areas that are too densely wooded to show individual trees are outlined and marked "woods," Figure 9-5. Removal of unwanted trees may require felling and stump removal, or may be accomplished with a bulldozer and dump truck. In either case, care must be exercised not to damage the trees that are to be saved.

GRADING

Grading refers to moving earth away from high areas and into low areas. Site grading is necessary to insure that water drains away from the building properly and does not puddle or run into the building. In some cases, grading may be necessary for access to

Figure 9-2   The winter sun passes through deciduous trees.

Figure 9-4   Typical note and symbol for individual tree.

Figure 9-3   The summer sun is shaded by deciduous trees.

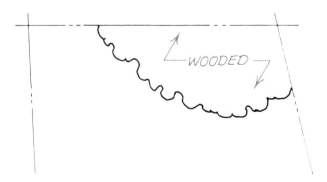

Figure 9-5   Typical note and symbol for wooded area.

the site. For example, if the site has a steep grade, it may be necessary to provide a more gradual slope for a driveway.

Grade is measured in vertical feet from sea level or from a fixed object such as a manhole cover. This vertical distance is called *elevation.* The term elevation to denote a vertical position should not be confused with elevation drawings that show the height of objects. The elevations of specific points are given as *spot elevations.* Spot elevations are used to establish points in a driveway, walk, or the slope of a terrace, Figure 9-6. Spot elevations are often given for trees that are to be saved.

The grade of a site is shown by *topographic contour lines.* These are lines following a particular elevation. The vertical difference between contour lines is the *vertical contour interval.* For plot plans this is usually 1 or 2 feet. When the land slopes steeply, the contour lines are closely spaced. When the slope is gradual, the contour lines are more widely spaced.

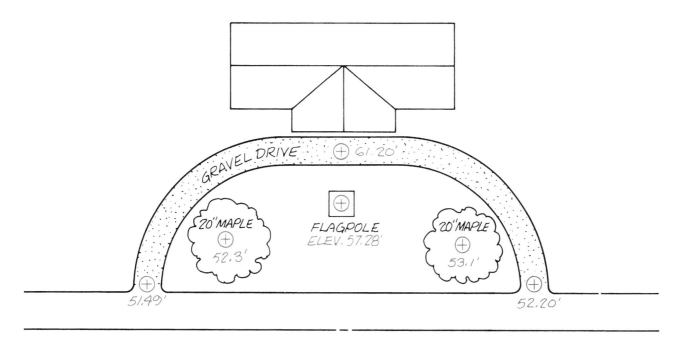

Figure 9-6 Spot elevations for specific locations.

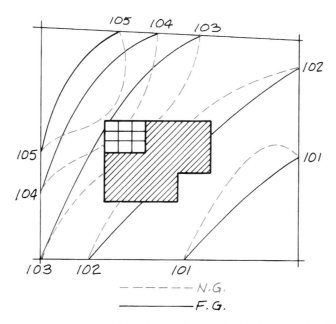

Figure 9-7 Two sets of contour lines show that this site will be graded to be more level in the area of the building.

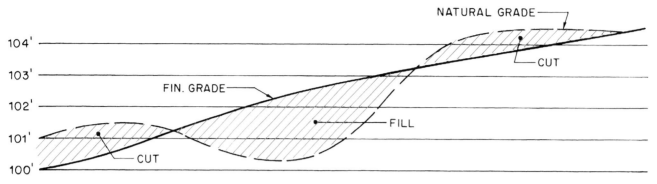

Figure 9-8   Cutting is required where NG is above FG. Fill is required where NG is below FG.

The builder must be concerned with not only the grade or contour of the existing site, but also that of the finished site. To show both contours, two sets of contour lines are included on the plot plan. Broken lines indicate natural grade (N.G.) and solid lines indicate finished grade (F.G.), Figure 9-7.

When the natural-grade elevation is higher than the finished-grade elevation, earth must be removed. This is referred to as *cut*. When the natural grade is at a lower elevation than the finished grade, *fill* is required. To determine the amount of cut or fill required at a given point, find the difference between natural grade and finished grade, Figure 9-8.

INTERPOLATING ELEVATIONS

Sometimes it is necessary to find an elevation that falls between two contour lines. This can be done by interpolation. *Interpolation* is a method of finding an unknown value by comparing it with known values.

Example: To interpolate the elevation of the tree at point A in Figure 9-9, follow the listed steps of the procedure using the information shown in the illustration and the numbers enclosed in parenthesis.

Step 1.   Scale the distance between the two adjacent contour lines (12 feet).
Step 2.   Scale the distance from the unknown point to the nearest contour line (4 feet).

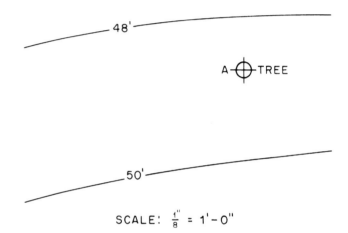

Figure 9-9   Interpolate the elevation of the tree.

Step 3.   Multiply the contour interval by the fraction of the distance between the contour lines to the unknown point. (Contour interval = 2 feet; fraction of distance between contour lines = $^4/_{12}$ = $^1/_3$. Therefore, 2 x $^1/_3$ = $^2/_3$ feet.)

Step 4.   If the nearest contour line is below the other one, add this to it. If the nearest contour lines is above the other one, subtract this amount. (Nearest contour = 48'. This is below the other contour line at 50', so $^2/_3$ is added to 48'. $^2/_3$' + 48' = 48.66'.)

✓ CHECK YOUR PROGRESS

Can you perform these tasks?

- ☐ Determine the bearings and lengths of property lines.
- ☐ Identify each tree to be removed from a site.
- ☐ Give the natural grades and finished grades of all points on a site.
- ☐ Explain which direction water will naturally run from any point on a site.

ASSIGNMENT

Refer to the site drawings of the Lake House (in the packet) to complete the assignment.

1. What are the lengths of the north, east, and west boundaries of the Lake House?

2. Which of the compass points shown in Figure 9-10 corresponds with the north boundary of the Lake House? Which compass point corresponds with the east boundary?

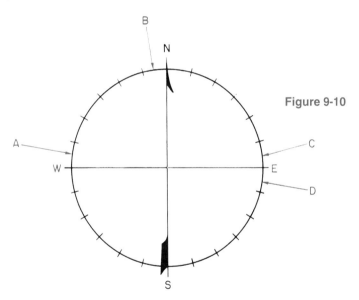

Figure 9-10

3. How many trees are indicated for removal?

4. How many trees are to remain on the site? (Do not include wooded areas.)

5. What is the finished grade elevation at the tree nearest the Lake House?

6. What was the natural-grade elevation of the most easterly tree to be saved?

7. What is the elevation of the tree to be saved nearest the Lake?

8. What is the natural-grade elevation at the southwest corner of the Lake House? Do not include the deck as part of the house.

9. What is the finished-grade elevation at the southwest corner of the house?

10. How much cut or fill is required at the entrance of the garage?

11. Is cut or fill required at the southwest corner of the house? How much?

12. What is the elevation at the northeast corner of the site?

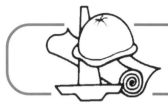

UNIT 10 Locating the Building

OBJECTIVES

After completing this unit, you will be able to perform the following tasks:

* Lay out building lines according to a site plan.

* Use the 6-8-10 or equal-diagonals method to check the squareness of corner.

* Use a leveling instrument to measure angles and depths of excavations.

LAYING OUT BUILDING LINES

The position of the building is shown on the site plan. Dimensions show the distance from the street (or lake) to the building and from the side boundaries to the building. The location of one corner can easily be found by measuring with a long (100' or 200') steel tape. The most efficient way to find all remaining corners is by the use of a leveling instrument, Figure 10-1. The functions of the parts of a leveling instrument are as follows:

* *Telescope* contains the lens, focusing adjustment, and cross has for sighting.

* *Telescope level* is a spirit level used for leveling the instrument prior to use.

* *Clamp screw* locks the instrument in position horizontally.

* *Fine adjusting screw* makes fine adjustments in a horizontal plane.

* *Leveling base* holds the leveling screws (usually four) for leveling the instrument prior to use.

* *Protractor* is a scale graduated in degrees and minutes for measuring horizontal angles.

Two accessories are required for most operations performed with a leveling instrument. The *tripod* is a three-legged stand that provides a stable base for the instrument. A *target rod* is a separate device with a

Figure 10-1 Builder's level
Courtesy of Keuffel & Esser Co.

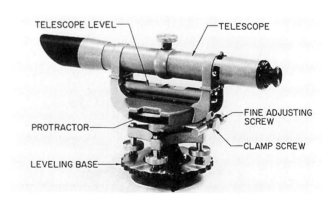

LOCATING THE BUILDING 67

scale graduated in feet and tenths of a foot. The telescope is focused on the target rod to measure elevations. The builder's level is a device for checking the difference in elevation between two points. It can also be used for measuring angles on a horizontal plane. Another instrument, a *transit,* can be tilted to measure angles in a vertical plane, Figure 10-2. The procedure described here for laying out square corners can be used with either instrument:

Step 1. Set the tripod up over a known corner. The exact position is determined by hanging a plumb bob from the tripod. The legs should be firmly set in the ground about three feet apart.

Step 2. Set the instrument on top of the tripod and hand tighten the clamp screw.

Step 3. Turn the leveling screws down, so they contact the tripod plate.

Step 4. Turn the telescope so that it is over one pair of leveling screws. Adjust these two screws so that the telescope is level.

Step 5. Rotate the telescope so that it is over the other pair of leveling screws. Adjust these two screws to level the telescope.

Step 6. Repeat this over each pair of leveling screws, until the telescope is level in all positions.

Step 7. Using a compass, carefully point the telescope to magnetic north. The north arrow on the drawing can point to either true north or magnetic north. The compass needle will point only to magnetic north. Because the true north pole of the earth is some distance from its magnetic north, the difference between bearings based on true north and those based on magnetic north can be several degrees. If the building is to be laid out according to true-north bearings, not magnetic compass bearings, contact the architect or surveyor to find out what correction should be made in your area.

Step 8. Set the protractor at zero degrees.

Step 9. Rotate the telescope to the bearing of one building line.

Step 10. Stretch a line (string) from the existing stake to several feet beyond the next corner.

Step 11. Have a partner hold the target rod plumb over the far end of this line. When the telescope cross hairs can be focused on the target rod, the bearing of the line is correct.

Step 12. Measure the length of this building line from the first stake and drive another stake.

Step 13. To lay out each of the remaining corners, set the tripod over the stake; line the telescope up on the existing line; and then use the protractor of the instrument to measure a 90° corner.

To check the squareness of the layout, measure the diagonals of a rectangle formed by the layout, Figure 10-3. When all four corners of the rectangle are 90°, the diagonals are equal.

Another method of checking a 90° angle is called the *6-8-10 method,* Figure 10-4. Measure 6 feet from the corner along one line. Measure 8 feet from the corner along the other line. These points should be 10 feet apart. (See Math Review 24).

The building lines can be saved, even after the corner stakes are removed for earthwork, by erecting

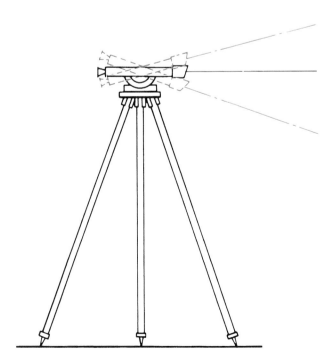

Figure 10-2 A transit is similar to a builder's level, but can be tilted to measure vertical angles.

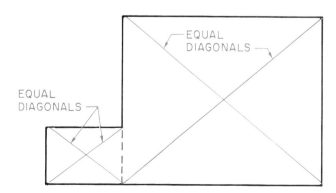

Figure 10-3 When the diagonals of a rectangle are the same length the corners are square.

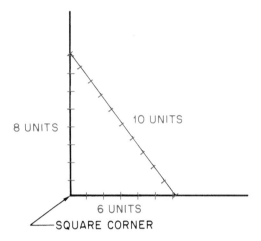

Figure 10-4   The 6-8-10 method of checking a square corner

batter boards, Figure 10-5. *Batter boards* are sturdy horizontal boards fastened between 2 x 4 stakes, at least 4 feet outside the building lines. The building lines are extended and marked on the batter boards.

EXCAVATING

Most buildings require some *excavation* (digging) to prepare the site for a foundation. The depth of the excavation is measured from a fixed bench mark. A bench mark can be any stationary object such as a surveyed point on a street or very large boulder. All elevations (vertical distances) are measured from this

benchmark. Only in the case of a real coincidence would the benchmark be at the same elevation as the surface of the ground where the excavation is to be done. The actual depth of the excavation is the difference between the elevation at the surface of the ground and the elevation at the bottom of the excavation, Figure 10-6.

Concrete footings are placed in the bottom of the excavation to support the entire weight of the building, Figure 10-7. These footings are placed on unexcavated earth to reduce the chance of the soil compacting under them. This means that the excavation contractor must measure the depth of the excavation accurately. The footings may be *stepped,* as in Figure 10-8, to accommodate a sloping site. This requires measuring the depth at each step of the footing. Information about the foot design is found on the foundation plan and building elevations. The layout of the foundation walls and their footings is shown on the foundation plan. The foundation walls are shown by two solid lines with dimensions to indicate their sizes, Figure 10-9. A dotted line on each side the foundation indicates the concrete footing. The size of the footing may be omitted when the plan was developed for use in several locations.

The depth of the foundation, including its footing, is shown on the elevations. To simplify calculating excavation depths, many architects indicate the elevation a key points along the footings, Figure 10-11. A section view through all or part of the building may show a typical depth, but it is wise to check all of the elevations for steps in the footing. The footings may be shown

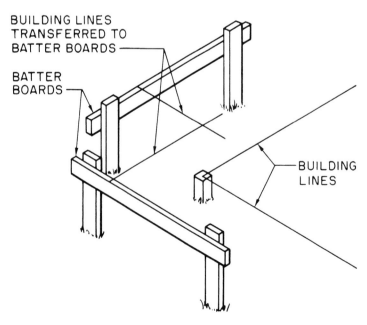

Figure 10-5   Batter boards

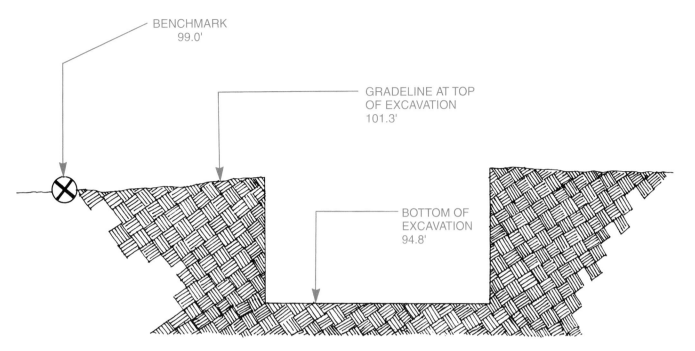

Figure 10-6   The top of this excavation is 2.3 feet above the benchmark. Its depth is 6.5 feet (101.3'–94.8').

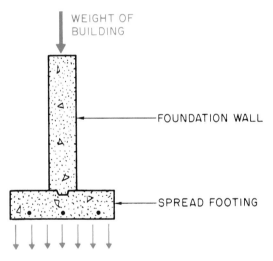

Figure 10-7   The footing spreads the weight of the building over a greater soil area.

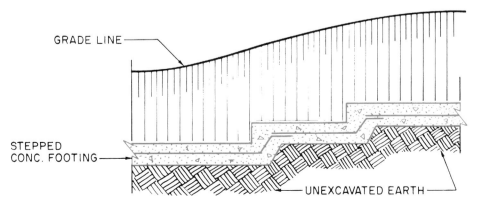

Figure 10-8   Footings can be stepped to accommodate a sloping site.

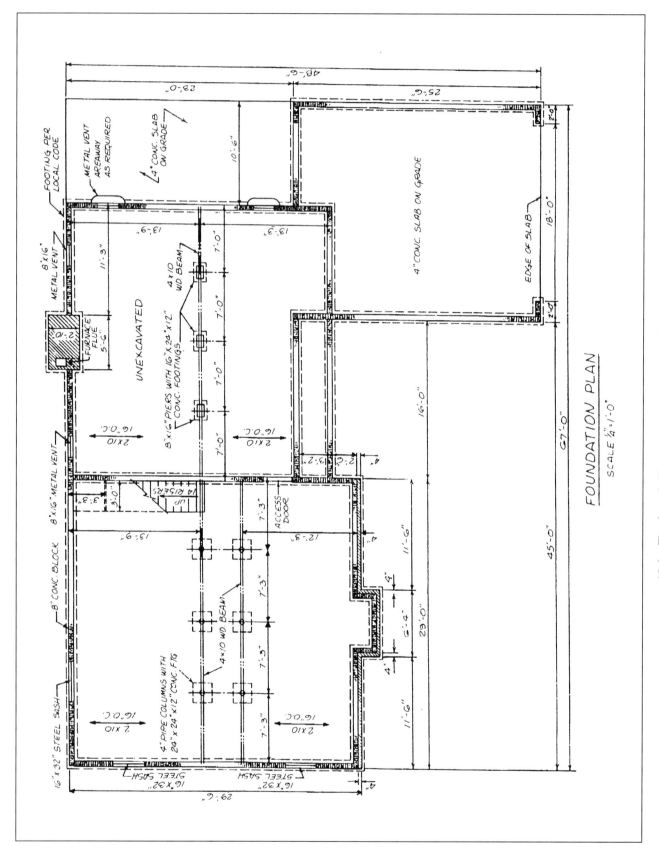

FOUNDATION PLAN
SCALE ½"=1'-0"

10-9   The foundation plan gives complete dimensions.

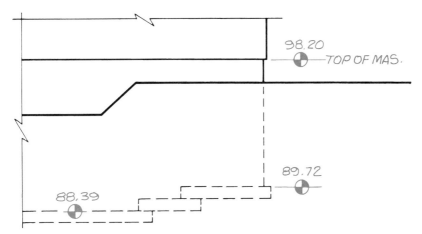

Figure 10-10   Spot elevations show key points on footing and foundation.

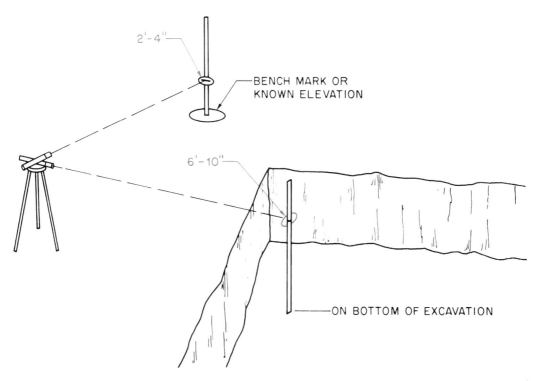

Figure 10-11   This excavation is 4'-6" deep—the difference in the two readings on the target rod.

on the elevation as a double or a single dotted line. In masonry foundations, steps in footings are usually in increments of 8 inches to conform to standard concrete block sizes.

Use the following procedure to measure differences in elevation, such as the depth of an excavation:

Step 1.   Set the instrument up on a tripod and level it in a convenient location.

Step 2.   Have a partner hold the target rod on a known elevation, such as a bench mark or ground of known elevation, while you focus the telescope and note the reading on the target rod where the cross hairs focus.

Step 3.   Take a similar reading with the target rod at the bottom of the excavation. The difference in the two readings is the depth of the excavation, Figure 10-11.

✓ CHECK YOUR PROGRESS ───

Can you perform these tasks?

☐ Properly set up a builder's level.

☐ Use a builder's level and tape measure to lay out building lines.

☐ Check the accuracy of right angles using a tape measure only.

☐ Check the squareness of a building layout using a tape measure only.

☐ Calculate the depth excavation required for footings.

☐ Measure the depth of an excavation using a building's level.

 ASSIGNMENT ───

Refer to the Lake House drawings (in the packet) to complete this assignment.

1. What is the distance from the Lake House to the nearest property boundary? (Do not treat the decks as part of the house for this question.)

2. What is the distance from the Lake House to the lake?

3. What is the distance from the north property line to the garage?

4. What is the area of the basement of the Lake House, including the foundation? Ignore slight irregularities in the shape of the foundation. For ease in calculating, divide the foundation into rectangles, Figure 10-12.

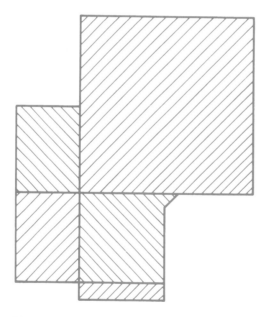

Figure 10-12   Use with assignment problem 4.

5. Find the highest and lowest natural grades meeting the house. Using the midpoint between the highest and lowest natural grades as a reference, determine the average depth of the excavation for the basement floor.

6. Approximately how many cubic yards of earth will have to be removed from the Lake House excavation? (See Math Review 28.)

7. Measuring from the natural grade, how deep is the excavation for the footing under the overhead garage door?

8. What is the elevation at the bottom of the deepest excavation for the Lake House? (Do not include the garage.)

9. Why would a row of large evergreen trees between the Lake House and the lake decrease the energy efficiency of the house?

10. What aspect of the location of a building is most often regulated by local ordinances?

UNIT 11 Site Utilities

After completing this unit, you will be able to perform the following tasks:

• Interpret symbols and notes used to describe site utilities.

• Explain the septic system indicated on a site plan.

• Determine the pitch of drain lines.

SEWER DRAINS

The *building sewer* carries the waste to the municipal sewer or septic system, Figure 11-1. Because sewer lines usually rely on gravity flow, they are large in diameter (4 inches, minimum) and are pitched to provide flow. Because water supply lines and gas lines are pressurized, pitch is not important in their installation. Therefore, the sewer is installed first and other piping is routed around it as necessary. The size, material, and pitch of drains are usually given in a note on the site plan, Figure 11-2. The pitch of a pipe is given in fractions

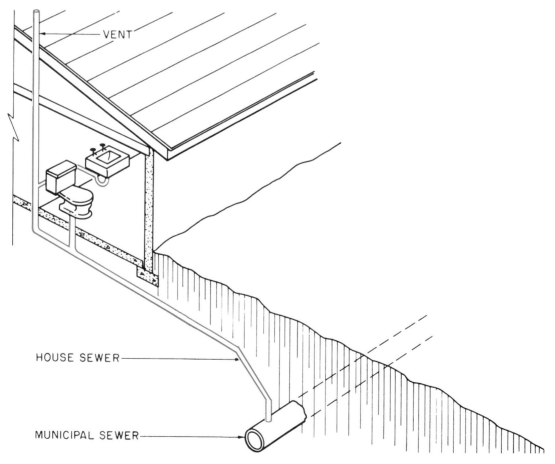

VENT

HOUSE SEWER

MUNICIPAL SEWER

Figure 11-1   The house sewer carries waste from the house to the municipal sewer or septic tank.

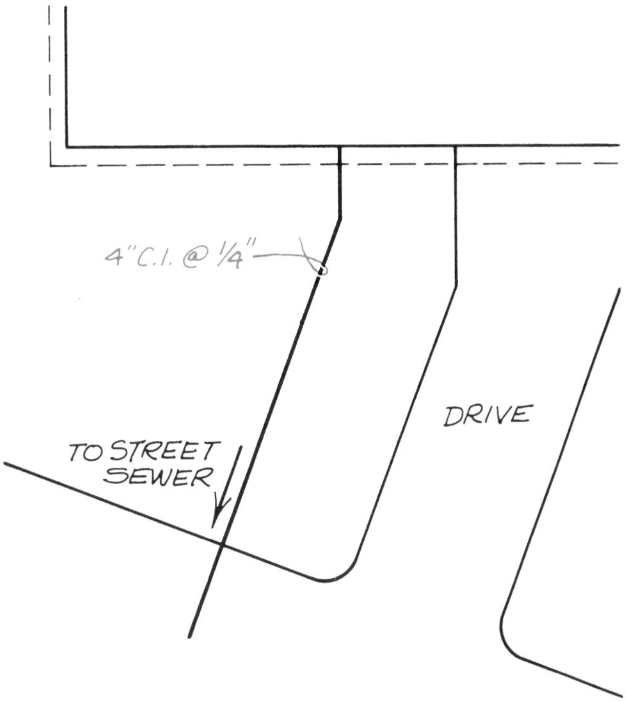

Figure 11-2 This note indicates cast iron pipe pitched ¼ inch for every foot of run.

of an inch per foot. A pitch of ¼ inch per foot means that for every horizontal foot, the pipe rises or falls ¼ inch.

In some cases, sewers may have to flow uphill. This is the case with the Lake House. Uphill flow is accomplished by a *grinder pump,* Figure 11-3.

BUILDING SEWER

Any plumbing that is to be concealed by concrete work must be installed without fixtures or *roughed in* before the concrete is placed. Because the plumbing contractor installs all plumbing inside the building lines, this phase of construction is discussed later with mechanical systems. However, employees of the general contractor may rough in the sewer from the building line to the street or septic system.

These workers must be able to determine the elevation at which the sewer passes through the foundation and the pitch of the line outside the building. The sewer line may be shown on plans as a solid or broken line. Although it is usually labeled, this is not always true. When the sewer is not labeled as such, it can still be recognized by its material, pitch, and ending place. In light construction, the sewer is usually the only 4-inch pipe to the building. Also, the sewer is the only line with the pitch indicated.

MUNICIPAL SEWERS AND SEPTIC SYSTEMS

In highly developed areas, the building sewer empties into a municipal sewer line in or near the street. The contractor for the new building is responsible for everything from the municipal sewer to the house.

In less developed areas, the sewer carries the sewage to a septic system. The *septic system* includes a septic tank and drain field, Figure 11-4. The septic tank holds the solid waste while it is decomposed by bacterial action. The liquids pass through the baffles and flow out of the tank to the distribution box. The distribution box (D.B.) diverts the liquid into *leach lines.* These perforated plastic or loose-fitting tile lines allow the liquid to be absorbed by the surrounding soil. The liquid gradually evaporates from or drains through the soil. The *drain field,* where the leach lines are laid, is usually made a layer of crushed stone.

The design of septic systems is closely regulated by most health and plumbing codes. These local codes should be checked before designing or installing any septic system. The building code or health department code often requires a percolation test before the system can be approved. In a *percolation test,* holes are dug. Then a measured amount of water is poured into each hole. The amount of time required for the water

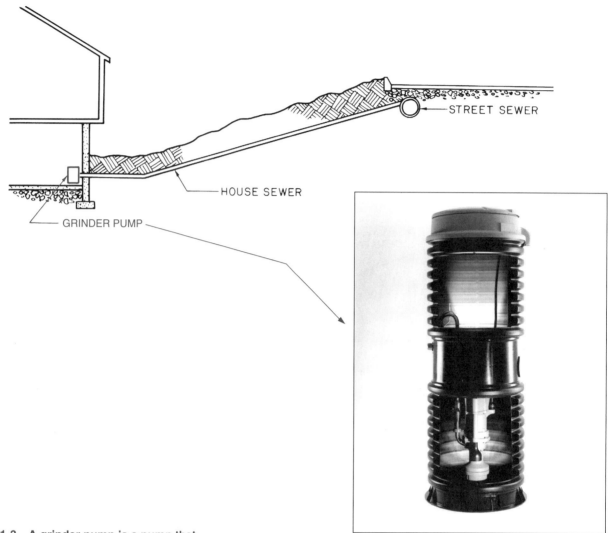

STREET SEWER

HOUSE SEWER

GRINDER PUMP

Figure 11-3   A grinder pump is a pump that moves sewage uphill.

Courtesy of Environment One

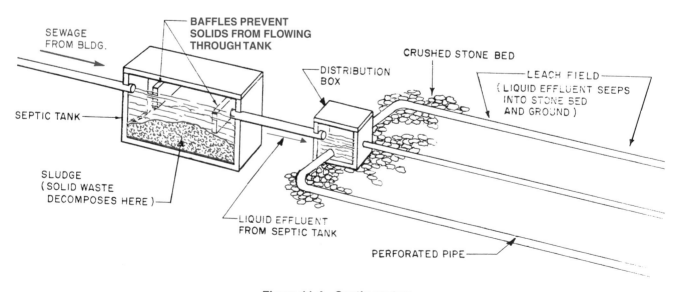

SEWAGE FROM BLDG.

BAFFLES PREVENT SOLIDS FROM FLOWING THROUGH TANK

DISTRIBUTION BOX

CRUSHED STONE BED

LEACH FIELD (LIQUID EFFLUENT SEEPS INTO STONE BED AND GROUND)

SEPTIC TANK

SLUDGE (SOLID WASTE DECOMPOSES HERE)

LIQUID EFFLUENT FROM SEPTIC TANK

PERFORATED PIPE

Figure 11-4   Septic system

to drain into the soil is an indication of how well the soil will accept water from the septic system. This ability to accept water is called *percolation.* The locations of key elements in the system are often shown on the site plan.

OTHER PIPING

Other utility piping, such as for water supply or gas, is shown on the site plan. If these lines will pass beneath the concrete footings or be concealed under a concrete slab, they must be roughed in before the concrete is placed. Water supply pipes follow the most direct route from the municipal water main or well to the main shut-off valve or pump. Gas lines run from the main to the gas meter. All supply lines on the plot plan should be labeled according to type and size of piping, Figure 11-5.

ELECTRICAL SERVICE

The electrical service is the wiring that brings electricity to the house. There are two types of electri-

cal service: overhead and underground (or buried). Overhead service involves a cable from the utility company transformer or pole to a weatherhead on the house, Figure 11-6. The *weatherhead* is a weathertight fitting on the *mast* (usually a pipe) which serves as a conduit to the meter receptacle. In an underground service the cable is buried, Figure 11-7. Although electrical service is a site utility, it is not usually installed until the building is enclosed.

The electrical service to residential and small commercial buildings is similar as previously explained and shown in Figures 11-6 and 11-7. This type of service has only one distribution point, the main electrical panel. Heavy commercial and industrial electrical services require multi-conductor feeders, and typically these are underground services from a utility transformer to a service entrance section in the designated electrical or utility room. Additional utility coordination will be required within the building, which will be discussed in Unit 14. These underground electrical utility line locations must be coordinated with the other site and building utilities prior to starting installation.

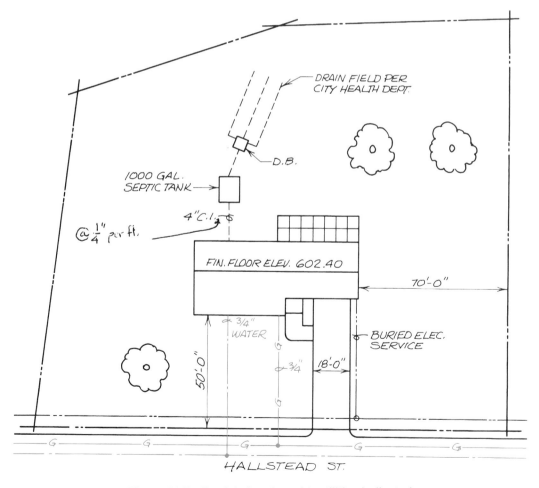

Figure 11-5   **Partial site plan with utilities indicated**

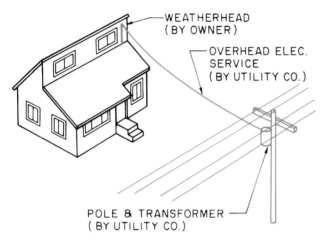

Figure 11-6   Overhead electrical service

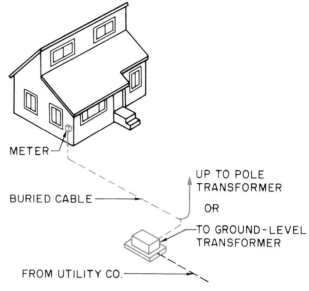

Figure 11-7   Underground electrical service

✓ CHECK YOUR PROGRESS

Can you perform these tasks?

☐ Identify electrical service, water, and sewage lines shown on a site plan.

☐ Explain the operation of a septic system.

☐ Tell at what elevation sewer lines are to pass through a foundation.

☐ Tell what the elevation should be at each end of a pitched sewer pipe.

ASSIGNMENT

Refer to the Lake House drawings (in the packet) to complete the assignment.

1.   What size is the sewer for the Lake House?

2.   How many lineal feet are required from the foundation wall to the septic tank?

3.   What is the rise of the sewer from the house to the septic tank?

4.   Where does the sewer pass through the foundation?

5.   How many lineal feet of perforated pipe are needed for the drain field?

6.   How many cubic yards of crushed stone are needed for the drain field?

7.   What is the location of the electrical service?

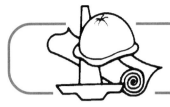

UNIT 12   Footings

After completing this unit, you will be able to perform the following tasks:

• Find all information on a set of drawings pertaining to footing design.

• Interpret drawings for stepped footings used to accommodate changes in elevation.

• Discuss applicable building codes pertaining to building design.

All soil can change shape under force. When the tremendous weight of a building is placed on soil, the soil tends to compress under the foundation walls and allow the building to settle. To prevent settling, concrete footings are used to spread the weight of the building over more area. The footings distribute the weight of the building, so that there is less force per square foot of area.

The simplest type of footing used in residential construction is referred to as *slab-on-grade.* In this system the main floor of the building is a single concrete slab, reinforced with steel to prevent cracking. This slab supports the weight of the building, Figure 12-1. Slab-on-grade foundations are common in warm climates. This type of construction is indicated on the floor plan by a note, Figure 12-2, and on section views of the construction, Figure 12-3. If excavation is involved in the construction where a slab-on-grade is to be placed, it is very important to thoroughly compact all loose fill before placing the concrete. Tamping the fill prevents the soil from compacting under the concrete later, causing the concrete to settle or crack.

A haunch is used to further strengthen the slab where concentrated weight, such as a wall, will be located. A *haunch* is an extra thick portion of the slab that

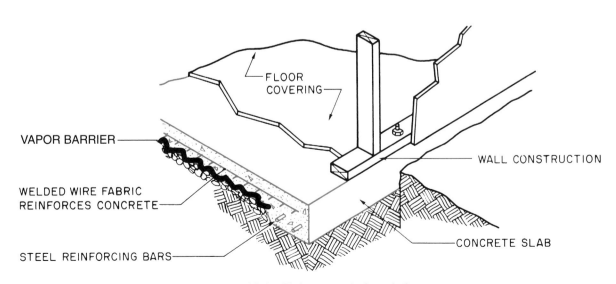

Figure 12-1   Slab-on-grade foundation

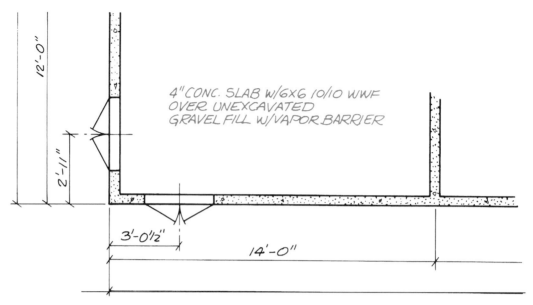

12'-0"

2'-11"

4" CONC. SLAB w/6x6 10/10 WWF
OVER UNEXCAVATED
GRAVEL FILL w/VAPOR BARRIER

3'-0½"

14'-0"

Figure 12-2   Note indicating slab-on-grade construction.

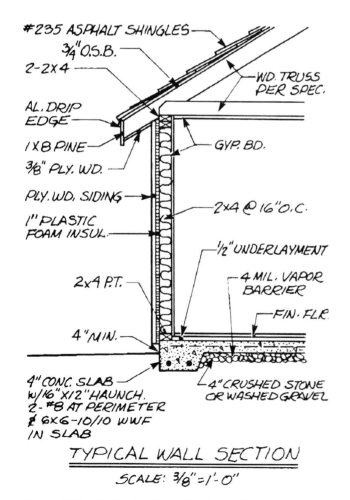

#235 ASPHALT SHINGLES
¾" O.S.B.
2-2x4
AL. DRIP EDGE
1x8 PINE
⅜" PLY. WD.
PLY. WD. SIDING
1" PLASTIC FOAM INSUL.
2x4 P.T.
4" MIN.
4" CONC. SLAB
w/16"x12" HAUNCH.
2-#8 AT PERIMETER
& 6x6-10/10 WWF
IN SLAB

WD. TRUSS PER SPEC.
GYP. BD.
2x4 @ 16" O.C.
½" UNDERLAYMENT
4 MIL. VAPOR BARRIER
FIN. FLR.
4" CRUSHED STONE OR WASHED GRAVEL

TYPICAL WALL SECTION
SCALE: ⅜"=1'-0"

Figure 12-3   Typical wall section for slab-on-grade construction.

is made by ditching the earth before the concrete is placed, Figure 12-4.

SPREAD FOOTINGS

In most sections of the country, the foundation of the house rests on a footing separate from the concrete floor. This separate concrete footing is called a *spread footing* because it spreads the force of the foundation wall over a wider area, Figure 12-5. Spread footings may be made by placing concrete inside wooden or metal forms, Figure 12-6, or placing the concrete in carefully measured ditches. In either case, the footing is shown on the foundation or basement plan by dotted lines outside the foundation wall lines, Figure 12-7.

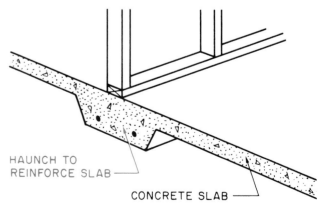

HAUNCH TO REINFORCE SLAB

CONCRETE SLAB

Figure 12-4   A *haunch* is a thickened part of a slab to reinforce it under a load-bearing wall.

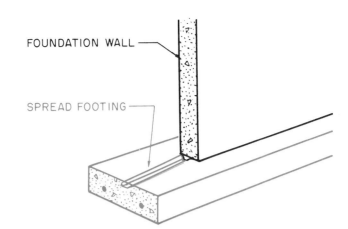

Figure 12-5 A spread footing is so named because it "spreads" the weight of the foundation wall over a wider soil area.

FOUNDATION WALL

SPREAD FOOTING

Figure 12-6 Footing forms

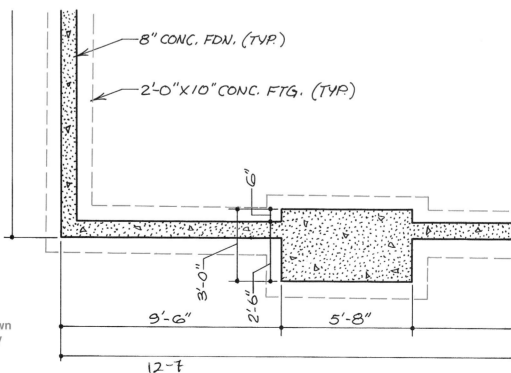

Figure 12-7 The footing lines are sometimes shown on the foundation plan by broken lines around the foundation.

8" CONC. FDN. (TYP.)

2'-0"X10" CONC. FTG. (TYP.)

6"

3'-0"

2'-6"

9'-6"

5'-8"

12-7

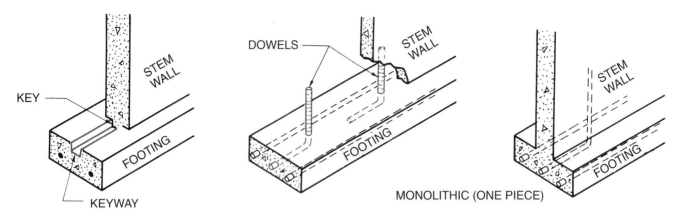

Figure 12-8   Anchoring the foundation wall to the footing

SECTION 1009.0 CONCRETE FOOTINGS

1009.1 Concrete strength: Concrete in footings shall have an ultimate compressive strength of not less than 2500 psi (1.76 kg/mm^2) at 28 days.

1009.2 Design: Concrete footings shall comply with Sections 1215.0 and 1216.0 and ACI 318 listed in Appendix A.

1009.3 Thickness: The thickness of concrete footings shall comply with Sections 1009.3.1 and 1009.3.2.

1009.3.1 Plain concrete: In plain concrete footings, the edge thickness shall be not less than 8 inches (203 mm) for footings on soil: except that for buildings of Use Group R-3 and buildings less than two stories in height of Type 4 construction, the edge thickness may be reduced to 6 inches (152 mm) provided the footing does not extend beyond 4 inches (102 mm) on either side of the supported wall.

1009.3.2 Reinforced concrete: In reinforced concrete footings the thickness at the edge above the bottom reinforcement shall be not less than 6 inches (152 mm) for footings on soil, nor less than 12 inches (305 mm) for footings on piles. The clear cover on reinforcement where the concrete is cast against the earth shall not be less than 3 inches (76 mm). Where concrete is exposed to soil after it has been cast, the clear cover shall be not less than 1 1/2 inches (38 mm) for reinforcement smaller than No. 5 bars of 5/8 inch (16 mm) diameter wire, nor 2 inches (51 mm) for larger reinforcement.

1009.4 Footings on piles and pile caps: Footings on piles and pile caps shall be of reinforced concrete. The soil immediately below the pile cap shall not be considered as carrying any vertical load. The top of all piles shall be embedded not less than 3 inches (76 mm) into pile caps and the caps shall extend at least 3 inches (76 mm) beyond the edge of all piles.

1009.5 Deposition: Concrete footings shall not be poured through water unless otherwise approved by the building official. When poured under or in the presence of water, the concrete shall be deposited by approved means which insure minimum segregation of the mix and negligible turbulence of the water.

1009.6 Protection of concrete: Concrete footings shall be protected from freezing during depositing and for a period of not less than 5 days thereafter. Water shall not be allowed to flow through the deposited concrete.

Figure 12-9   Building code section on concrete footings

The dimensions of the footings can be determined from the dimensions shown for the foundation. The foundation rests on the center of the footing unless otherwise specified. Therefore, if an 8-inch foundation rests on a 16-inch footing, the footing projects four inches beyond the foundation on each side. To lay out these footing lines, measure four inches from the building lines marked on the batter boards. Where footing lines cross to form a corner, suspend a plumb bob. Drive a stake under the plumb bob. Then, drive a nail in the stake to accurately mark the corner. When all corners are located in this manner, stretch a line between the nails to locate the inside of the footing forms.

A complete set of construction drawings also includes sections that show the spread footing in greater detail. However, these drawings are often superseded by local building codes. Building codes for footings include such things as minimum permissible depth of footing, required strength of concrete for footings, the width of footings, and the use of key or dowels made of reinforcing steel, Figure 12-8. The drawings for the Lake House are drawn to satisfy the building codes that are enforced in the community where it is to be built. If the drawings were done for a plan catalog where the locality is not known in advance, they might refer the builder to local building codes and omit many dimensions. Figure 12-9 shows an example of a building code section on footings.

When reading the foundation plan for footing dimensions, pay particular attention to special features like fireplaces and pilasters. (*Pilasters* are thickened sections of the foundation wall that add strength to the wall.) The footing in these areas will probably be wider and may be deeper than under straight sections of foundation wall. There are times when a footing or foundation wall must be crossed or penetrated by a utility line. Care must be taken when penetrating a footing or foundation wall. Prior approval is most often required. Items of concern include expansion and contraction between the utility line and footing or foundation wall, reduction in structural strength, and breaking the moisture barrier.

COLUMN PADS

Where steel columns, masonry piers, and wooden posts are used in the construction, a special concrete pad is indicated on the foundation plan, Figure 12-10. As with other footings, building codes for the area should be consulted for the design of these pads. In the absence of any applicable building code, the foundation plan normally includes enough information to construct the pads. These pads, as with all other footings, should rest on unexcavated or well-tamped earth.

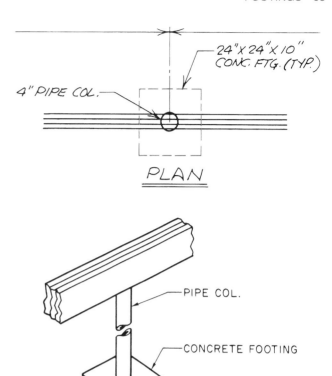

Figure 12-10   Column footings appear as a rectangle of broken lines on the plan.

REINFORCEMENT BARS	
Size Designation	Diameter in Inches
3	.375
4	.500
5	.625
6	.750
7	.875
8	1.000
9	1.128
10	1.270
11	1.410
14	1.693
18	2.257

Figure 12-11   Standard sizes of rebars

REINFORCEMENT

Footings, column pads, and other structural concrete frequently include steel reinforcement. Footing reinforcement is normally in the form of steel reinforcement bars, commonly called *rebars.* Reinforcement bars are designated by their diameters in eighths of an inch, Figure 12-11.

DEPTH OF FOOTINGS

In many sections of North America, the moisture in the surface of the earth freezes in the winter. As this frost forms, it causes the earth to expand. The force of this expansion is so great that if the earth under the footing of a building is allowed to freeze, it either cracks the footing or moves the building. To eliminate this problem, the footing is always placed below the depth of any possible freezing. This depth is called the *frostline,* Figure 12-12.

Two methods are commonly used to indicate the elevation, or depth, of the bottom of the footings. The easiest to interpret is with the elevation drawings. Where elevations are given in this manner, the top of the footing forms are leveled with a leveling instrument, using a bench mark for reference.

The most frequently used method is to dimension the bottom of the footing from a point of known elevation. This may be the finished floor or the top of the masonry foundation, for example. These dimensions are given on the building elevations. Footings and other features marked with a reference symbol (⊕ , called a *datum* symbol) are to be used as reference points for other dimensions.

For an example, see the South Elevation 3/3 of the Lake House (included in the packet). The left end of this view shows a footing with its bottom at an elevation of 334.83 feet. This is a variation of the usual practice of showing the elevation of the top of the footing. The top of this footing is 10 inches higher or 335.66 feet. (See Math Review 16). What room of the Lake House is this footing under?

At the right end of the South Elevation 3/3, the footing is shown to be 5'-4" below the finished floor and masonry. This is the basement floor, which the Site Plan

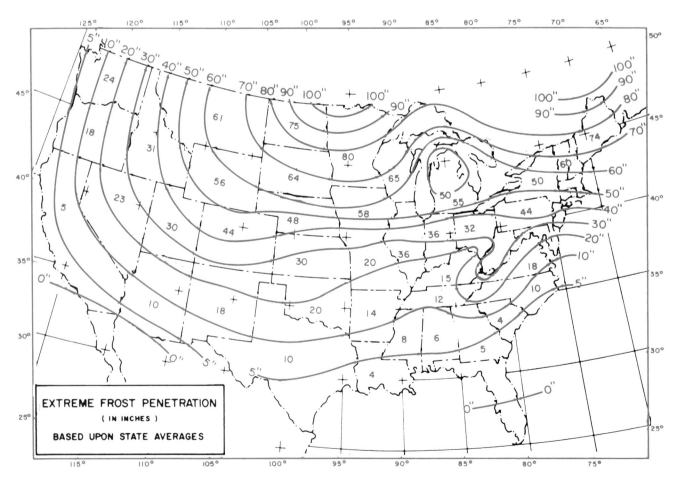

Figure 12-12   Average frost depths in the United States.

shows to be at 337.0 feet. Therefore, the top of this footing is at 331'-8" or 331.66'.

STEPPED FOOTINGS

On sloping building sites, it is necessary to change the depth of the footings to accommodate the slope. This is done by *stepping* the footings. When concrete blocks are to be used for the foundation walls, these steps are normally in increments of eight inches. This allows the concrete blocks to be laid so that the top of each footing step is even with a masonry course. Some buildings require several steps in the footing to accommodate steeply sloping sites. Steps in the footings are shown on the elevation drawings and on the foundation plan by a single line across the footing.

The Lake House has several steps in the footing. For example, see the east side of the garage. This is shown in the East Elevation 2/3. This step is also shown on the Foundation Plan. It is 8'-0" from the north end of the garage. Notice that the two levels of the footing overlap one another. These are built in one overlapping section as shown in Figure 12-13.

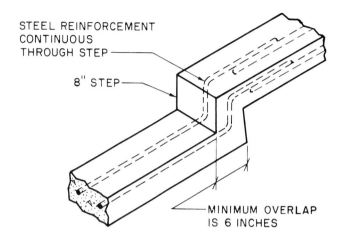

Figure 12-13   Stepped footing

✓ CHECK YOUR PROGRESS

Can you perform these tasks?

- ☐ List the dimensions of a slab-on-grade including footings, when this type of foundation is shown on construction drawings.

- ☐ List the thickness, width, and reinforcement to be used for spread footings.

- ☐ Identify steps in footings, and give the dimensions of each step.

- ☐ Give the locations and dimensions of footings for columns, posts, and other features.

- ☐ Give the dimensions and reinforcement to be used for any thickened haunches in a concrete slab-on-grade.

ASSIGNMENT

Refer to drawings of the Lake House drawings (included in the packet) to complete this assignment.

1. What is the typical width and depth of the concrete footings for the Lake House?

2. What is the total length and width (outside dimensions) of the concrete footings for the garage of the Lake House? (Remember to allow for the footings to project beyond the foundation wall.)

3. How many concrete pads are shown for footings under columns or piers in the Lake House?

4. What are the dimensions of these pads?

5. What reinforcement is indicated for these pads?

6. What reinforcement is indicated for the spread footings under the Lake House?

7. What is indicated by the 2-inch dimension between the 12-inch round concrete footings?

8. What is the elevation of the top of the footing under the garage door?

9. What are the elevations of the tops of each section of concrete footing shown on the East Elevation 2/3?

10. How far outside the foundation walls are the typical footings?

11. Refer to the building code in your community (or the model code section shown in Figure 12-9) and list the specific differences between the Lake House footings and the minimum code requirements.

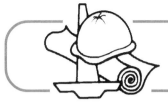

UNIT 13   Foundation Walls

OBJECTIVES

After completing this unit, you will be able to perform the following tasks:

- Determine the locations and dimensions of foundation walls indicated on a set of drawings.

- Describe special features indicated for the foundation on a set of drawings.

LAYING OUT THE FOUNDATION

When the concrete for the footings has hardened and the forms are removed, carpenters can begin erecting forms for concrete foundations or masons can begin laying blocks or bricks for masonry foundations. Although the material differs, the drawings and their interpretation for each type of foundation are similar.

In Unit 12 you referred to the dimensions on the foundation plan to lay out the footings. The same dimensions are used to lay out the foundation walls. The

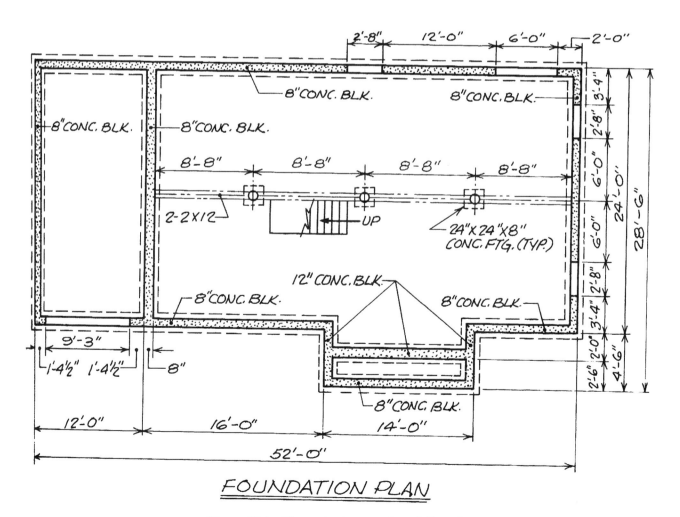

FOUNDATION PLAN

Figure 13-1   Dimensioning on a foundation plan

layout process is also similar. The outside surface of the foundation wall is laid out using previously constructed batter boards. Then the forms are erected or the masonry units are laid to these lines. The foundation plan includes overall dimensions, dimensions to interior corners and special constructions, and dimensions of special smaller features. It is customary to place the smallest dimensions closest to the drawing. The overall dimensions are placed around the outside of the drawing, Figure 13-1.

All drawing sets include, at least, a wall section showing how the foundation is built, how it is secured to the footings, and any special construction at the top of the foundation wall, Figure 13-2. Although a typical wall section may indicate the thickness of the foundation wall, you should carefully check around the entire wall on the foundation plan to find any notes that indicate varying thicknesses of the foundation wall. For example, the wall may be 12 inches thick where it has to support brick veneer above, while it is only 8 inches thick on the back of the building where there is no brick veneer. A careful check of the foundation plan for the Lake House shows that the house foundation calls for 12-inch, 10-inch, and 6-inch concrete block and that the garage foundation calls for 8-inch block.

The details for a masonry foundation may call for horizontal reinforcement in every second or third course. This is usually prefabricated wire reinforcement to be embedded in the mortar joints. Prefabricated wire reinforcement is available in varying sizes for different sizes of concrete blocks.

The height of the foundation wall is dimensioned on the building elevations. These are the same dimensions as those used to determine the depth of the footings in the preceding unit. Just as the footing was stepped to accommodate a sloping building site, the top of the foundation wall may be stepped to accommodate varying floor levels in the *superstructure* (construction above the foundation), Figure 13-3.

The top of a masonry foundation may be built with smaller concrete blocks to form a ledge upon which later brickwork will be built. It is also common practice to use one course of 4-inch solid block as the top course of the foundation wall.

In concrete foundations, anchor bolts, are usually placed in the top of the foundation, Figure 13-4. These bolts are left protruding out of the top of the foundation so that the wood superstructure can be fastened in place later. Anchor bolts are not normally shown on the foundation plan, but a note on the wall section indicates their center-to-center or on-center spacing. On masonry walls, anchor bolts can be placed in the hollow cores of the concrete blocks. They are held in place by filling the core with mortar grout. *Grout* is a portland cement mixture which has high strength.

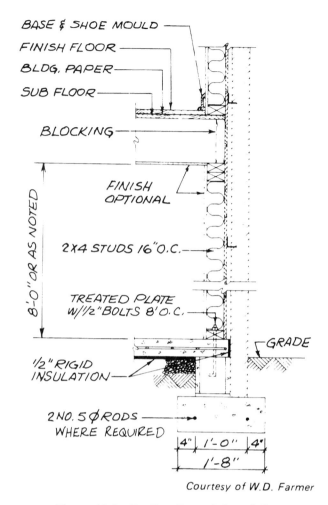

Courtesy of W.D. Farmer

Figure 13-2   Section through foundation

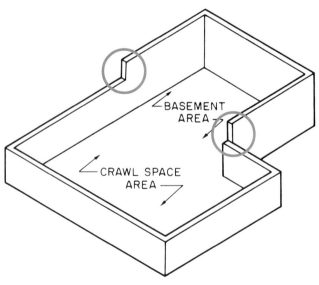

Figure 13-3   The top of the foundation may be stepped to allow for a partial basement or varying floor levels.

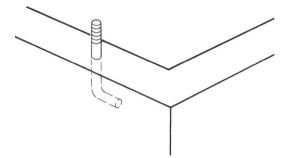

Figure 13-4 Anchor bolt

In areas where there is a threat of extremely high winds or earthquakes, additional hold-down straps may be called for, Figure 13-5. These hold-downs are normally only used with concrete foundations.

SPECIAL FEATURES

Many foundations include steel or wooden beams which act as girders to support the floor framing over long spans, Figure 13-6. When the girder is steel, it is indicated by a single line with a note specifying the size

Figure 13-5 Anchor bolts and hold-down strap are used in an earthquake zone.

Figure 13-6 The girder supports the floor framing.

Courtesy of Trus Joist Corporation

and type of structural steel. A wood girder is usually indicated by two or more lines and a note specifying the number of pieces of wood and their sizes in a built-up girder, Figure 13-7.

If the top of the girder is to be flush with the top of the foundation beam, pockets must be provided in the foundation, Figure 13-8. Beam pockets are usually shown on the details and sections of the construction drawings. The locations of these beam pockets are dimensioned on the foundation plans.

If windows are to be included in the foundation, the form carpenter or mason must provide rough openings of the proper size. The locations of windows should be dimensioned on the foundation plan. The sizes of the windows may be shown by a note or given on the window schedule. Window sizes and window schedules are discussed later in Unit 26. It is important, however, to get the masonry opening size from the window manufacturer before forming the opening in the foundation wall. The *masonry opening* is the size of the opening required in the foundation wall to accommodate the window. This size may be different from the nominal size given in a note on the foundation plan.

The foundation may include pilasters for extra support. A *pilaster* is a thickened section of the foundation which helps it to resist the pressure exerted by the earth on the outside. The location and size of pilasters is shown on the foundation plan.

The Lake House drawing includes a special feature not commonly found on foundation plans for houses.

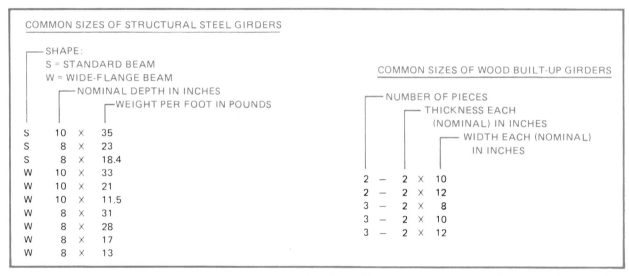

Figure 13-7   Typical specifications for structural steel and wood built-up girders—other types are described in the project specifications.

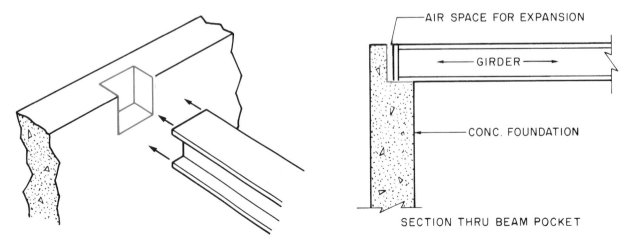

SECTION THRU BEAM POCKET

Figure 13-8   A beam pocket is a recess in the wall to hold the girder.

There are four notes which read 3½" ⧄ STD. WT. STL. COL. W/8 × 8 × ½ B.PL. These notes indicate a 3½-inch square, standard-weight, steel column with 8-inch-by-8-inch-by-½-inch thick base plates. This structural steel work is explained in more detail later, but to completely understand the foundation plan, it is necessary to know that the steel will be erected. The stress that this steel work will place on the foundation is the reason for another note on the foundation plan in the area near the 18-inch round concrete footings. That note says FILL CORE SOLID W/ GROUT 1:1:6 = PORTLAND:MAS. CEMENT: AGGREGATE. The cores in the concrete block foundation where the steelwork will be erected are filled with this grout to provide the extra strength necessary to support the steel.

PERMANENT WOOD FOUNDATION

Foundations are usually constructed of concrete or concrete block. However, a type of specially treated wood foundation is sometimes used. Figure 13-9. These *permanent wood foundations* do not use concrete footings.

Instead, they are built on 2×8s or 2×10s laid on gravel fill below the frostline. The foundation walls are framed with lumber that has been pressure treated to make it rot resistant and insect resistant. The framing is covered with a plywood skin and the plywood is covered with polyethylene film for complete moisture proofing, Figure 13-10. All of the nails in a wood foundation are stainless steel, to prevent rusting.

Courtesy of National Forest Products Association

Figure 13-9   Permanent wood foundation

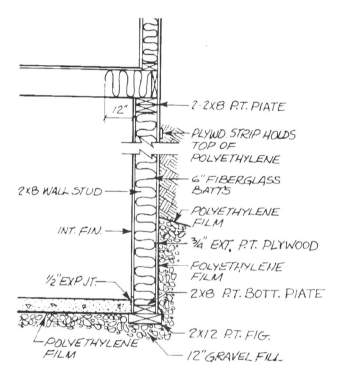

Figure 13-10   Section of wood foundation

✓ CHECK YOUR PROGRESS

Can you perform these tasks?

☐ Give the length, width, and any offsets in a foundation wall.

☐ Name the material to be used for the foundation walls, including any reinforcement.

☐ Describe the locations and dimensions of all window openings, door openings, beam pockets, and other openings in the foundation walls.

☐ Tell what the elevation is at any point on the foundation wall.

☐ Describe anchor bolts and other tie-downs to be embedded in the foundation.

ASSIGNMENT ——————————————————————————————

Refer to the Lake House drawings (in the packet) to complete the assignment.

1.   What is the typical thickness of the concrete block foundation for the Lake House?

2.   How many lineal feet of block wall are included in the foundations of the Lake House and garage?

3.   How thick is the south foundation of the fireplace?

4.   What is the elevation of the top of the north end of the east foundation wall of the Lake House?

5.   What is the highest elevation on the Lake House foundation?

6.   How many courses of block are required at the highest elevation of the Lake House foundation?

7.   What size anchor bolts are indicated at the top of the Lake House foundation?

8.   What spacing is indicated for the anchor bolts.

9.   How close (in courses) is horizontal reinforcement to be placed in the Lake House foundation?

10.   What is the elevation of the top of the concrete block wall at the southwest 3^1/$_2$" ▱ steel column?

11.   In how many places are the concrete block walls to be filled with grout?

UNIT 14 Drainage, Insulation, and Concrete Slabs

OBJECTIVES

After completing this unit, you will be able to perform the following tasks:

- Locate and explain information for control of ground water as shown on a set of drawings.

- Locate and describe subsurface insulation.

- Determine the dimensions of concrete slabs and the reinforcement to be used in concrete slabs.

DRAINAGE

After the foundation walls are erected and before the excavation outside the walls is *backfilled* (filled with earth to the finished gradeline), footing drains, if indicated, must be installed. *Footing drains* are usually perforated plastic pipe placed around the footings in a bed of crushed stone, Figure 14-1. If the site has a natural slope, the footing drains can be run around the foundation wall to the lowest point, then away from the building to drain by gravity. In areas where there is no natural drainage, the drain is run to a dry well or municipal storm drain.

At one time, clay drain tile was the most common type of pipe for this purpose. However, perforated plastic pipe is used in most new construction. Plastic drain pipe is manufactured in 10-foot lengths of rigid pipe and in 250-foot rolls of flexible pipe, Figure 14-2. An assortment of plastic fittings is available for joining rigid plastic pipe. When footing drains are to be included, they are shown on a wall section or footing detail, Figure 14-3. A note on the drawing indicates the size and material of the pipe.

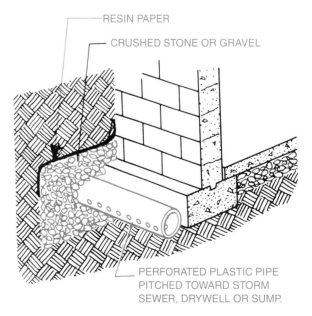

RESIN PAPER

CRUSHED STONE OR GRAVEL

PERFORATED PLASTIC PIPE PITCHED TOWARD STORM SEWER, DRYWELL OR SUMP.

Figure 14-1 Footing drain

Figure 14-2   Plastic drain pipe

If the floor drains are to be included in concrete slab floors, they are indicated by a symbol on the appropriate floor plan, Figure 14-4. If these floor drains run under a concrete footing, the piping had to be installed before the footing was placed (See Unit 11). The *riser* (vertical part through the floor) and drain basin are usually set at the proper elevation just prior to placing the concrete floor.

The floor plan may include a spot elevation for the finished drain, or it may be necessary to calculate it from information given for the pitch of the concrete slab. Pitch of concrete slabs is discussed later in this unit.

VAPOR BARRIERS

Another technique often used to prevent ground-water from seeping through the foundation is coating the foundation wall. In residential construction, the most common foundation moisture-control system is a combination of parging and coating with asphalt foundation coating. *Parging* consists of plastering the outside of the masonry wall below the grade line with a portland cement and lime plaster. The parging is them covered with asphalt-based foundation coating, Figure 14-5.

At this point, subsurface work outside the foundation wall is completed, but backfilling should not be done until the superstructure is framed. The weight and rigidity of the floor on the foundation wall helps the wall resist the

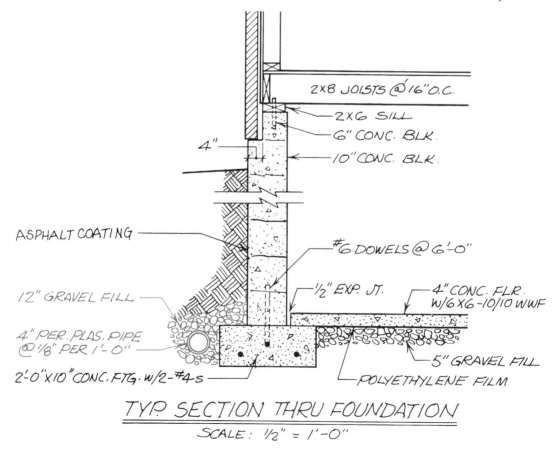

Figure 14-3   Footing drains are shown outside the footing.

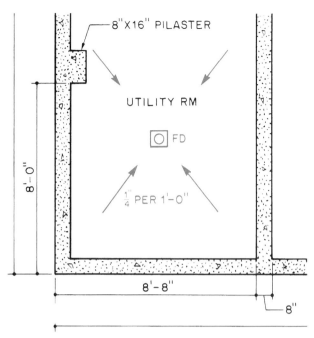

Figure 14-4 The floor drain is shown by a symbol. (Notice that the floor is pitched toward the drain.)

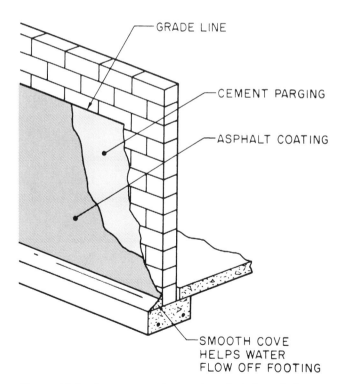

Figure 14-5 Cement parging and asphalt foundation coating are used for moisture proofing.

pressure of the backfill. If the backfilling must be done before the framing, the foundation walls should be braced. To retard the flow of moisture from the earth through the concrete-slab floor, the drawings may call for a layer of gravel over the entire area before the concrete is placed. A polyethylene vapor barrier is laid over the gravel underfill. The thickness of polyethylene sheeting is measured in mils. One mil equals $1/1000$ of an inch. For vapor barriers, 4-mil or 6-mil polyethylene is generally used.

INSULATION

In cold climates, it is desirable to insulate the foundation and concrete slab. This insulation is usually rigid plastic foam boards placed against the foundation wall or laid over the gravel underfill, Figure 14-6.

Like all materials, concrete and masonry expand and contract slightly with changes in temperature. To allow for this slight expansion and contraction, the joint between the concrete slab and foundation wall should include a compressible expansion joint material. Expansion joints can be made from any compressible material such as neoprene or composition sheathing material. This expansion joint filler is as wide as the slab is thick and is simply placed against the foundation wall before the concrete is placed.

CONCRETE SLABS

When the house has a basement, the floor is a concrete slab-on-grade. The areas to be covered with concrete are indicated on the foundation plan or basement floor plan. This is usually done by an area not giving the thickness of the concrete slab and any reinforcing steel to be used. To help the concrete resist minor stresses, it is usually reinforced with welded wire fabric. The specifications for welded wire fabric are explained in Figure 14-7. Where the slab must support bearing walls or masonry partitions, it may be haunched as discussed in Unit 12.

When floor drains are included or where water must be allowed to run off, the slab is *pitched* (sloped slightly). A note on drawings indicates the amount of pitch. One-quarter inch per foot is common. When there is any possibility of confusion about which way the slab is to be pitched, bold arrows are drawn to show the direction the water will run, Figure 14-8.

When floor drains or forms are set for pitched floors, it is necessary to find the total pitch of the slab. This is done by multiplying the pitch-per-foot by the number of feet over which the slab is pitched. (See Math Review 8.) For example, if the note on a concrete apron in front of a garage door indicates a pitch of $1/2$-inch per foot and the apron is 4 feet wide, the total pitch is 2 inches. The proper elevation for the form at the

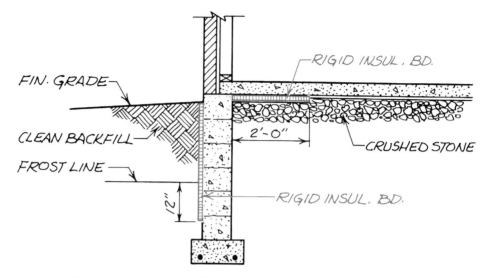

Figure 14-6   Rigid plastic-foam insulation may be laid under the perimeter of the floor or against the foundation wall.

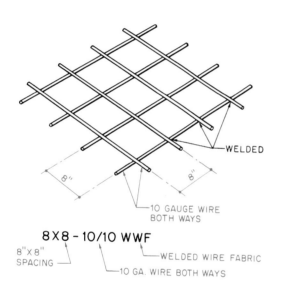

Figure 14-7   The callout for welded wire fabric explains the size and spacing of the wires.

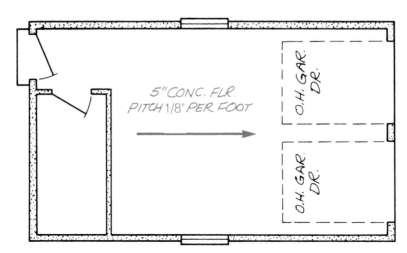

Figure 14-8   A bold arrow indicates the way that water will run off a pitched surface.

outer edge of the apron is 2 inches less than the finished floor elevation.

Commercial and industrial installations may require electrical raceways to be installed in or just below the concrete slab in the crushed or gravel fill. These installations must be made so that they do not reduce the structural integrity of the concrete slab. An oversized raceway in the concrete slab may cause the slab to crack and settle unevenly. An electrician should be present during concrete placement to observe and correct any damage to these electrical raceways.

All underground utility systems must be coordinated prior to starting installation especially in commercial and industrial buildings as previously discussed in Unit 11. The service entrance section may have up to six sub-distribution sections or panels located throughout the building. Electrical feeders are required from the electrical service entrance section to these subdistribution sections or panels. This requires detailed coordination between the electrical installer and the other utility installers prior to starting the installation of the utility systems. The electrical instal-

lation must be laid out around the other utilities with the other utility layouts normally having priority. The other utilities to be coordinated may include plumbing, fire sprinkler, heating and air conditioning, and specialty systems. This coordination must be done for both the underground and above ground systems.

Some of the concrete work in the Lake House is of particular interest, because it is part of the passive solar heating system the Lake House uses. Section 1/4 and the lower level floor plan 1/2 indicate that the area under the living room and dining room floors is a hear sink. A *heat sink* is a mass of dense material which absorbs the energy of the sun during the day and radiates it at night. The living room and dining room are on the south side of the Lake House. In the winter, when the leaves are off the deciduous trees, the sun shines in the large areas of glass in these rooms and warms the heat sink. At night, this heat is radiated into the house to provide additional heat when it is needed most. The floor over the heat sink is a concrete slab similar to that used in the playroom. Detail 2/5 helps explain this concrete slab.

✓ CHECK YOUR PROGRESS

Can you perform these tasks?

- ☐ Describe the footing drains shown on a set of construction drawings.
- ☐ Name the material and its thickness when foundation insulation is shown.
- ☐ Describe any vapor barriers or foundation coating to be applied to the foundation or under concrete slabs.
- ☐ List the dimensions of any concrete slab floors shown on the drawings.
- ☐ Describe the reinforcement to be used in concrete slabs.
- ☐ Explain the pitch of a concrete slab to provide for drainage.

ASSIGNMENT

Refer to the Lake House drawings (in the packet) to complete this assignment.

1. What is the thickness of the concrete slab over the heat sink in the Lake House? What is the largest raceway (conduit) that could be installed in this slab?

2. Describe the reinforcement used in the concrete slab in the playroom of the Lake House.

3. How many square feet of two-inch rigid insulation are needed for the Lake House heat sink?

4. What prevents moisture from seeping through the concrete slab floor in the Lake House?

5. What is the finished floor elevation of the Lake House garage?

6. What is the elevation of the floor drain in the utility room of the Lake House?

7. What is the purpose of the eight-inch-thick concrete haunch in the middle of the Lake House slab?

8. How many cubic yards of concrete are required for the garage floor? The basement floor?

UNIT 15 Framing Systems

OBJECTIVES

After completing this unit, you will be able to identify each of the following types of framing on construction drawings:

- Platform
- Balloon
- Post-and-beam
- Energy-saving

PLATFORM FRAMING

Platform framing, also called *western framing,* is the type of framing used in most houses built in the last 30 years, Figure 15-1. It is called platform framing because as the rough floor is built at each level, it forms a platform on which to work while erecting the next level, Figure 15-2.

A characteristic of platform framing is that all wall *studs,* the main framing members in walls, extend only the height of one story. Interior walls, called partitions, are the same as exterior walls. The bottoms of the studs are held in position by a *bottom* (or *sole*) plate. The tops

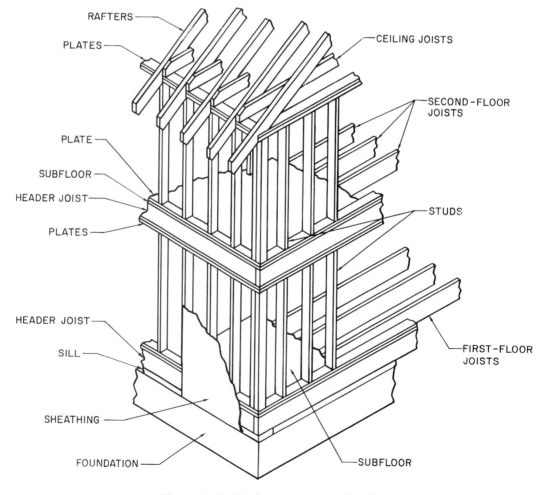

RAFTERS

PLATES

CEILING JOISTS

SECOND-FLOOR JOISTS

PLATE

SUBFLOOR

HEADER JOIST

PLATES

STUDS

HEADER JOIST

SILL

FIRST-FLOOR JOISTS

SHEATHING

FOUNDATION

SUBFLOOR

Figure 15-1   Platform or western framing

Figure 15-2   Platform or western framing provides a convenient work surface during construction.

of the studs are held in position by a *top plate.* Usually, a *double top plate* is overlapped at the corners to tie intersecting walls and partitions together, Figure 15-3. In some construction, the second top plate is not used. Instead, metal framing clips are used to tie intersecting walls together. Upper floors rest on the top plate of the walls beneath. The framing members of the upper floors or roof are positioned over the studs of the wall that supports them.

Platform construction can be recognized on wall sections, Figure 15-4. Notice that the studs extend only from one floor to the next.

BALLOON FRAMING

Although it is not widely used now, balloon framing was once the most common framing system. In *balloon framing,* the exterior wall studs are continuous from the foundation to the top of the wall, Figure 15-5. Floor framing at intermediate levels is supported by *let-in ribbon boards.* This is a board that fits into a notch in each joist and forms a support for the joists.

In both platform-frame and balloon-frame construction, the structural frame of the walls is covered with sheathing. Sheathing encloses the structure and, if a structural grade is used, prevents wracking of the wall. *Wracking* is the tendency of all of the studs to

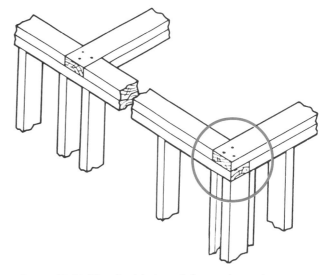
Figure 15-3   The double top plate overlaps at corners.

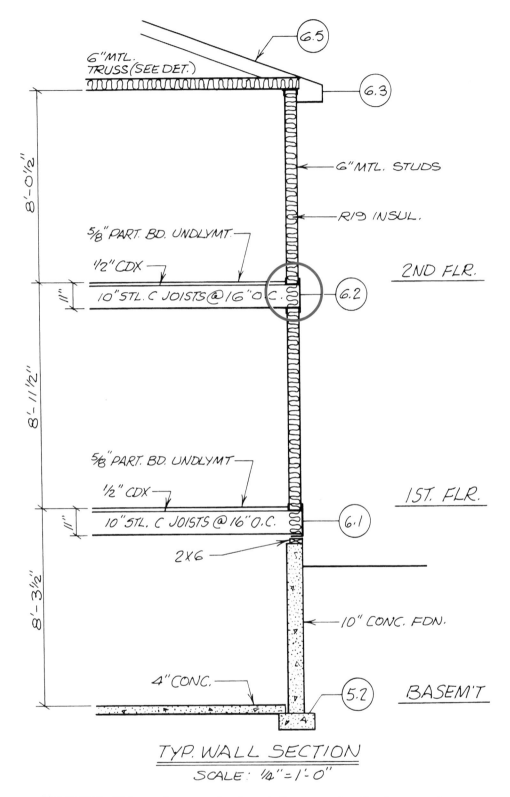

6.5

6"MTL.
TRUSS (SEE DET.)

6.3

6"MTL. STUDS

R19 INSUL.

8'-0½"

5/8" PART. BD. UNDLYMT.

½" CDX

2ND FLR.

10" STL. C JOISTS @ 16" O.C.

6.2

11"

8'-11½"

5/8" PART. BD. UNDLYMT.

½" CDX

1ST. FLR.

10" STL. C JOISTS @ 16" O.C.

6.1

2X6

10" CONC. FDN.

8'-3½"

11"

4" CONC.

BASEM'T

5.2

TYP. WALL SECTION
SCALE: ¼"=1'-0"

Figure 15-4   This can be recognized as platform construction because the studs extend only from one floor to the next.

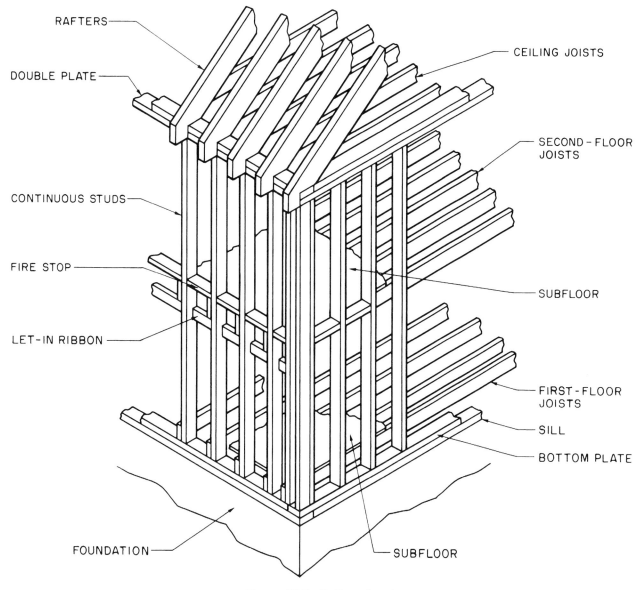

RAFTERS

DOUBLE PLATE

CONTINUOUS STUDS

FIRE STOP

LET-IN RIBBON

FOUNDATION

CEILING JOISTS

SECOND-FLOOR JOISTS

SUBFLOOR

FIRST-FLOOR JOISTS

SILL

BOTTOM PLATE

SUBFLOOR

Figure 15-5   Balloon framing

move, as in a parallelogram, allowing the wall to collapse to the side, Figure 15-6. There are two ways to prevent wracking. Plywood or other structural sheathing at the corners of the building prevents this movement. Also, diagonal braces can be attached to the wall framing at the corners to prevent wracking, Figure 15-7.

Seismic codes (building codes for earthquake protection) often call for interior shear walls. These are framed walls covered with structural sheathing, Figure 15-8. Interior shear walls make the building stronger.

POST-AND-BEAM FRAMING

Platform framing and balloon framing are characterized by closely spaced, lightweight framing members,

Figure 15-9. *Post-and-beam* framing uses heavier framing members spaced further apart, Figure 15-10. These heavy timbers are joined or fastened with special hardware, Figure 15-11. Because post-and-beam framing uses fewer pieces of material, it can be erected more quickly. Also, although the framing members are large (ranging from 3 inches by 6 inches to 5 inches by 8 inches), their wider spacing results in a savings of material. However, to span this wider spacing, floor and roof decking must be heavier. Post-and-beam framing is sometimes left exposed to create special architectural effects, Figure 15-12.

The structural core of the Lake House uses posts and beams, detail 6/6. However, this is not purely post-and-beam construction. The posts are $3^{1}/_{2}$-inch square steel tubing and the beams are plywood box beams.

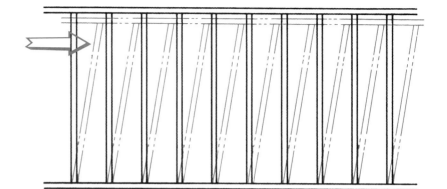

Figure 15-6   Wracking

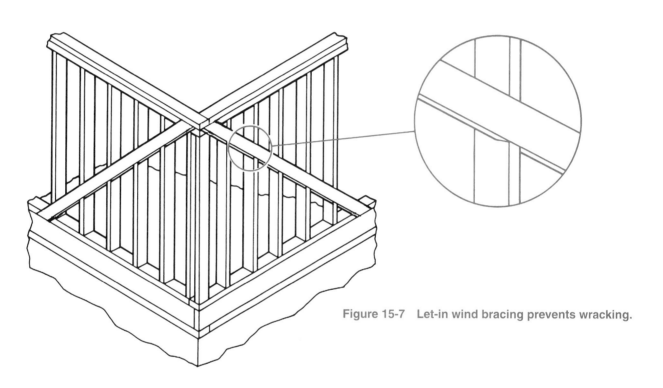

Figure 15-7   Let-in wind bracing prevents wracking.

Figure 15-8   Interior shear walls
may be used in earthquake zones.

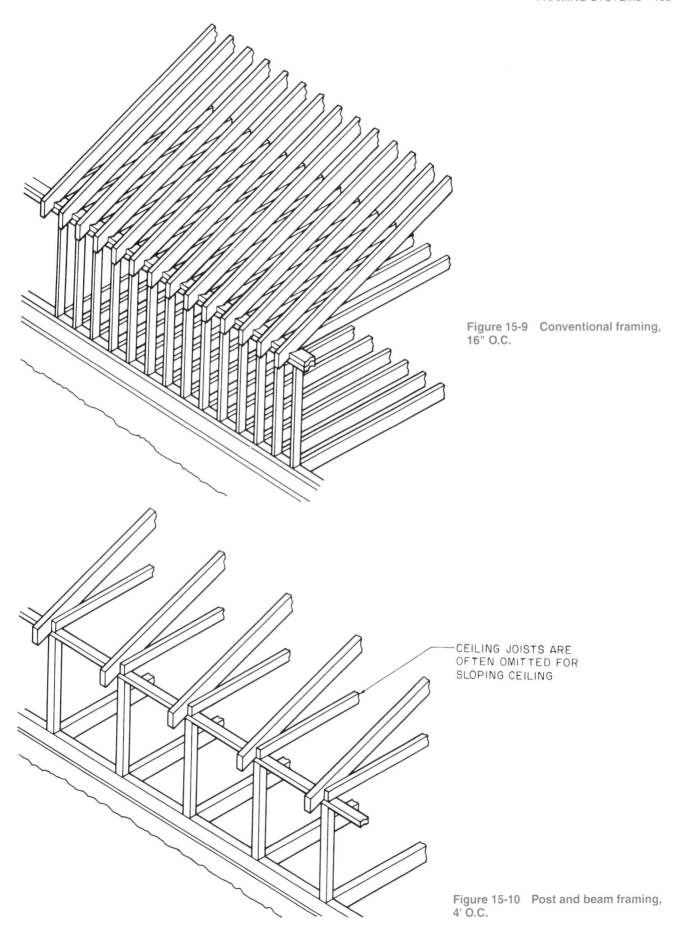

Figure 15-9   Conventional framing, 16" O.C.

CEILING JOISTS ARE OFTEN OMITTED FOR SLOPING CEILING

Figure 15-10   Post and beam framing, 4' O.C.

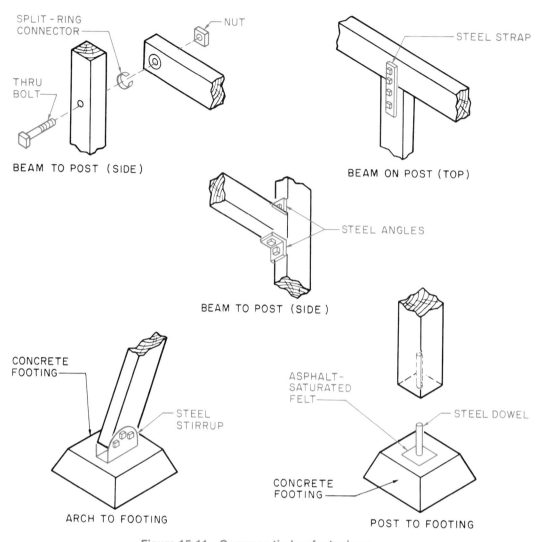

Figure 15-11   Common timber fastenings

Figure 15-12   The framing in this cottage is exposed to create a special architectural effect.

These are properly called *beams* because they carry a load without continuous support from below. These beams are supported only in the beam pockets on the steel posts, Figure 15-13. In the Lake House the post-and-beam construction does not include exterior walls. In pure post-and-beam construction, the exterior walls have widely spaced posts and the space between is filled in with nonload-bearing curtain walls. These curtain walls are merely panels that fill in the space between the structural elements—the posts and beams.

In areas where hurricanes and high tides are a threat, some houses are built as pole structures. Pole buildings are a variation of post-and-beam construction. Poles, which are treated to be insect and rot resistant, are set several feet in the ground and 8 to 12 feet apart. A *band joist,* or *header,* is then bolted to these poles, Figure 15-14. The floors, walls, and roof are framed within the pole structure, Figure 15-15. Pole buildings are strong; they resist severe winds. Pole construction allows buildings to be kept above damaging flood waters.

METAL FRAMING

Steel and aluminum metal framing is usually specified in commercial applications. Metal framing in residential applications is increasing as the price of wood framing increases and fire insurance decreases on metal-framed structures. Metal framing members are manufactured in the same standard lengths as wood framing members. Metal framing can be pre-assembled just as you would pre-assemble wood framing. The metal framing is fastened to the floor with powder actuated fasteners. The wallboard material is fastened to

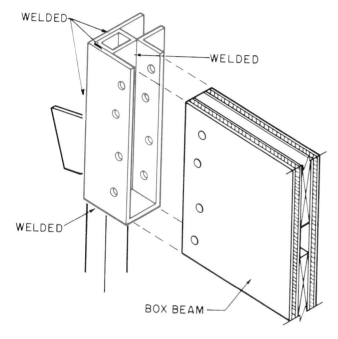

Figure 15-13   Beam pocket for Lake House

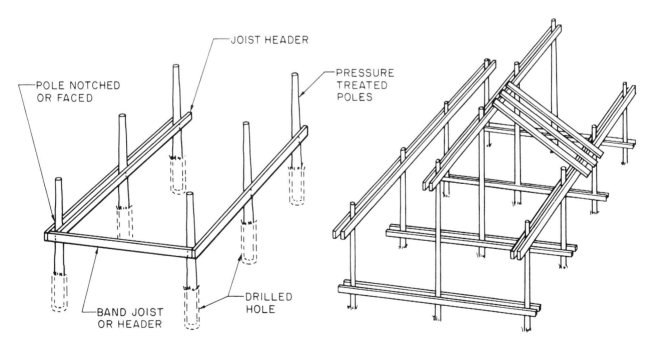

EMBEDMENT AND ALIGNMENT OF POLES. DEPTH OF EMBEDMENT DEPENDS ON SPACING AND SIZE OF POLES, WIND LOADS, AND SO FORTH, AND MAY VARY FROM 5 TO 8 FEET.

Courtesy of American Wood Preservers Institute

Figure 15-14   Basic elements of a pole building

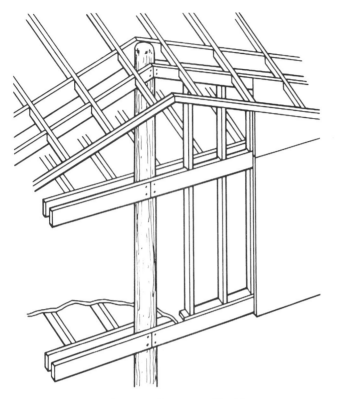

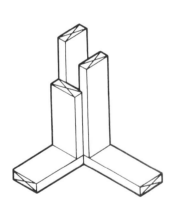

Courtesy of American Wood Preservers Institute

Figure 15-15   Framing in a pole building

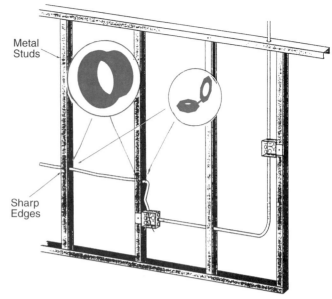

Figure 15-16   Nonmetallic-sheathed cable installed through metal framing

Metal Studs

Sharp Edges

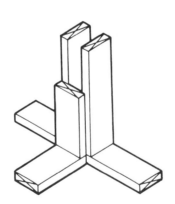

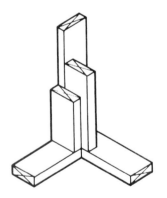

Figure 15-17   Conventional corner posts for 2 × 4 framing

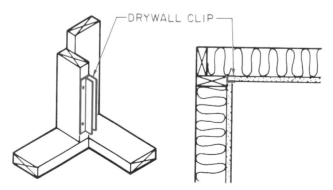

DRYWALL CLIP

Figure 15-18   Corner construction for Arkansas Energy Saving System

the metal studs using an electric screw gun with sheet metal screws. Sub-flooring material is fastened to the metal floor joists similarly with power-driven screws or nails. The construction drawings and specifications will indicate the type of metal framing to be used. Metal framing comes in various gages (metal thickness). The gage must be taken into consideration before using certain tools (stud punches, power actuated tools, etc.) and special mounting devices (box hangers, conduit brackets, etc.).

Where electrical nonmetallic-sheathed cable is installed through metal framing, a bushing or grommet must be installed prior to installing the cable (Figure 15-16).

ENERGY-SAVING TECHNIQUES

When wall framing is done with 2 × 4s spaced 16 inches on centers, up to 25 percent of the wall is solid wood. Wood conducts heat out of the building. Only the space between the solid wood framing can be filled with insulation. By using 2 × 6 studs spaced 24 inches on centers, the area of solid wood is reduced to less than 20 percent of the wall. The amount of framing material is the same. Not only does this reduce the amount of wood exposed to the surface of the wall, but it also allows for 2 inches more insulation.

The area of exposed wood is further reduced by special corner construction. In conventional framing, three pieces are used to frame the corner of a wall, Figure 15-17. To reduce heat loss through the wall only two pieces are used, Figure 15-18. The third piece, which normally provides a nailing surface for the interior drywall, is replaced by metal clips.

Examination of the first floor plan and the detail drawings of the Lake House shows that the walls are framed for maximum efficiency. Several details indicate that the studs are 2 × 6s © 24 O.C. This allows room for more insulation in the wall. Also notice that the house is sheathed with 3/4-inch insulation sheathing. Interior partitions do not need insulation, so they are framed with 2 × 4s.

When installing nonmetallic-sheathing cable through wood wall studs and floor joists, Unit 17, the cable must be installed 1¼ inches from the nearest edge. Where this 1¼ inches cannot be met, a 1/16-inch thick metal plate must be installed to protect the cable from nails and screws, Figure 15-19.

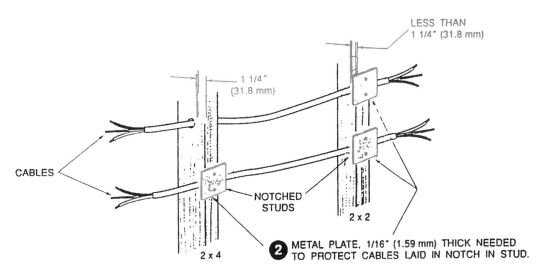

Figure 15-19   Nonmetallic-sheathed cable installed in wood framing

✓ CHECK YOUR PROGRESS

Can you perform these tasks?

☐ Identify platform framing shown on drawings and describe how it differs from other types of framing.

☐ Identify balloon framing shown on drawings and describe how it differs from other types of framing.

☐ Identify post-and-beam framing shown on drawings and describe how it differs from other types.

☐ Identify the following framing members on a wall or building section: joists, bottom or sole plate, top plate, stud, ribbon (to support joists), posts, beams, and sheathing.

☐ Explain what prevents a building frame from wracking.

☐ Describe at least three techniques for reducing the amount of heat lost through solid wood in a building frame.

ASSIGNMENT

1. Identify *a* through *h* in Figure 15-20.

2. What kind of framing is shown in Figure 15-18?

3. Identify *a* through *c* in Figure 15-21.

4. What kind of framing is shown in Figure 15-21?

5. Identify *a* through *c* in Figure 15-22.

6. Sketch a plan view of a conventional corner detail. Include drywall and sheathing.

7. Sketch a plan view of an energy-efficient corner detail. Include drywall and sheathing.

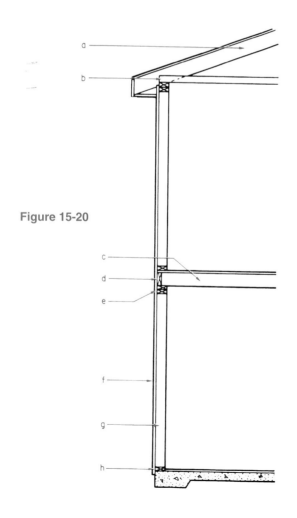

Figure 15-20

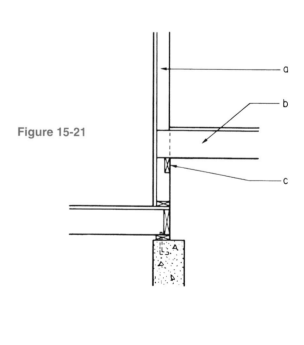

Figure 15-21

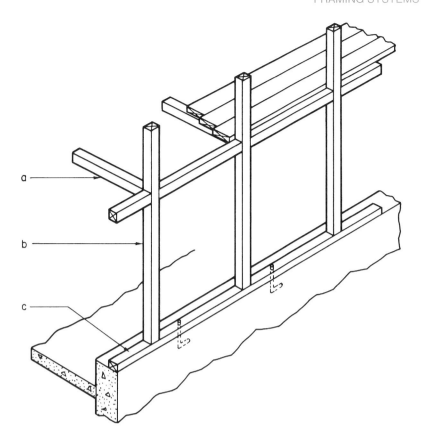

Figure 15-22

8. What two materials are used for framing?

9. What framing material requires bushings or grommets for nonmetallic-sheathed cable installations?

Refer to Lake House drawings (in the packet) to complete the rest of the assignment.

10. Which of the types of framing discussed in this topic is used for the Lake House?

11. What supports the west ends of the floor joists in bedroom #1?

12. What supports the east ends of the kitchen rafters?

13. How are the box beams fastened to the $3^1/_2$-inch square posts?

14. How are the steel posts anchored?

15. What does the north-west square steel post rest on?

16. What supports the north edge of the living room floor?

17. List the dimensions (thickness $\times$ depth $\times$ length) of all box beams. (Note: The length can be found by subtracting the outside dimension of the posts from the centerline spacing shown on the plan views.)

18. List the length of each piece of $3^1/_2$-inch steel.

UNIT 16   Columns, Piers, and Girders

OBJECTIVES

After completing this unit, you will be able to perform the following tasks:

- Locate columns and piers and describe each from drawings.

- Locate and describe the girders which support floor framing.

- Determine the lengths of columns and the heights of piers.

The most common system of floor framing in light construction involves the use of joists and girders. *Joists* are parallel beams used in the floor framing, Figure 16-1. Usually buildings are too wide for continuous joists to span the full width. In this case, the joists are supported by one or more *girders* (beams) running the length of the building. The girder is supported at regular intervals by wood or metal posts or by masonry or concrete piers.

COLUMNS AND PIERS

Metal posts called *pipe columns* are the most common supports for girders. However, masonry or concrete piers may be specified. The locations of columns, posts, or piers are given by dimensions to their centerlines. When metal or wooden posts are indicated, the only description may be a note on the foundation plan, Figure 16-2. This note may give the size and material of the posts only, or it may also specify the kind of bearing plates to be used at the top and bottom of the post, Figure 16-3. A *bearing plate* is a steel plate

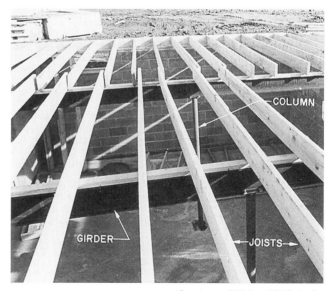

Courtesy of Richard T. Kreh, Sr.

Figure 16-1   Joist-and-girder floor framing

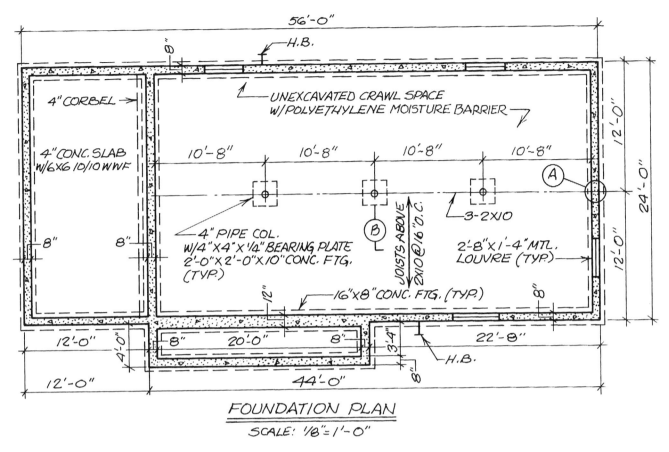

Figure 16-2 Foundation plan with note for column and footing

that provides a flat surface at the top or bottom of the column or post. If the girder is supported by masonry or concrete piers, a special detail may be included to give dimensions and reinforcement details.

GIRDERS

Before the total length of the columns or height of the piers can be calculated, it is necessary to determine the size of the girder and its relationship to the floor joist. There are three types of girders commonly used in residential construction. Steel beams are often used where strength is a critical factor. Built-up wood girders are constructed on the site. These consist of two, three, or more pieces of 2-inch lumber nailed together with staggered joints to form larger beams. The sizes and specifications for built-up wood girders and structural steel girders are discussed in Unit 15. Laminated veneer lumber (LVL), Figure 16-4, is becoming increasingly popular for use as girders. LVL is an engineered lumber product made by gluing layers of veneer together. These beams can be manufactured in almost

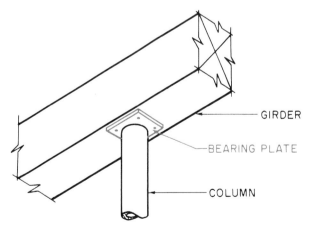

Figure 16-3 Bearing plate

Figure 16-4   LVL laminated beams are popular, because they are strong and can be manufactured in any length and depth. (Trus Joist Macmillan, A limited Partnership, Boise, ID)

any size and they are very strong, because lumber defects can be eliminated and all of the grain runs in the best possible direction. The notes commonly found on a foundation plan to indicate the type of girder to be used are shown in Figure 16-1.

DETERMINING THE HEIGHTS OF COLUMNS AND PIERS

The length of the columns or height of the piers depends on how the joists will be attached to the girder. The floor joists may rest directly on top of the girder or may be butted against the girder so that the top surface of the floor joist is flush with the top surface of the girder, Figure 16-5.

To find the height of the columns or piers, first determine the dimension from the basement floor to the finished first floor. Then subtract from this dimension the depth of the first floor including all of the framing and

the girder. Then add the distance from the top of the basement floor to the bottom of the column. The result equals the height of the column or pier. (See Math Reviews 5 and 6.) For example, the following shows the calculation of the height of the steel column in Figure 16-6:

- Dimension from finished basement floor to finished first floor = 8'-10$\frac{1}{2}$"

  Allowance for finished floor = 1"

  Nominal 2 × 8 joists = 7$\frac{1}{4}$"

  2 × 4 bearing surface on girder = 1$\frac{1}{2}$"

  W8 × 31 = 8"

  Total floor framing = 17$\frac{3}{4}$" or 1'-5$\frac{3}{4}$"

- 8'-10$\frac{1}{2}$" minus 1'-5$\frac{3}{4}$" = 7'-4$\frac{3}{4}$"

  Add thickness of concrete slab

- 7'-4$\frac{3}{4}$" plus 4" = 7'-8$\frac{3}{4}$"

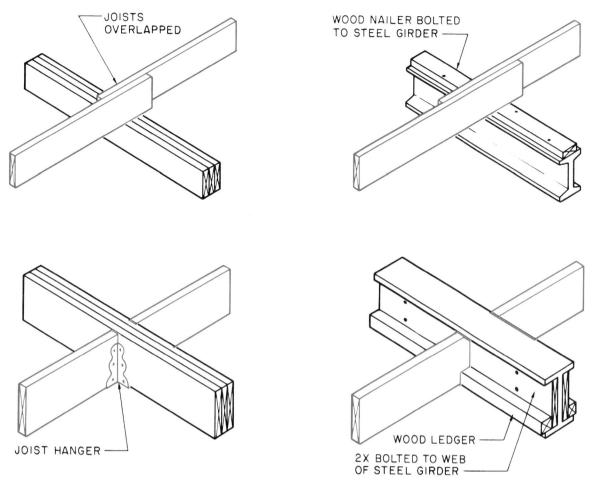

Figure 16-5   Several methods of attaching joists to girders

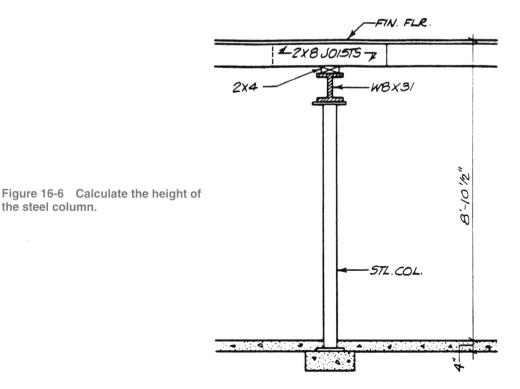

Figure 16-6   Calculate the height of the steel column.

✓ CHECK YOUR PROGRESS

Can you perform these tasks?

- ☐ Give the dimensions to locate columns or piers to support floor framing.
- ☐ Describe columns, including their material, diameter, and bearing plates.
- ☐ Calculate the length or height of columns and piers.
- ☐ Describe the girders to be used in a building, including their material, depth, weight or thickness, and length.
- ☐ Explain how the floor joists bear on the girders.

ASSIGNMENT

Questions 1 through 5 refer to Figure 16-7.

1. What is the length of the girder?

2. Describe the material used to build the girder including the size of material.

3. What supports the girder? (Include material and cross sectional size.)

4. How many posts, columns, or pier support the girder?

5. What is the height of the column or piers supporting the girder including bearing plates?

Questions 6 through 10 refer to Figure 16-8.

6. What is the length of the girder?

7. Describe the material used to build the girder including the size of material.

8. What supports the girder? (Include material and cross sectional size.)

9. How many posts, columns, or piers support the girder?

10. What is the height of the column or pier supporting the girder, including bearing plates?

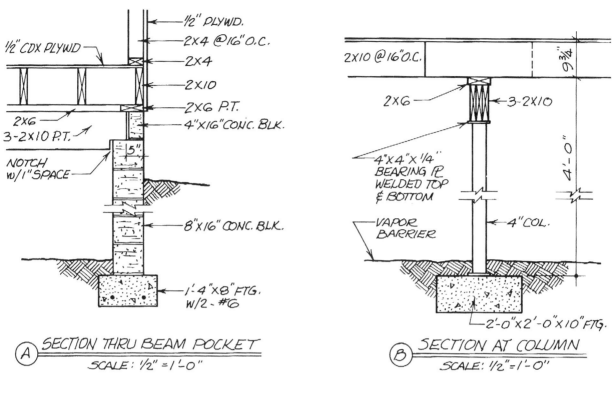

½" CDX PLYWD

2x6
3-2x10 P.T.

NOTCH
w/1" SPACE

½" PLYWD.
2x4 @16"O.C.
2x4
2x10
2x6 P.T.
4"x16" CONC. BLK.

5"

8"x16" CONC. BLK.

1'-4"x8" FTG.
W/2-#6

(A) SECTION THRU BEAM POCKET
SCALE: ½"=1'-0"

2X10 @16"O.C.

2x6

9¾"

3-2x10

4"x4"x¼"
BEARING ℙ
WELDED TOP
& BOTTOM

VAPOR
BARRIER

4" COL.

4'-0"

2'-0"x2'-0"x10" FTG.

(B) SECTION AT COLUMN
SCALE: ½"=1'-0"

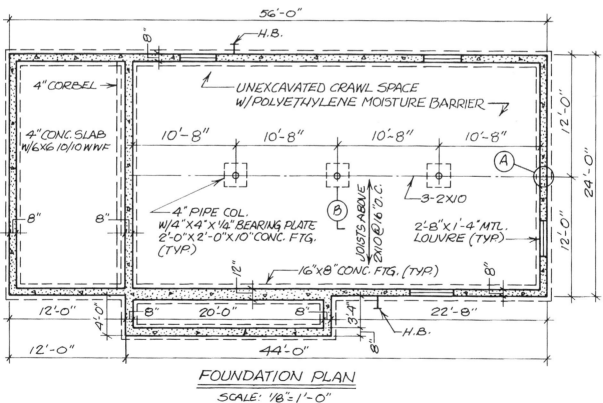

56'-0"

8"

H.B.

4" CORBEL

UNEXCAVATED CRAWL SPACE
W/POLYETHYLENE MOISTURE BARRIER

4" CONC. SLAB
W/6X6 10/10 WWF

10'-8"   10'-8"   10'-8"   10'-8"

12'-0"

A

24'-0"

8"   8"

4" PIPE COL.
W/4"X4"X¼" BEARING PLATE
2'-0"X2'-0"X10" CONC. FTG.
(TYP.)

B

JOISTS ABOVE
2X10 @ 6"O.C.

3-2X10

2'-8"X1'-4" MTL.
LOUVRE (TYP.)

12'-0"

12"

16"X8" CONC. FTG. (TYP.)

8"

4'-0"

8"   20'-0"   8"

3'-4"

22'-8"

H.B.

8"

12'-0"

44'-0"

FOUNDATION PLAN
SCALE: ⅛"=1'-0"

Figure 16-7   Use with assignment questions 1–5.

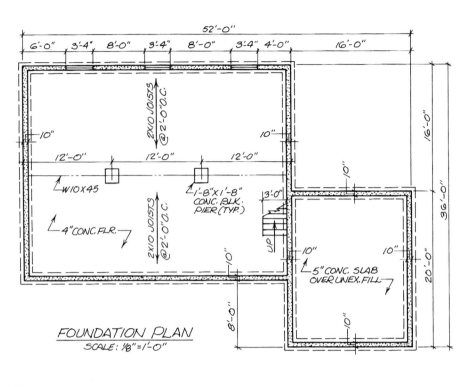

FOUNDATION PLAN
SCALE: 1/8"=1'-0"

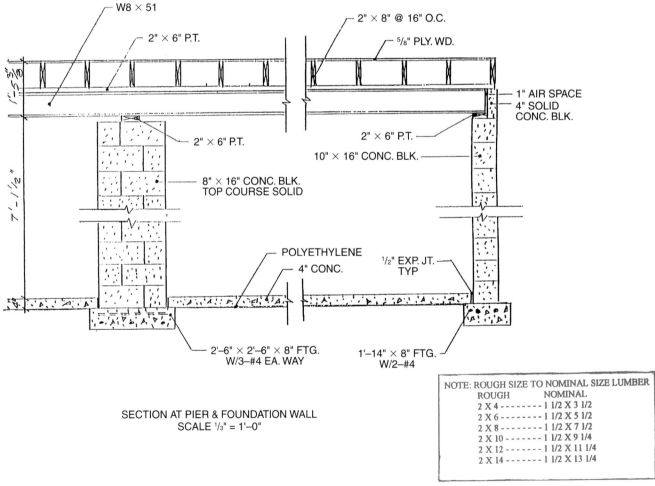

SECTION AT PIER & FOUNDATION WALL
SCALE 1/3" = 1'-0"

NOTE: ROUGH SIZE TO NOMINAL SIZE LUMBER
ROUGH	NOMINAL
2 X 4	1 1/2 X 3 1/2
2 X 6	1 1/2 X 5 1/2
2 X 8	1 1/2 X 7 1/2
2 X 10	1 1/2 X 9 1/4
2 X 12	1 1/2 X 11 1/4
2 X 14	1 1/2 X 13 1/4

Figure 16-8   Use with assignment questions 6–10.

UNIT 17   Floor Framing

OBJECTIVES

After completing this unit, you will be able to perform the following tasks:

- Describe the sill construction shown on a set of drawings.

- Identify the size, direction, and spacing of floor joists according to a set of drawings.

- Describe the floor framing around openings in a floor.

SILL CONSTRUCTION

Where the framing rests on concrete or masonry foundation walls, the piece in contact with the foundation is called the *sill plate,* Figure 17-1. The sill plate is the piece through which the anchor bolts pass to secure the floor in place. To prevent the sill plate from coming in direct contact with the foundation, and to seal any small gaps, a *sill sealer* is often included. This is a compressible, fiberous material that acts like a gasket in the sill construction.

The entire construction of the floor frame at the top of the foundation is called *sill construction* or the *box sill.* The box sill is made up of the sill sealer, sill plate, joist, and joist header, Figure 17-2. The sizes of materials are given on a wall section or sill detail. For areas where termites are present, a termite shield is included in the sill construction, Figure 17-3. A *termite shield* is a continuous metal shield that prevents termites from getting to the wood superstructure.

Figure 17-1   Sill plate

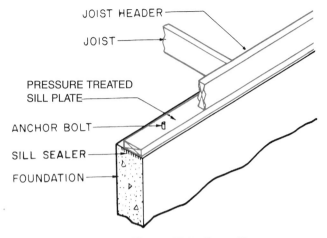

Figure 17-2 Box sill

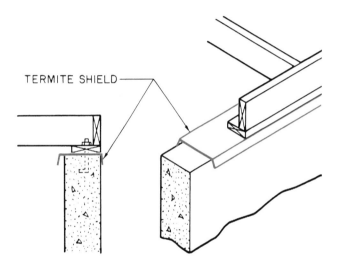

Figure 17-3 Termite shield

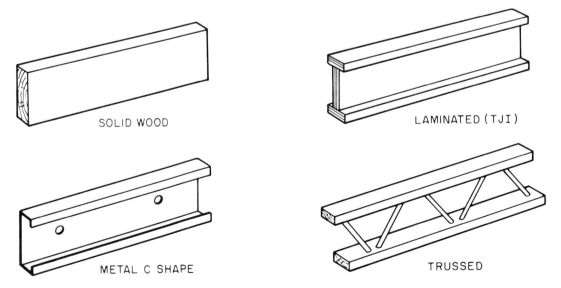

Figure 17-4 Several styles of joists

FLOOR JOISTS

Floor joists are the parallel framing members that make up most of the floor framing. Until recently, joists in residential construction were 2-inch framing lumber. However, recent advances in the use of materials have produced several types of engineered joists, Figure 17-

4. Although the materials in each type are different, their use is essentially the same.

Metal floor joists are often used in commercial and industrial construction. Where nonmetallic-sheathed cable is installed through metal studs or floor joists, a bushing or grommet must be installed prior to installing the cable. (Figure 15-16).

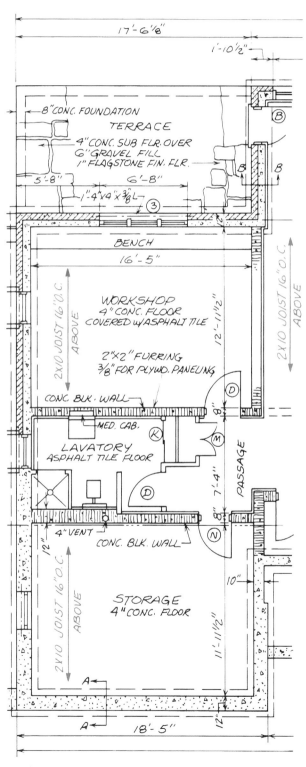

Figure 17-5   Joist callouts for the first floor are shown on the foundation plan.

Notes on the floor plans indicate the size, direction, and spacing of the joists in the floor above, Figure 17-5. For example, notes on the foundation plan give the information for the first floor framing. When the arrangement of framing members is complicated, a framing plan may be included, Figure 17-6. On most framing plans, each member is represented by a single line.

In the simplest building, all joists run in the same direction and are supported between the foundation walls by a single girder. However, irregularities in building shapes require that joists run in different directions, Figure 17-7. As the building design becomes more complex, more variations in floor framing are necessary. In a building such as the Lake House, which has floors at varying levels, the joists are supported by a combination of girders, beams, and load-bearing walls.

Lake House Floor Framing

The floor framing for the Lake House is shown on Framing Plan 1/6 (included in the packet). Notice that the framing plan is made up of a simplified floor plan and single lines to represent floor joists, beams, and joist headers. A more elaborate type of framing plan uses double lines to represent the thickness of each member, Figure 17-8. The framing plan shows the location and direction of framing members, but for more detail it will be necessary to refer to Floor Plan 2/2 also.

The floor framing in the Lake House can be studied most easily if it is viewed as having four parts: the kitchen, bedroom #1, bedroom #2 and bathrooms, and the loft. There is not floor framing for the living room and dining room. These two areas form the heat sink

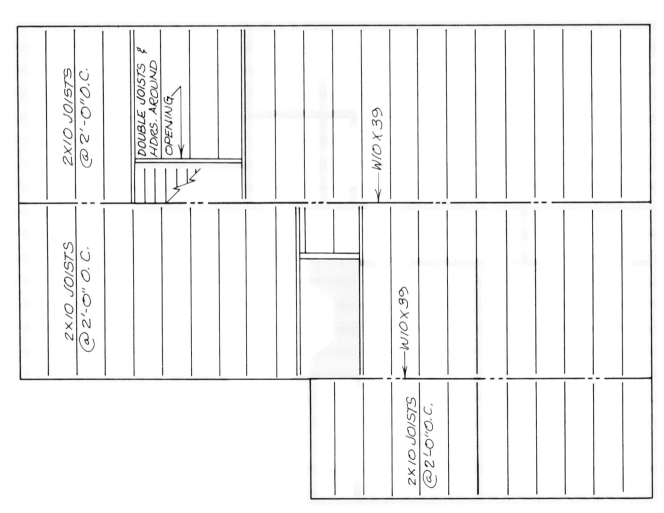

FLOOR FRAMING PLAN
SCALE: 3/16" = 1'- 0"

Figure 17-6   A simple floor framing plan

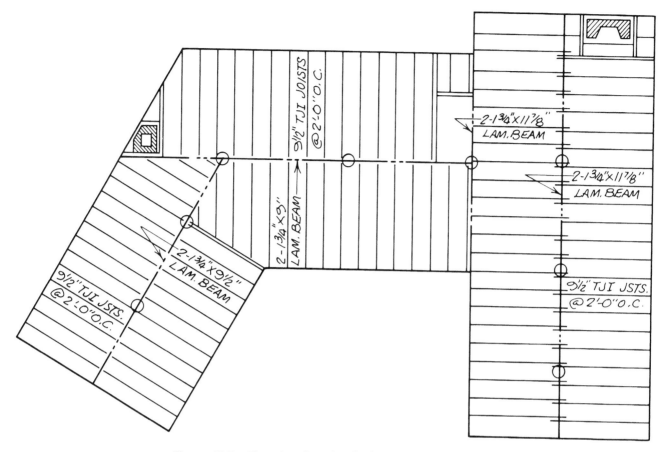

Figure 17-7   Floor framing plan for irregular-shaped house

for the passive-solar features discussed in earlier units. The floors in these rooms are concrete to absorb and radiate solar heat. As each area is framed, the carpenter must identify the following:

- Joist headers (locate the outer ends of joists)
- Bearing for inner ends of joists (beams, walls, etc.)
- Size and type of framing materials
- Length of joists
- Spacing
- Framing at openings

The joists headers are easily identified on the framing plan, but their exact position should be checked on the detail drawings. They may be set back to create a brick ledge, to accommodate wall finish, or to be flush with the foundation, Figure 17-9.

The *bearing* (support) for the inner ends of some of the joists in the Lake House is the structural-steel-and-box-beam core, Detail 6/6. This core consists of

four $3^{1}/_{2}$" square steel posts with wood box beams and steel-channel beams. The positions of the posts are shown on all plan views of the house. The C8 × 11.5 steel acts as a beam to support the floor in bedroom #2. Two 2 × 10s support the loft. The plywood box beams form part of the roof framing and are discussed in Unit 21. The double 2 × 10 beam rests in beam pockets that are welded to the posts. The structural steel channel (C8 × 11.5) is bolted directly to the posts.

The inner ends of the remaining floor joists in the Lake House are supported by bearing walls in the lower level. For example, some of the kitchen floor joists are supported by the west wall of the playroom, near the fireplace; and some, by the 2 × 10 header that spans the distance from the playroom wall to the foundation in the crawl space.

The size and spacing of material to be used are given on the framing plan by a note. They can also be found on the wall section. Lengths of framing members are usually not included on framing plans. However, these lengths can be found easily by referring to the floor plan that shows the location of the walls or beams

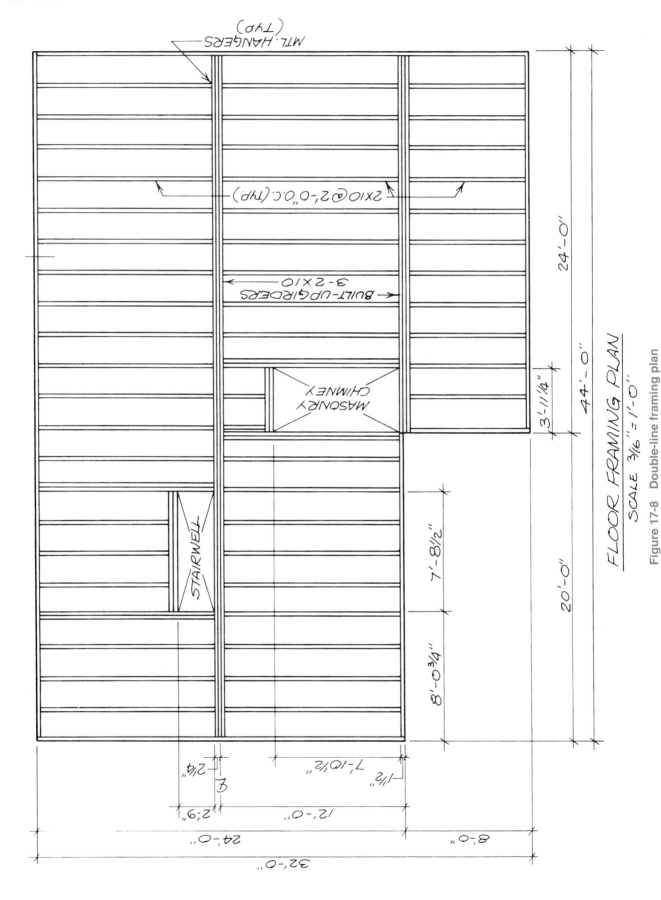

FLOOR FRAMING PLAN
SCALE 3/16" = 1'-0"

Figure 17-8 Double-line framing plan

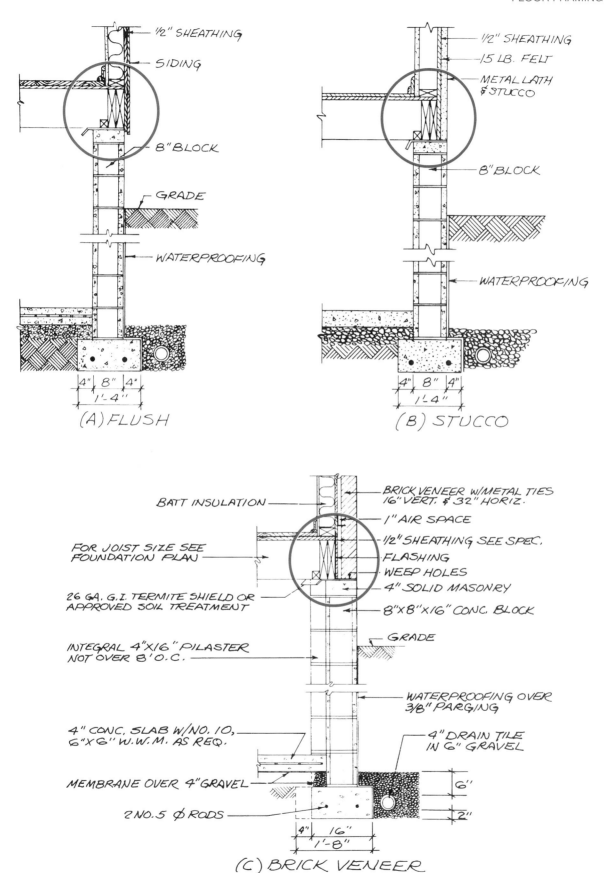

1/2" SHEATHING

SIDING

8"BLOCK

GRADE

WATERPROOFING

4" 8" 4"

1'-4"

(A) FLUSH

1/2" SHEATHING

15 LB. FELT

METAL LATH & STUCCO

8" BLOCK

WATERPROOFING

4" 8" 4"

1'-4"

(B) STUCCO

BATT INSULATION

FOR JOIST SIZE SEE FOUNDATION PLAN

26 GA. G.I. TERMITE SHIELD OR APPROVED SOIL TREATMENT

INTEGRAL 4"X16" PILASTER NOT OVER 8' O.C.

4" CONC. SLAB W/NO. 10, 6"X6" W.W.M. AS REQ.

MEMBRANE OVER 4" GRAVEL

2 NO. 5 Ø RODS

BRICK VENEER W/METAL TIES 16" VERT. & 32" HORIZ.

1" AIR SPACE

1/2" SHEATHING SEE SPEC.

FLASHING

WEEP HOLES

4" SOLID MASONRY

8"X8"X16" CONC. BLOCK

GRADE

WATERPROOFING OVER 3/8" PARGING

4" DRAIN TILE IN 6" GRAVEL

6"

2"

4" 16"

1'-8"

(C) BRICK VENEER

Courtesy of W. D. Farmer

Figure 17-9   The joist headers are positioned according to the exterior finish to be used.

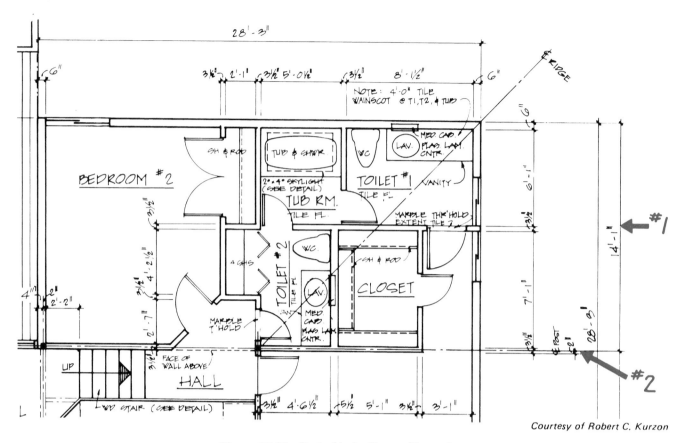

Courtesy of Robert C. Kurzon

Figure 17-10   Part of Lake House Floor plan

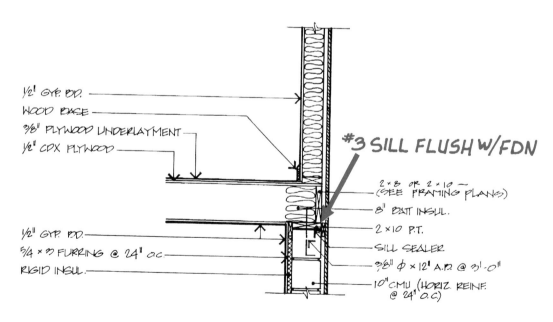

Courtesy of Robert C. Kurzon

Figure 17-11   Part of Lake House wall section

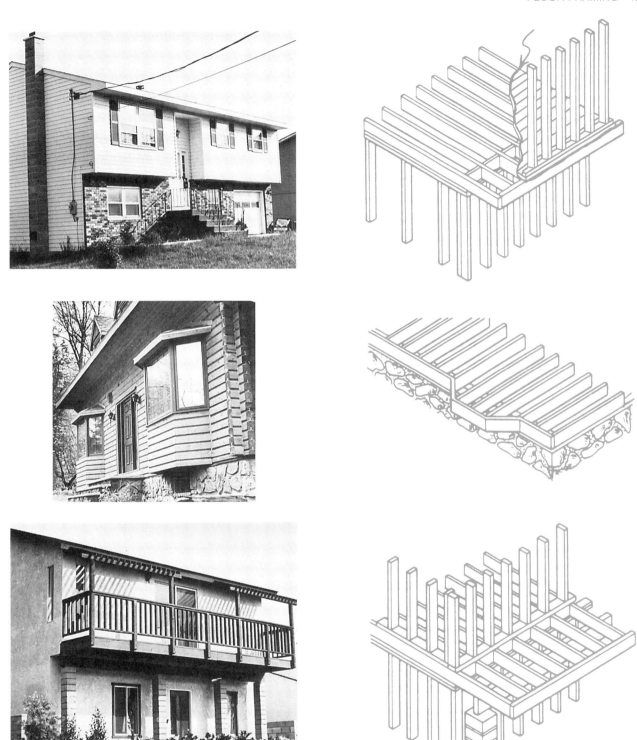

Figure 17-12   Typical uses of cantilevered framing

on which the joists rest. For example, refer to Figures 17-10 and 17-11 and find length of the floor joists in bedroom #2 as follows:

1. The dimension from the outside of the north wall to the centerline of the 3½" ⊠ post is 14'-1".

2. The dimension from the centerline of the 3½" ⊠ post is 2", so the overall dimension of the bedroom floor is 14'-3".

3. According to Wall Section 3/4, the joist header is flush with the north foundation wall, so subtract 1½" (the thickness of the joist header) from each end.

4. 14'-3" minus 3" (1½" at each end) equals 14'-0" (the length of the joists).

Some floor framing is cantilevered to create a seemingly unsupported deck. *Cantilevered* framing consists of joists that project beyond the bearing surface to create a wide overhang. This technique is used extensively for balconies, under bay windows, and for garrison-style houses, Figure 17-12.

FRAMING AT OPENINGS

Where stairs and chimneys pass through the floor frame, some of the joists must be cut out to form an opening. The ends of these joists are supported by headers made of two or more members. The full joists at the sides of the opening have to carry the extra load of the shortened joists and headers, so they are also doubled or tripled, Figure 17-13. The number of joists and headers required around openings of various sizes is often spelled out in building codes and shown on framing plans.

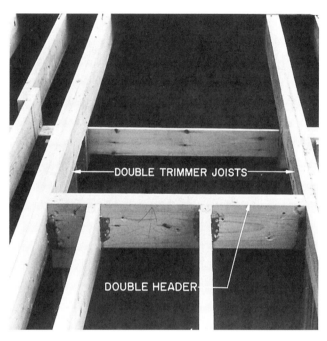

Figure 17-13   Framing around opening in floor

✓ CHECK YOUR PROGRESS ─────────────────────

Can you perform these tasks?

☐ Describe the box sill shown on a set of drawings, including size of materials, anchoring, and setback from the face of the foundation.

☐ List the sizes of all the floor joists in a building.

☐ Describe the direction and spacing of all of the floor joists in a building.

☐ Describe the framing around openings in the floor.

ASSIGNMENT ──────────────────────────────

Refer to the Lake House drawings to complete this assignment.

1.   What size lumber is used for the floor framing in bedroom #2?

2.   What size lumber is used for the floor joists in the loft?

3.   What size lumber is used for the joist headers in the loft?

4. How long are the joists in the loft?

5. How long are the floor joists in bedroom #1?

6. What supports the west end of the floor joists in bedroom #1?

7. What supports the south end of the floor joists in bedroom #2?

8. How long are the floor joists in bedroom #2?

9. How many floor joists are needed for bedroom #2 and the adjacent closets?

10. What size material is used for the sill?

11. What does the double 2 × 10 beam in the structural core of the Lake House support?

12. How long are the headers that support the loft floor joists?

13. Is the box sill of the Lake House flush with the foundation wall or set back?

14. When an opening in a floor is framed, why are the joists at the sides of the opening doubled?

UNIT 18 Laying Out Walls and Partitions

OBJECTIVES

After completing this unit, you will be able to perform the following tasks:

- Describe the layout of a house from its floor plans.
- Find specific dimensions given on floor plans.

When the deck (framing and subfloor or concrete slab) is completed, the framing carpenter lays out the location of walls and partitions. The size and location of each wall is indicated on the floor plans. Drawings 1/2 and 2/2 are the floor plans for the Lake House.

Of all the sheets in a set of construction drawings, the floor plans often contain the most information. Before looking for specific details on floor plans, it may help to mentally walk through the house. Start at the main entrance to the lowest floor and visualize each room as if you were walking through the house.

VISUALIZING THE LAYOUT OF WALLS AND PARTITIONS

The lowest floor with frame walls in the Lake House is the basement. Start at the $6^0 \times 6^8$ SGD (sliding glass door) on the east wall of the playroom. This large L-shaped room covers most of the basement floor.

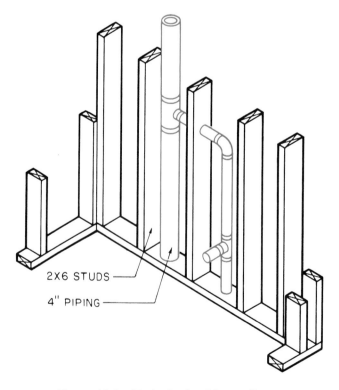

2X6 STUDS ——

4" PIPING ——

Figure 18-1   Six-inch plumbing wall.

Some plans list overall dimensions for each room. The plans for the Lake House give this information by conventional dimension lines only.

The north-south dimension of the playroom is 28'-3". The east-west dimension of the north part of the playroom is 14'-1". The section of the playroom with the fireplace is 13'-8" by 13'-10". It may be helpful to notice the overall dimensions of each room as you visualize its shape and relationship to other rooms. It will be necessary to refer to these dimensions many more times. At the north end of the playroom is a small closet. South of the closet is a hall leading to the bathroom and utility room.

The bathroom has a shower, water closet (toilet), and lavatory. Notice that although most interior partitions are $3\frac{1}{2}$ inches thick, the wall behind the water closet and lavatory is $5\frac{1}{2}$ inches thick. This thicker wall is called a *plumbing wall*. Its extra thickness allows room for plumbing, Figure 18-1.

The purpose of the utility room beyond the bathroom is to house mechanical equipment such as water heater, furnace or boiler, and water pump. In the southwest corner of the utility room is a small opening into the crawl space on the west side of the house. Notice that the concrete ends here.

Walking back through the playroom to the stairs, you will see some interesting features, Figure 18-2. A broken line in this area indicates the edge of a floor above. These floors can be seen more easily on the first floor plan. Also notice the location of the fireplace which will extend up through the upper floor. The L-shaped stairs lead up to the first floor. It is obvious that his is not a full story higher because only four steps are shown. This can be seen more clearly in the Sections on sheet 4. The L-shaped stairs lead up to the living room, Figure 18-3. Against the west and south walls is a plywood platform or built-in bench.

Another set of stairs in the northwest corner of the living room leads to the kitchen and dining room. These rooms are separated only by a peninsula of kitchen cabinets. The kitchen is separated from a hall by a free-standing closet and enclosure for the refrigerator and oven, Figure 18-4. This closet and enclosure cannot be recognized as free-standing (meaning it does not reach the ceiling) on the floor plan. However, this can be seen on Section 2/4. On the north side of the kitchen is a door to the deck outside. The east side of the hall has a railing that continues up another set of stairs to the upper hall. Beyond this railing the floor is open to the playroom below.

Above the open area of the playroom is a loft that provides storage or extra sleeping space, Figure 18-5. Access to the loft is by a ladder in the upper hall. The loft is shown on a separate plan on Sheet 4. The loft is suspended on two beams built-up of three 2 × 10s each. The north and south sides of the loft are enclosed by a wall 2'-6" high. The east and west ends have a railing.

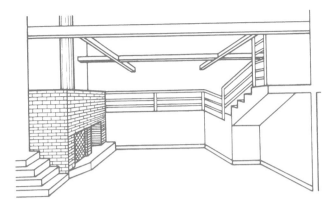

Figure 18-2   Lake House playroom

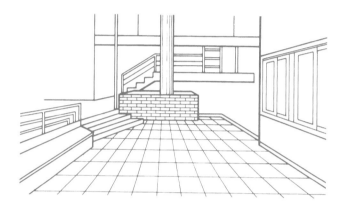

Figure 18-3   Lake House living room

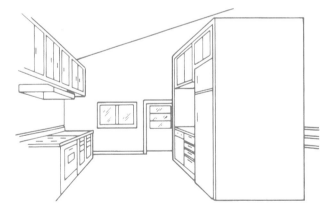

Figure 18-4   Lake House kitchen

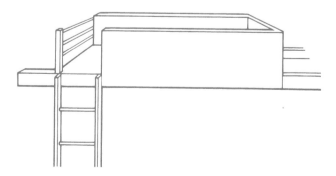

Figure 18-5   Lake House loft

One of the doors in the upper hall opens into bedroom #1. On the east side of this bedroom is a hall that leads past a large (7'1" × 5'-1") closet to toilet #1 in the northeast corner of the house. This toilet room has only a lavatory and water closet. A tub in the next room also serves toilet #2 to the east. Toilet #2 has a closet with shelves for linen storage. This toilet room can also be entered from the upper hall. The remaining room is bedroom #2.

The garage is attached to the house only by the wood deck and the roof. The garage is a rectangular building with 4-inch walls, an overhead door, and a walk-through door.

FINDING DIMENSIONS

When you understand the relationships of the rooms to one another, you are ready to look for more detailed information. Frame walls are dimensioned in one of three ways, Figure 18-6. Exterior walls are usually dimensioned to the face of the studs or the face of the sheathing. Interior walls may be dimensioned either to the face of the studs or to their centerlines.

When walls are dimensioned to their centerlines, one half of the wall thickness must be subtracted to find the face of the studs. For example, in Figure 18-7, the end walls are 12'-4" on centers. However, the plates for the side walls are 12'-0$\frac{1}{2}$" long (12'-4" minus 3$\frac{1}{2}$", the width of the studs, equals 12'-0$\frac{1}{2}$").

Dimensions are usually given in a continuous string where practical, Figure 18-8. Overall dimensions and major wall locations are given outside the view. Minor partitions and more detailed features are dimensioned either on or off the view, whichever is most practical.

The Lake House includes an angled wall. The length of such walls can be accurately found by trigonometry. However, the accuracy obtained by measuring with an architects' scale should be adequate for normal estimating.

As each wall is laid out, the plates are cut to length and the positions of all openings and intersecting walls are marked. The wall frame is usually assembled flat on the deck. Its position is marked on the deck with a chalkline, then the assembly is tipped up and slid into place.

Commercial buildings may require a thicker wall where the electrical service panel is flush mounted and the panel can is deeper than the standard 3$\frac{1}{2}$-inch wall. Many times this thicker wall requirement is not shown on the floor plan or drawing details.

The utilities that are to be stubbed up from a concrete floor slab into a wall are installed prior to floor slab concrete placement and, therefore, prior to the wall locations being determined by the framing contractor. The utility installer must be able to accurately predetermine the specific wall locations.

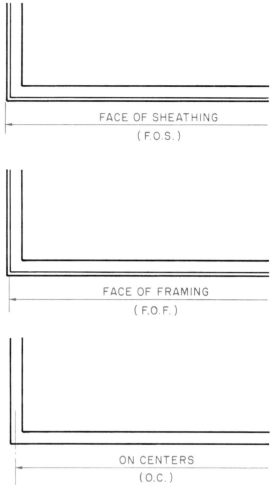

Figure 18-6   Three types of dimensioning

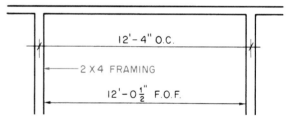

Figure 18-7   It is important to consider the size of the framing material when working with O.C. dimensions.

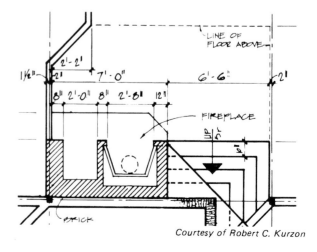

Figure 18-8   Dimensions are placed in neat strings as much as possible.

Courtesy of Robert C. Kurzon

✓ CHECK YOUR PROGRESS ———————————————————————

Can you perform these tasks?

- ☐ List dimensions for locating all of the walls and partitions in a house.
- ☐ Explain whether wall dimensions are to face of sheathing, face of framing, or on centers.
- ☐ Identify plumbing walls shown on floor plans.

ASSIGNMENTS ———————————————————————

Refer to the Lake House drawings (in the packet) to complete the following assignment.

1. What are the inside dimensions of each room? Disregard slight irregularities in room shape, but remember to allow for wall thicknesses.

Room	N-S dimension	E-W dimension
Utility	X	
Basement Bath	X	
Living Room	X	
Bedroom #1	X	
Bedroom #2	X	
Toilet #1	X	
Tub	X	
Toilet #2	X	
Loft	X	

2. How many lineal feet of 2 × 4 frame wall are there in the basement?

3. How many lineal feet of 2 × 6 frame wall are there in the basement?

4. What is between the dining room and the living room?

5. What is directly below the loft?

6. What is directly below the tub room?

7. How many lineal feet of 2 × 6 frame wall are there on the Upper Level Floor Plan?

8. What utility must be roughed-in prior to concrete slab placement?

9. Where would you find the utility concrete slab rough-in requirements?

UNIT 19 *Framing Openings in Walls*

After completing this unit, you will be able to perform the following tasks:

• Describe typical rough openings.

• Locate and interpret specific information for framing openings.

Two types of dimensions must be known before window and door openings can be framed: location and size. Opening locations are given by dimensioning to their centerlines in a string of dimensions outside the floor plan, Figure 19-1. Such dimensions are usually given from the face of the studs. One-half the rough opening is than allowed on each side of the centerline.

Courtesy of Robert C. Kurzon

Figure 19-1   The locations of openings in framed walls are usually given to their centerlines. The ± on the sliding doors allows the builder to place the doors next to the corner post.

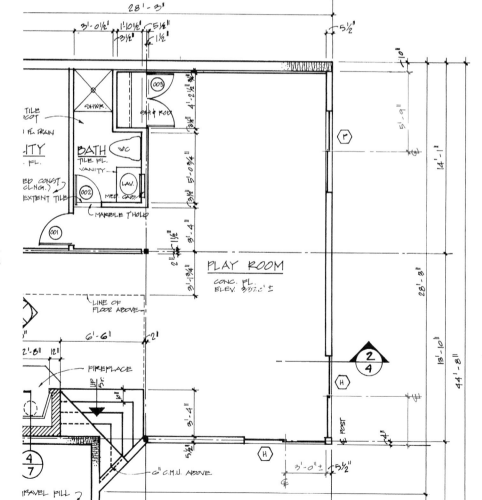

DIMENSIONS OF ROUGH OPENINGS

The size of the *rough opening* (R.O.), or opening in the framing, is often listed on the door and window schedule. This is a list of all doors and windows in the house, Figure 19-2. Doors and windows are identified on the floor plan by a *mark*—a letter or number. All doors or windows of a certain size and type have the same mark. Each mark is listed on the schedule with the information for the doors or windows of that type and size.

Occasionally the rough opening dimensions are not given on the drawings. In this case they should be

SCHEDULE OF WINDOWS

⬡	SIZE	TYPE	GLASS	COMMENTS
A	3'-0" × 3'-0"	THERMAL BREAK ALUMINUM	3/8 INSULATED	
B	3'-0" × 5'-0"	D O	D O.	
C	4'-0" × 4'-0"	D O.	D O	
D	4'-0" × 5'-0"	D O.	D O.	
E	5'-0" × 5'-0"	D O.	D O	
F	6'-0" × 4'-0"	D O	1" INSULATED	
G	6'-0" × 5'-0"	D O.	1" INSULATED	
H	6'-0" × 6'-8"	D.O.	1" TEMP. INSUL.	SLIDING GLASS DOORS

SCHEDULE OF DOORS

◯	SIZE		THK	DOOR MAT'L	DOOR TYPE	FRAME MAT'L	FRAME TYPE	GLASS	LOCKSET	COMMENTS
001	2'-8" × 6'-8"		1 3/8	WD	C	WD.	A		CLASSROOM	
002	2'-0" × 6'-8"		1 3/8	WD	C	WD.	A		PRIVACY	
003	PR 2'-0" × 6'-8"		1 3/8	WD	D	WD.	A		PASSAGE R.H. DUMMY TRIM L.H.	
101	3'-0" × 6'-8"		1 3/4	H.M.	A	WD.	A	1" INSUL	ENTRY	1'-8" SIDELIGHT
102	3'-0" × 6'-8"		1 3/4	H.M.	B	WD.	A		ENTRY	
103	PR 2'-4" × 6'-8"		1 3/8	WD	D	WD.	B		NONE	BI-PASSING HDWARE
104	2'-6" × 6'-8"		1 3/8	WD.	C	WD.	B		PRIVACY	
105	PR 2'-6" × 6'-8"		1 3/8	WD	D	WD.	B		PASSAGE R.H. DUMMY TRIM L.H.	
106	2'-0" × 6'-8"		1 3/8	WD	C	WD.	B		PRIVACY	
107	(4) 1'-0" × 6'-8"		1 3/8	WD	E	WD.	B		NONE	BI-FOLDING HDWARE
108	2'-0" × 6'-8"		1 3/8	WD	C	WD.	B		PRIVACY	
109	2'-0" × 6'-8"		1 3/8	WD	C	WD.	B		PRIVACY	
110	2'-6" × 6'-8"		1 3/8	WD	C	WD.	B		PRIVACY	
111	2'-0" × 6'-8"		1 3/8	WD	C	WD.	B		PASSAGE	
112	2'-6" × 6'-8"		1 3/8	WD	C	WD.	B		PRIVACY	

Courtesy of Robert C. Kurzon

Figure 19-2   Window and door schedules list all of the windows and doors with their sizes.

Andersen® Windowalls® — Windows·Gliding Doors

Perma-Shield® Casement Windows In Terratone Color

BASIC UNIT INCLUDES:

(1) **FRAME:** Treated wood frame, preformed rigid vinyl sheath and flashing over all exterior parts.
(2) **SASH:** Treated wood core sash members covered with rigid vinyl, glazed double-pane insulating glass.
(3) **HARDWARE:** Ventilating sash with roto-operator, sash lock and stainless steel sliding hinges applied.
(4) **HINGING:** Specify left, right or stationary on single sash units. Hinging on multiple units only as indicated.
(5) **JAMB WIDTHS: Basic jamb 2⅞".** Pine head, sill and side extension jambs available for 5¼" wall (included in basic price, not applied). Contact your distributor for other thicknesses.
(6) **MISC.:** Pine stops tacked to frame.

Vent Layout: LH, RH

UNIT DIM.	1'5"	2'0 1/8"	2'4 3/8"	3'4 3/4"	4'0"	4'8 1/2"	5'11 7/8"	2'11 15/16"	4'0"	5'11 7/8"
RGH. OPG.	1'5 1/2"	2'0 5/8"	2'4 7/8"	3'5 1/4"	4'0 1/2"	4'9"	6'0 3/8"	3'0 1/2"	4'0 1/2"	6'0 3/8"
GLASS*	12 5/8"	19 3/4"	24"	16 1/8"	19 3/4"	24"	19 3/4"	31 9/16"	43 5/8"	67 1/2"

CR3, C3 — Height: 2'-11 15/16" unit dim., 3'-0 1/2" rgh. opg., 31 1/8" glass

Basic Unit	CR13T	C13T	C23T	C33T	CP23T	CP33T
DBL.-PANE INSUL. GLA., w/scrns.	$146.85	$158.16	$306.21	$423.80	—	—
DBL.-PANE INSUL. GLA., no scrns.	138.45	149.21	288.31	405.90	—	—
DBL.-PANE INSUL. GLA., sta.	119.79	130.55	—	—	$188.58	$275.79
Deduct for no pine ext. jambs (not applied)	10.02	11.10	15.98	21.90	15.98	21.90
ADD for set div. light grilles★	7.60	8.03	16.06	24.09	—	—
DBL.-PANE INSUL. GREY or BRONZE GLA., w/scrns.†	176.82	190.38	370.65	520.46	256.12	379.91
ADD for Removable Glazing pnls.★	15.62	18.78	37.56	56.34	47.05	69.30

CN35, C35, CX35 — Height: 3'-4 13/16" unit dim., 3'-5 3/8" rgh. opg., 36" glass

Basic Unit	C135T	CX135T	CN235T	C235T	CX235T	C335T	CP235T	CP335T
DBL.-PANE INSUL. GLA., w/scrns.	$172.62	$192.25	$290.94	$331.78	$370.76	$463.71	—	—
DBL.-PANE INSUL. GLA., no scrns.	162.60	181.29	271.70	311.74	348.84	443.67	—	—
DBL.-PANE INSUL GLA., sta.	143.94	162.63	—	—	—	—	$221.30	$312.49
Deduct for no pine ext. jambs (not applied)	12.30	13.06	15.84	17.18	19.50	23.10	17.18	23.10
ADD for set div. light grilles★	8.65	10.50	16.24	17.30	21.00	25.95	—	—
DBL.-PANE INSUL. GREY or BRONZE GLA., w/scrns.†	207.64	237.15	333.44	401.82	460.56	568.77	295.29	427.93
ADD for Removable Glazing pnls.★	20.35	21.93	36.48	40.70	43.86	61.05	53.91	70.83

CR4, C4, CX4 — Height: 4'-0" unit dim., 4'0 1/2" rgh. opg., 43 3/16" glass

Basic Unit	CR14T	C14T	CX14T	C24T	CX24T	C34T	CP24T	CP34T
DBL.-PANE INSUL. GLA., w/scrns.	$179.09	$189.14	$216.86	$363.97	$417.27	$503.42	—	—
DBL.-PANE INSUL. GLA., no scrns.	169.32	178.18	204.83	342.05	393.21	481.50	—	—
DBL.-PANE INSUL. GLA., sta.	148.12	156.98	183.63	—	—	—	$252.85	$353.16
Deduct for no pine ext. jambs (not applied)	12.56	13.64	14.40	18.52	20.84	24.44	18.52	24.44
ADD for set div. light grilles★	9.49	10.67	11.48	21.34	22.96	32.01	—	—
DBL.-PANE INSUL. GREY or BRONZE GLA., w/scrns.†	210.48	232.83	261.01	451.35	505.57	634.49	346.56	514.80
ADD for Removable Glazing pnls.★	18.60	22.67	25.19	45.34	50.38	68.01	58.20	88.31

C5, CX5 — Height: 4'-11 7/8" unit dim., 5'-0 3/8" rgh. opg., 55 1/16" glass

GLASS* 65 7/16 x 53 1/16 (CP35T)

Basic Unit	C15T	CX15T	C25T	CX25T	C35T	CP305T	CP25T	CP35T
DBL.-PANE INSUL. GLA., w/scrns	$219.13	$251.52	$421.19	$484.49	$573.48	—	—	—
DBL.-PANE INSUL. GLA., no scrns.	206.02	237.26	394.97	455.97	547.26	—	—	—
DBL.-PANE INSUL. GLA., sta.	184.82	216.06	—	—	—	$236.51	$286.79	$472.00
Deduct for no pine ext. jambs (not applied)	16.11	16.87	20.99	23.31	26.91	18.71	20.99	26.91
ADD for set div. light grilles★	12.29	12.91	24.58	25.82	36.87	—	—	—
DBL.-PANE INSUL. GREY or BRONZE GLA., w/scrns.†	271.75	296.83	526.43	575.11	731.34	305.42	399.86	554.12
ADD for Removable Glazing pnls.★	33.15	38.03	66.30	76.06	99.45	58.24	70.49	—

C6 — Height: 5'-11 7/8" unit dim., 6'-0 3/8" rgh. opg., 67 1/16" glass

*Unobstructed glass size shown in inches.

Basic Unit	C16T	C26T	CP26T
DBL.-PANE INSUL. GLA., w/scrns.	$259.29	$483.17	—
DBL.-PANE INSUL. GLA., no scrns.	242.60	449.79	—
DBL.-PANE INSUL. GLA., sta.	221.40	—	$352.46
Deduct for no pine ext. jambs (not applied)	18.86	23.74	23.74
ADD for set. div. light grilles★	17.44	34.88	—
DBL.-PANE INSUL. GREY or BRONZE GLA., w/scrns.†	320.94	606.47	514.10
ADD for Removable Glazing pnls.★	38.72	77.44	86.97

For hinging and combinations other than shown, contact your distributor.
†Check with your distributor for shipping schedule.
Picture Units stationary — no screen.
★ NOT INSTALLED
SEE PAGE 19 FOR PRICES ON ASSEMBLY AND PARTS
NOTE: See Page 19 for miscellanous parts and optional equipment.

Courtesy of Iroquois Millwork Corporation

Figure 19-3 Typical page from a window catalog

obtained from the manufacturer. Window manufacturers list glass size, sash opening, rough opening, and unit dimensions in their catalogs, Figure 19-3. If the finished doorway, the *jamb,* is to be built on the site by a carpenter, the rough opening size will not be available. In this case the rough opening for swing doors can be built 2 inches wider and 2 inches higher than the door. If the door is another type, you must first determine what the finished opening is to be. Sliding door panels should overlap 1 inch so the finished opening is 1 inch narrower than the width of two panels added together. Sliding doors, bifold doors, and accordian doors all need an extra inch at the top for hardware; build these rough openings 2 1/2 inches higher than the door itself.

Sizes of doors and windows are given with the width first and height second. To further simplify dimensioning, they are often listed as feet/inches. For example, a $2^6 \times 6^8$ door is 2 feet 6 inches wide by 6 feet 8 inches high.

The information in this unit is intended only to help you determine rough opening sizes. Electrical utility installers should familiarize themselves with the locations and sizes of these rough openings prior to installing the electrical wiring. More information about windows and doors can be found in Unit 24.

FRAMING OPENINGS

In stud-wall construction, it is usually necessary to cut off or completely eliminate one or more studs where windows and doors are installed. The load normally carried by these studs must be transferred to the sides of the opening. The construction over an opening which transfers this load to the sides of the opening is called the *header.* Additional studs are installed at the sides to carry this load, Figure 19-4.

A set of construction drawings includes details showing the type of rough opening construction intended. The simplest type of wood construction uses two 2 × 10s or 2 × 12s with plywood spacers between to form the header, Figure 19-5. A flat 2 × 4 may be nailed to the bottom of the header to reduce the height of the top of the opening.

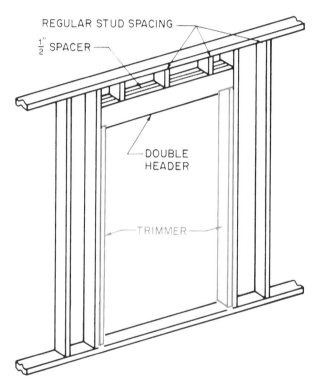

Figure 19-4   The trimmer studs support the header.

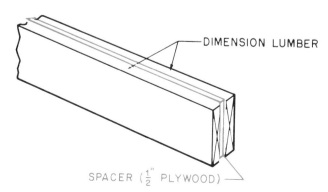

Figure 19-5   Solid wood header

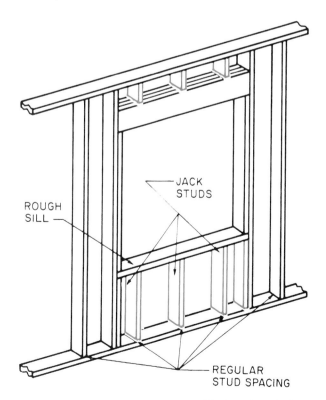

Figure 19-6   The spaces above and below the window opening are framed with cripple or jack studs.

To conserve lumber in nonbearing walls, the header may be two 2 × 4s. The area over the header is framed with cripple studs. The area below a window, also, is framed with *cripple studs,* Figure 19-6. These cripple studs are installed on the normal spacing for wall studs. This may require a cripple within a few inches of a side trimmer, but is necessary to provide a nailing surface for the sheathing.

Several systems of metal framing are available for light construction. The basic elements of light-gauge metal framing are the same as wood framing, Figure 19-7. These systems use top and bottom wall plates, studs, and floor joists. To make light-gauge framing compatible with wood, the metal members are made in common sizes for wood framing. The greatest difference is that metal framing is joined with screws instead of nails.

Headers over openings in light-gauge metal framing are usually very similar to those in wood framing. However, the system designed by the metal framing manufacturer should always be followed.

Courtesy of Zinc Institute, Inc.

Figure 19-7   The parts of this steel frame are the same as a wood frame. Notice the window header and cripple studs.

✓ CHECK YOUR PROGRESS

- [ ] Identify each door or window listed on a schedule of doors and windows according to its location on the floor plan.
- [ ] Determine the rough opening for a window from the information in a manufacturer's catalog.
- [ ] Describe the framing for a window opening.
- [ ] Describe the framing for a door opening.
- [ ] Determine the rough opening for a door.
- [ ] Explain the difference between a solid wood header and a trussed header

ASSIGNMENT

Refer to the Lake House drawings (in the packet) to complete this assignment.

1. What type of header should be used over the door from the hall to bedroom #1?

2. What type of header should be used over the door from the deck to the kitchen?

3. Why should these two headers be made differently?

4. What is the length of the header over the garage overhead door? Allow for two trimmers at each side and $1^1/_4$" for jambs at each side.

5. What are the R.O. dimensions for the door from the deck into the garage?

6. Name the location and give the R.O. dimensions for each interior door on the Upper Level Floor Plan.

7. According to Figure 19-3, what are the R.O. dimensions for the windows in bedroom #2?

8. How long is the header over the window in bedroom #2?

9. How many cripple studs are needed beneath the windows in the south wall of the living room?

10. What is the length of the cripple studs beneath the window in bedroom #2? (Assume the bottom of the header is 6'-8$^1/_2$" from the top of the subfloor.)

UNIT 20 Roof Construction Terms

OBJECTIVES

After completing this unit, you will be able to perform the following tasks:

• Identify common roof types.

• Define the terms used in laying out and constructing a roof.

• State roof pitches as a ratio of rise to run or as a fraction.

TYPES OF ROOFS

Several types of roofs are commonly used in residential construction, Figure 20-1. Variations of these roof types may be used to create certain architectural styles.

The *gable roof* is one of the most common types used on houses. The gable roof consists of two sloping sides which meet at the ridge. The triangle formed at the ends of the house between the top plates of the wall and roof is called the *gable*.

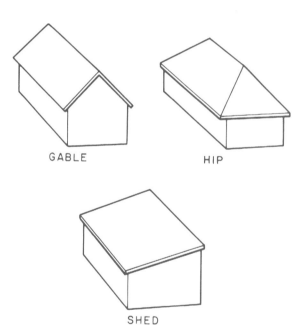

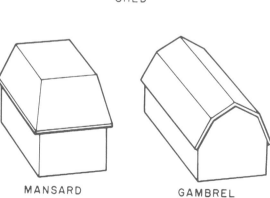

Figure 20-1   Common roof types

The *gambrel roof* is similar to the gable roof. On this roof, the sides slope very steeply from the walls to a point about halfway up to roof. Above this point, they have a more gradual slope.

The *hip roof* slopes on all four sides. The hip roof has no exposed wall above the top plates. This means that all four sides of the house are equally protected from the weather.

The *mansard roof* is similar to the hip roof, except the lower half of the roof has a very steep slope and the top half is more gradual. This roof style is used extensively in commercial construction—on stores, for example.

The *shed roof* is a simple sloped roof with no ridge. A shed roof is much like one side of a gable roof. This type of roof is used in modern architecture and for additions to existing buildings.

ROOF CONSTRUCTION TERMS

The roof construction terms defined in the following list are illustrated in Figure 20-2.

- *Span* is the distance between the outsides of the walls covered by a roof.

- *Run* is the horizontal distance covered by one rafter. Run does not include any part of the rafter that extends beyond the wall. On a common two-sided roof, the run is one-half the span.

- *Rise* is the vertical distance form the top of the wall to the measuring line of the ridge board.

- The *measuring line* is an imaginary line along which all roof dimensions are taken. The measuring line of a rafter is a line parallel to its edges and passing through the deepest part of the bird's mouth.

- A *bird's mouth* is a notch cut in the lower edge of the rafter to fit around the wall plate.

- The *ridge board* is the horizontal framing member to which the upper ends of the rafters are connected.

- The *overhang* is the horizontal distance covered by the roof outside the walls.

- The *tail* is the portion of the rafter that is outside the walls. The rafter tail is measured along the measuring line and forms the overhang.

- The *pitch* of a roof is a way of indicating how steep the roof is. Pitch is given as a fraction. The steepness of a roof can also be given as the number of inches of rise per foot of run. This is called *slope* or *unit rise.*

- *Collar ties* prevent the weight on the roof from spreading the walls.

In the fractional method of indicating pitch the rise is the numerator (top of the fraction) and the span the denominator (bottom of the fraction), Figure 20-3. This

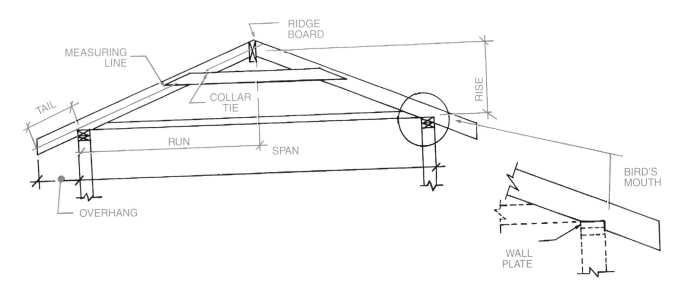

Figure 20-2   Common roof terms

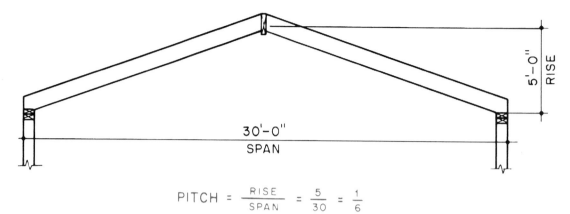

$$PITCH = \frac{RISE}{SPAN} = \frac{5}{30} = \frac{1}{6}$$

Figure 20-3 Fractional pitch

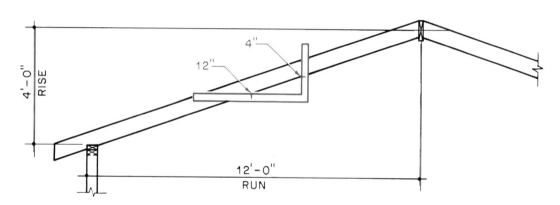

Figure 20-4 Rise per foot of run: This roof has 4 in 12 slope.

fraction is always reduced to the lowest possible terms. (See Math Review 1.)

In the rise-per-foot-of-run method, the slope is given as the number of inches of rise for every twelve inches of run, Figure 20-4. The rise per foot of run is given on drawings with the symbol ▱ . The horizontal leg of this triangle represents the run. The vertical leg represents the rise, Figure 20-5. Notice that Figure 20-5 shows two roof pitches. These pitches are written as 10 in 12 and 5 in 12.

When the dimensions are given for a run of other than 12 feet, the rise per foot of run can be calculated: Divide 12 by the actual run to find the proper ratio of rise. Multiply this result by the actual rise to find the rise per foot of run. For example, the run is 14'–6" and the rise if 5'0"; what is the rise per foot of run? 12 = 14'–6" (14.5') = 0.83 ÷ 5 = 4.15. The result (4.15) is close enough to 4 that rafter calculations can be based on 4 in 12.

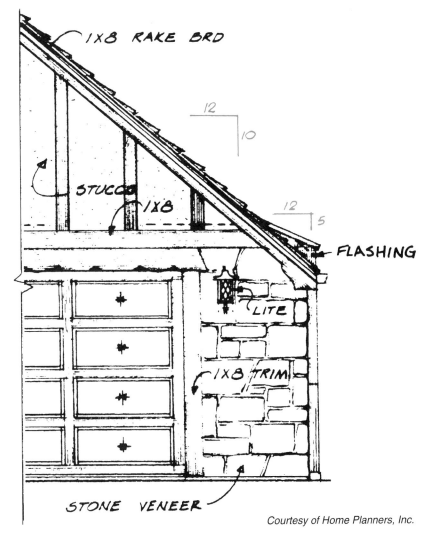

1X8 RAKE BRD

12
10

STUCCO

1X8

12
5

FLASHING

LITE

1X8 TRIM

STONE VENEER

Courtesy of Home Planners, Inc.

Figure 20-5 Roof slope is usually shown on building elevations.

✓ CHECK YOUR PROGRESS

Can you perform these tasks?

☐ Identify the following roof types from information given on construction drawings: gable, gambrel, hip mansard, and shed.

☐ List each of the following from information given on construction drawings: span, run, rise, overhang, rafter tails, slope.

ASSIGNMENT

1.   Give the following information for the roof shown in Figure 20-6:

   a. Span

   b. Run

   c. Rise

   d. Overhang

   e. Length of the rafter tails

   f. Fractional pitch

   g. Rise per foot of run

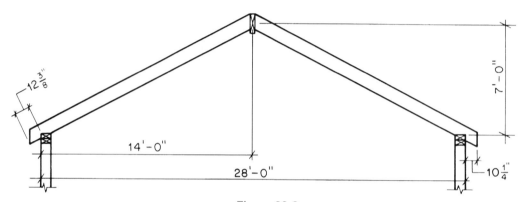

Figure 20-6

Refer to Lake House drawings (in the packet) to complete rest of assignment.

2.   What style roof is used for most of the Lake House?

3.   What is the span of the rafters over the Lake House garage?

4.   What is the rise per foot of run of the rafters over the Lake House garage?

5.   What is the run of the rafters over the Lake House bedroom #1?

6.   What is the rise per foot of run of the rafters over the Lake House bedroom #1.

7.   What is the overhang of the rafters over the Lake House bedroom #1?

UNIT 21　Common Roof Framing

After completing this unit, you will be able to perform the following tasks:

- Find information about roof construction on drawings.
- Calculate the length of common rafters.
- Calculate the length of rafter tails when overhang is given.
- Describe framing.

ROOF CONSTRUCTION

The roof is designed to support weight and give protection from the weather. In a common frame roof, the rafters and ridge board are the structural members. They are sized and spaced to support the weight of the roof itself plus any snow that can be expected. The protection from weather is provided by sheathing (or roof decking) and roofing material, Figure 21-1.

The size and spacing of the rafters varies depending on their length, pitch, and *load* (weight they must support). If the rafters span a great distance or are spaced far apart, they must be deep to support their load. (The vertical dimension of rafters and joists is called *depth*). The size and spacing of the rafters is shown on a section view, Figure 21-2. The ridge board is also usually shown on a section view. The ridge board should be made of stock two inches deeper than the rafters. The greater depth is needed because an angled

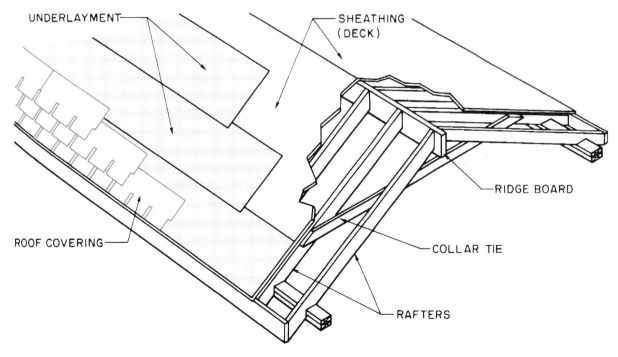

Figure 21-1　Elements of a roof

143

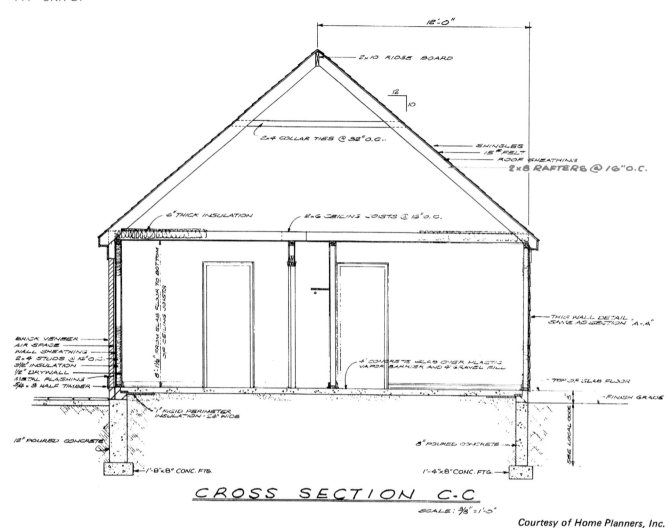

CROSS SECTION C-C

SCALE: 3/8"=1'-0"

Courtesy of Home Planners, Inc.

Figure 21-2 A section view of the roof shows the size and spacing of rafters.

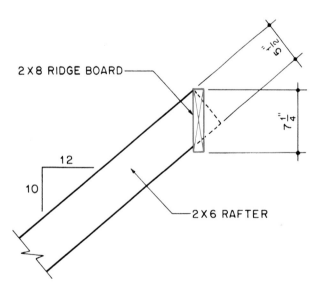

Figure 21-3 The ridge board must be made of wider lumber than the rafters to allow for the angled plumb cut.

cut across the rafters is longer than a square cut across the rafters, Figure 21-3.

When the roof framing is complicated, a separate roof framing plan may be included, Figure 21-4. Roof framing plans are used only to show the general arrangement of the framing. Therefore, unless specific dimensions are included, framing plans should not be relied upon for the lengths of the framing members.

The roof frame may also include *collar beams.* Collar beams are usually made on one-inch (nominal thickness) lumber. They are normally included on every second or third pair of rafters. This information should be shown on the roof section if collar beams are planned.

The slope of the roof may be shown on any section view that shows the rafter size. Slope is usually also shown on the building elevations.

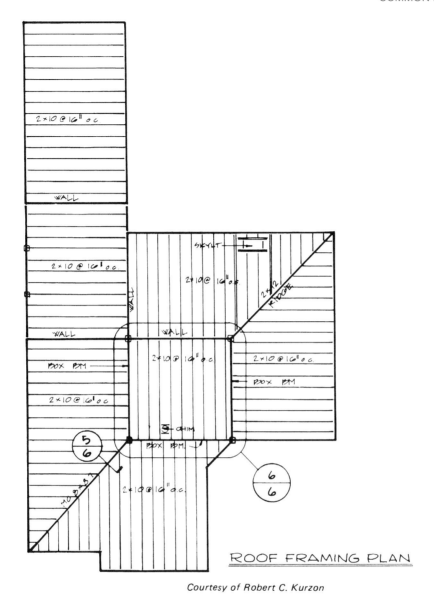

Courtesy of Robert C. Kurzon

Figure 21-4   Roof framing plans show the arrangement of rafters.

GABLE AND RAKE FRAMING

The triangle formed by the top plate of the end walls and the rafters in a gable roof is the *gable.* The gable is framed like the lower parts of the house walls. Unless a detail drawing shows differently, the gable is framed with the same size and spacing of studs as the rest of the walls, Figure 21-5.

The end of a gable roof is called the *rake.* If the rake is *tight* (does not overhang), the gable studs are notched to fit the rake rafters, Figure 21-6. If there is a rake overhang, the gable framing includes a plate. The rake rafters are then supported by *lookouts* nailed to

the last rafter and the gable plate, Figure 21-7. The overhang of the lookouts should not be more than their length inside the gable frame.

ROOF COVERING

The most common roof decking materials for houses are plywood and oriented-strand board (OSB). However, in post-and-beam construction where the extra strength is needed to span the distance between the rafters, dimensional lumber is used for the roof deck.

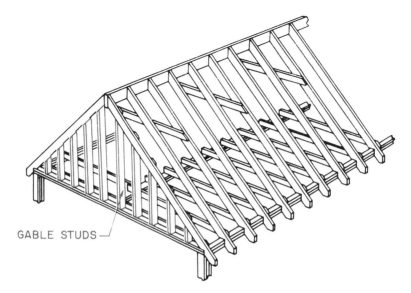

Figure 21-5   The gable is framed with studs.

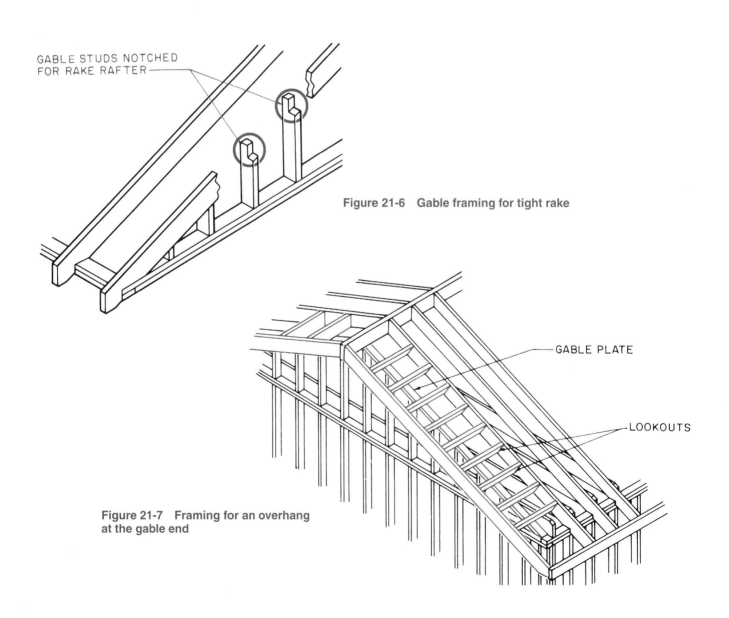

GABLE STUDS NOTCHED
FOR RAKE RAFTER

Figure 21-6   Gable framing for tight rake

GABLE PLATE

LOOKOUTS

**Figure 21-7   Framing for an overhang
at the gable end**

The material used to provide weather protection is often chosen for its architectural style, Figure 21-8. Some of the most common materials are asphalt shingles, wood shingles, and terra-cotta tiles. The material to be used is shown on the section view of the roof and usually on the building elevations. The roofing should be applied over a layer of asphalt-saturated building paper—sometimes called *slater's felt.*

Asphalt roofing materials, including felt and shingles, are sold by their weight per hundred square feet of coverage. One hundred square feet of roof is called a *square.* If enough shingles to cover one square weigh 235 pounds, they are called 235-lb shingles. The felt used under roofing is typically 15-lb weight.

Figure 21-8   Tile roofs are popular in some parts of the country.

FINDING THE LENGTH OF COMMON RAFTERS

Carpenters use a rafter table to find the length of rafters. These tables are available in handbooks and are printed on the face of a framing square, Figure 21-9. To find the length of a common rafter, you must know the run of the rafter and its rise per foot of run. The run can be found on the floor plan of the building. The rise per foot of run is shown on the building elevations. The length of the common rafter is then found by following these steps:

Step 1.   Find the number of inches of rise per foot of run at the top of the table. These numbers are the regular graduations on the square.

Step 2.   Under this number, find the length of the rafter per foot of run. A space between the numbers indicates a decimal point.

Step 3.   Multiply the length of the common rafter per foot of run (the number found in Step 2) by the number of feet of run.

Step 4.   Add the length of the tail and subtract one-half the thickness of the ridge board. The result is the length of the common rafter as measured along the measuring line.

Note: If the overhang is given on the working drawings, it can be added to the run of the rafter instead of adding the length of the tail.

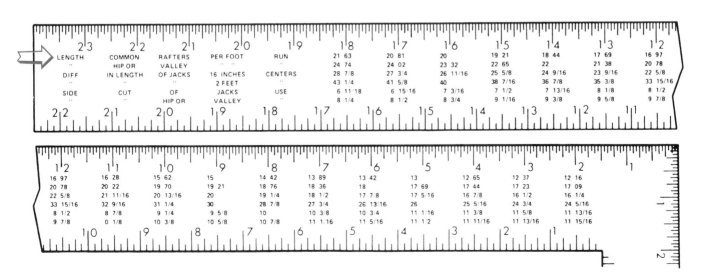

Figure 21-9   Rafter table on the face of a square: The top line is the length of common rafters per foot of run.

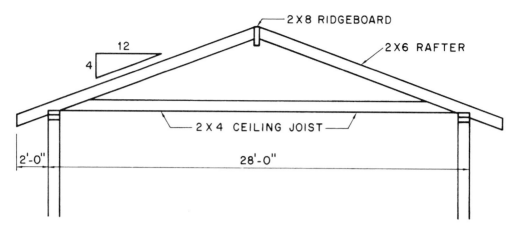

Figure 21-10   Find the length of a common rafter.

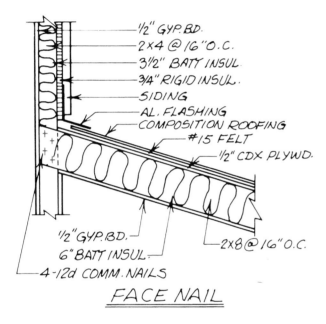

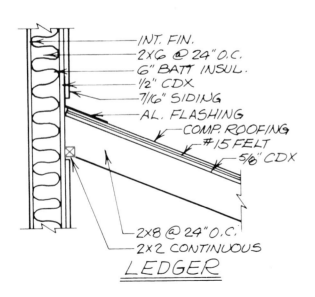

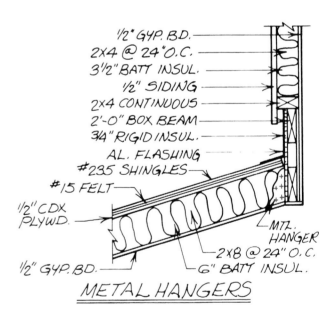

Figure 21-11   Details of shed roof to wall

Example: Find the length of a common rafter for the roof in Figure 21-10.

1. Rise per foot of run = 4"

2. Length of common rafter per foot of run = 12.65"

3. Run of one rafter including overhang = 16'0"

4. 16 × 12.65" = 202.40" (round off to 202½")

5. Subtract ½ the thickness of the ridge board: 202½"-¾" = 201¾"

SHED ROOF FRAMING

Most of the Lake House has shed roof construction. That is, the common rafters cover the full span of the area they cover in a single slope. To find the length of shed rafters, treat the entire span of the rafters as run. For example, the total width of the garage is 13'9". There is no overhang on either side of the garage, so the span of the garage rafters is 13'-9". Use 13'-9", or 13.75 feet, as the run in calculating the length of common rafters. (See Math Review 20.)

Where shed rafters butt against a wall or other vertical surface, the drawings should include a detail to show how they are fastened. Figure 21-11 shows three methods of fastening rafters to a vertical surface.

ROOF OPENINGS

It is often necessary to frame openings in the roof. Where chimneys, skylights, or other features require openings through the rafters; headers and double framing members are used. This method of framing openings is similar to that used in floor framing.

✓ CHECK YOUR PROGRESS

Can you perform these tasks?

☐ List the size and spacing of rafters, ridge board, and collar beams.

☐ Describe the framing of a gable and rake overhang.

☐ List the materials to be used for roof covering.

☐ Use a rafter table to find the length of a common rafter including the rafter tail.

 ASSIGNMENT

Give the following information for the rafters of the Lake House:

Rafter Location	Thickness × Depth	Run	Rise per Foot Run	Length	O.C. Spacing
1. Kitchen					
2. Bedroom #2 and Loft					
3. Living room					
4. Bedroom #1					
5. Garage					

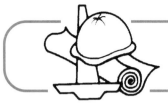

UNIT 22   Hip and Valley Framing

OBJECTIVES

After completing this unit, you will be able to perform the following tasks:

- Calculate the length of hip rafters.
- Calculate the length of valley rafters.
- Calculate the length of hip and valley jack rafters.

HIP RAFTERS

Hip rafters run from the corner of the building to the ridge at a 45° angle, Figure 22-1. The length of hip rafters can be found by using a table found on most framing squares, Figure 22-2. The second line of this table is used to calculate the length of hip rafters. This table is based on the unit-run-and-rise method for finding the length of common rafters, explained earlier in Unit 21.

To calculate the length of a hip rafter, you must know the run of the common rafters in the roof and the unit rise of the roof, Figure 22-3. The length of the hip rafters is then found by using the table for length of hip and valley rafters in the same way the table for the length of common rafters was used in Unit 21.

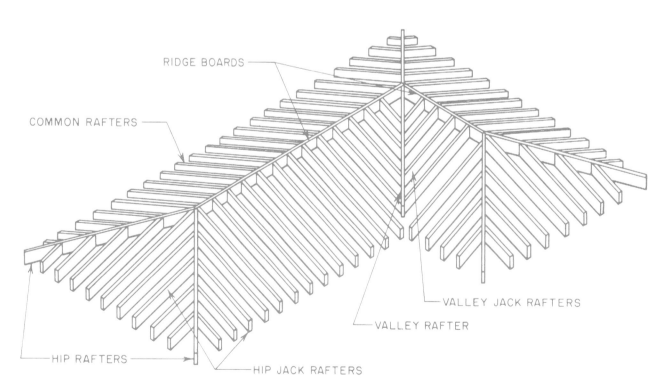

Figure 22-1   Parts of a hip and valley roof frame

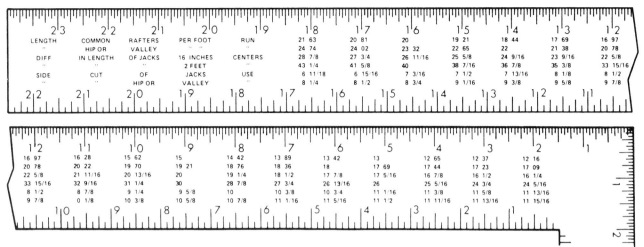

Figure 22-2   Rafter table on the face of a square

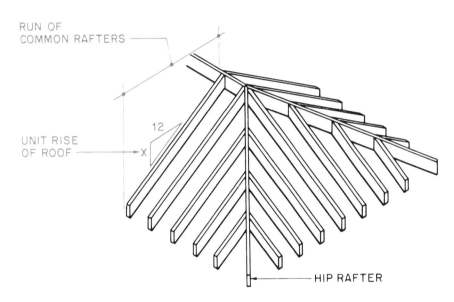

Figure 22-3   To use the table for the length of hip and valley rafters use the run of the common rafters in the roof.

Step 1.   Find the unit rise (number of inches of rise per foot of run) at the top of the table. These numbers are the regular graduations on the square.

Step 2.   Under this number, find the length of the hip rafter per foot of run of the common rafters.

Step 3.   Multiply the length of the hip rafter per foot of common-rafter run (the number found in Step 2) by the number of feet of run of the common rafters (½ the width of the building).

Step 4.   Subtract the ridge allowance. Because the hip rafter meets the ridge board at a 45° angle, the ridge allowance is one half the 45° thickness of the ridge board, Figure 22-4. The 45° *thickness* is the length of a 45° line across the thickness of the ridge board. The 45° thickness of a 1½-inch (2-inch nominal) ridge board is 2⅛ inches. Therefore, the ridge allowance for a hip rafter on a 1½-inch ridge board is 1¹⁄₁₆ inches.

Note: If the hip rafter includes an overhang, add the overhang of the common rafters to the run of the common rafters.

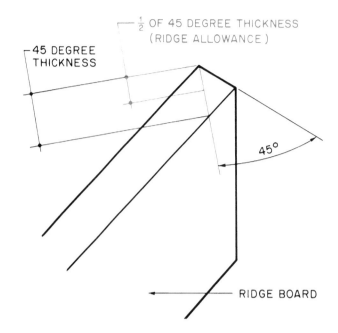

Figure 22-4   Use one-half of the 45° thickness of
the ridge as a ridge allowance for hip rafters.

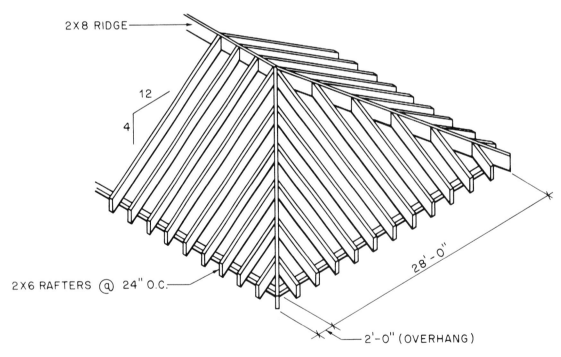

Figure 22-5   Find the length of a hip rafter.

Example: Find the length of a hip rafter for the roof shown in Figure 22-5.

1. Rise per foot of run = 4"

2. Length of hip rafter per foot of common-rafter run = 17.44"

3. Common-rafter run including overhang = 16'-0"

4. 16 × 17.44" = 279.04" (round off to 279$^1/_{16}$")

5. Subtract 1$^1/_{16}$" ridge allowance: 279$^1/_{16}$"- 1$^1/_{16}$" = 278"

The rafters that butt against the hip rafter (called *hip jack rafters*) are cut at an angle, Figure 22-6. This angled cut produces a surface which is longer than the width of the lumber from which the rafter is cut. Therefore, the hip rafters are made of wider lumber than the common rafters and jack rafters.

VALLEY RAFTERS

The line where two pitched roofs meet is called a *valley.* The rafter that follows the valley is a *valley rafter,* as shown in Figure 22-6. It is most common for both roofs to have the same pitch. This results in the valley rafter being at a 45° angle with both ridges—the same angle as a hip rafter. Because the angles of the hip and valley rafters are the same and the pitch is the same, the same table can be used to compute their lengths.

All steps of the procedure given earlier for hip rafters can be followed to find the length of valley rafters. However, the valley rafters often have no tail even though the roof has an overhang, Figure 22-7. In this case, the total length of the valley rafters is computed on the basis of the run of the common rafters excluding the overhang.

When both roofs have the same span and rise, the valley extends from the eave to the ridge, Figure 22-8. When one roof has a greater span than the other, the valley does not reach the ridge, Figure 22-9. In this case, the valley is framed in one of two ways. One valley rafter can extend to the ridge, and the other can butt against the first, Figure 22-10. The other method is to install a common rafter on the wider roof in line with the ridge of the narrower roof. Both valley rafters can then butt against this common rafter, Figure 22-11.

The length of the valley rafters is based on the run of the common rafters that have their upper (ridge) ends at the same level as the valley rafter. In other words, when the long and short valley rafters are used, the length of the long valley rafter is based on the run of the common rafters in the wider roof. The length of the short valley rafter is based on the run of the common rafters in the narrower roof. When both valley rafters butt against a common rafter of a higher roof, their lengths are based on the run of the common rafters in the narrower roof.

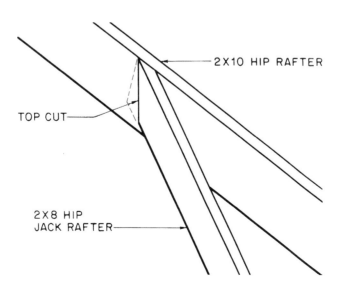

Figure 22-6 Because the top of the hip jack is cut on an angle, the hip rafter must be wider.

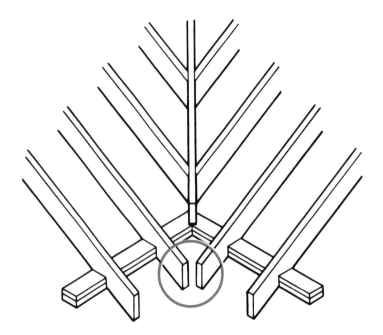

Figure 22-7   On some roofs the common rafter tails are close enough so no tail is needed on the valley rafters.

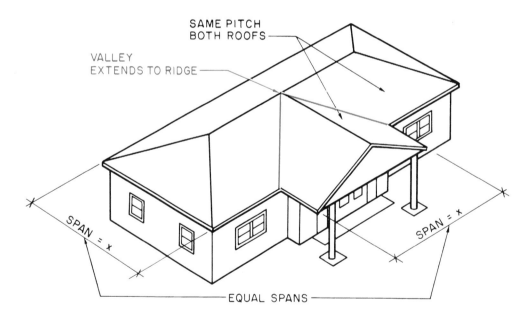

Figure 22-8   When both spans are the same distance and the same pitch, the valley goes all the way to the ridge.

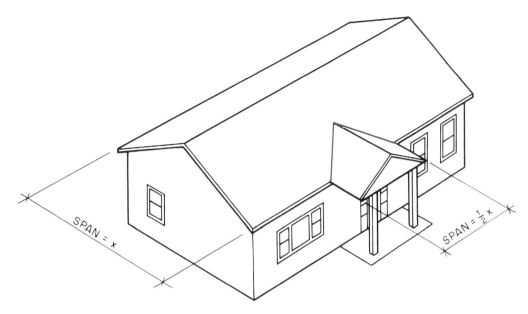

Figure 22-9   When the spans are not equal, the valley does not reach the ridge.

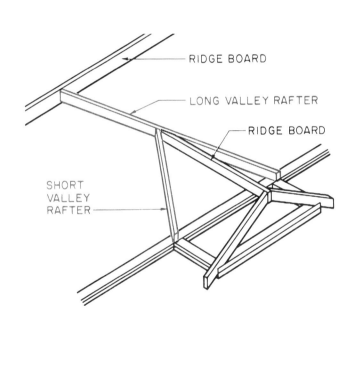

Figure 22-10   Framing valleys with long and short valley rafters

Figure 22-11   A valley framed against a common rafter.

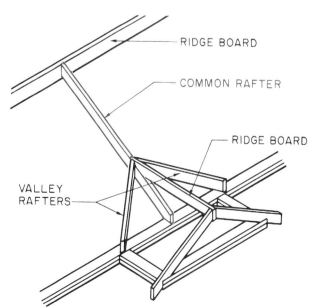

JACK RAFTERS

Rafters that extend from the wall plate to a hip rafter are called *hip jack rafters,* as was shown in Figure 22-1. Those that extend from a valley rafter to the ridge board are called *valley jack rafters,* also shown in Figure 22-1.

The third and fourth lines of the rafter table on most framing squares, shown in Figure 22-2, are used to calculate the length of jack rafters. The length of each jack rafter in a roof varies from the length of the one next to it by the same amount, Figure 22-12. The amount of this variance depends on the spacing of the rafters and the pitch of the roof. The third line of the rafter table is used when the rafters are spaced 16 inches O.C.; the fourth line is used when they are spaced 24 inches O.C. As with the other lines of the rafter table, the inch numerals at the top of the square are used to indicate the unit rise of the roof. For example, if a roof has a 6 in 12 slope and the roof framing is 16 inches O.C., the difference in the length of jack rafters is $17^7/_8$ inches.

The first jack rafter should be a full space (16 inches or 24 inches) from the bottom of the hip or the top of the valley. Therefore, the first jack rafter should be the length shown on the rafter table.

The length of the tail must be added to the theoretical length from the table. The length of the tail on hip jacks is the same as the length of the tail on common rafters. Therefore, the length of the tail can be found by using the table for the length of common rafters. Simply treat the tail as a very short common rafter. If the overhang is not in even feet, divide the length of common rafter per foot of run on the table by twelve. This gives you the length of the common rafter (or jack rafter tail) per inch of run.

Example: Find the length of the tails of the jack rafters in Figure 22-13.

1. Unit rise (rise per foot of run) = 4"

2. Length of common rafter per foot of run = 12.65"

3. Length of common rafter per inch of run =
$$\frac{12.65"}{12} = 1.05"$$

4. Run of rafter tail (overhang) = 8"

5. Length of rafter tail = 8 × 1.05" = 8.4", approximately $8^7/_{16}"$.

Hip jack rafters are also shortened at the top to allow for the thickness of the hip rafter they butt against. This allowance is one-half the 45° thickness of the hip rafter. Valley jack rafters are shortened at the bottom to allow for the thickness of the valley rafter. This allowance is one-half the 45° thickness of the valley rafter. The valley jack rafters are also shortened at the top to allow for the thickness of the ridge board. The ridge board allowance for valley jacks is the same as the ridge board allowance for common rafters—one-half the actual thickness of the ridge board.

Example: Find the length of the hip jack rafters (A, B, and C) and the valley jack rafters (D, E, and F) in Figure 22-13.

Hip Jack Rafters

1. Rise per foot of run = 4"

2. Spacing of rafters = 24" O.C.

3. Difference in the length of jacks = $25^5/_{16}"$

4. Theoretical length of hip jack rafter A = 0 + $25^5/_{16}" = 25^5/_{16}"$

5. Add tail as found in earlier example: $25^5/_{16}"$ + $8^7/_{16}" = 33^3/_4"$

6. Subtract one half the 45° thickness of the hip rafter $33^3/_4" - 1^1/_{16}" = 32^{11}/_{16}"$ (actual length of A)

7. Actual length of hip jack B—$32^{11}/_{16}"$ (length of A) + $25^5/_{16}$ (from rafter table) = 58"

8. Actual length of hip jack C = 58" (length of B) + $25^5/_{16}" = 83^5/_{16}"$

Valley Jack Rafters

9. Theoretical length of valley jack rafter D = 0 + $25^5/_{16}" = 25^5/_{16}"$

10. Subtract one-half the 45° thickness of the valley rafter: $25^5/_{16}" - 1^1/_{16}" = 24^1/_4"$

11. Subtract one-half the actual thickness of the ridge board: $24^1/_4" - ^3/_4" = 23^1/_2"$ (actual length of D)

12. Actual length of valley jack E = $23^1/_2"$ (length of D) + $25^5/_{16}"$ (from rafter table) = $48^{13}/_{16}"$

13. Actual length of valley jack F = $48^{13}/_{16}"$ (length of E) + $25^5/_{16}" = 74^1/_8"$

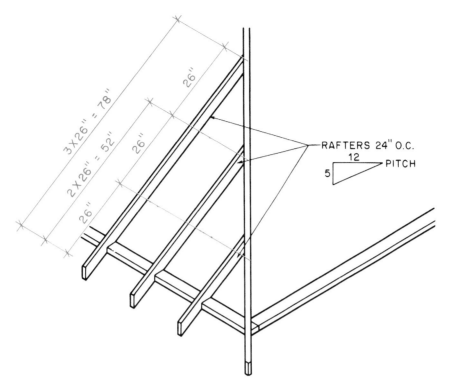

Figure 22-12   Each jack rafter in a string varies from the next by the same amount.

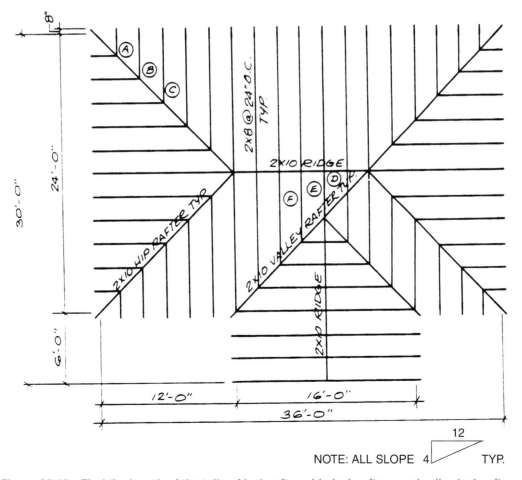

Figure 22-13   Find the length of the tails of jack rafters, hip jack rafters, and valley jack rafters.

✓ CHECK YOUR PROGRESS ——————————————————————————

Can you perform these tasks?

☐ Use a rafter table to find the length of hip rafters.

☐ Use a rafter table to find the length of valley rafters.

☐ Use a rafter table to find the length of hip jack rafters and valley jack rafters.

ASSIGNMENT ——————————————————————————————————

A. Refer to Figure 22-14 to complete questions 1-11.

1. What is the run of the common rafters at A?

2. How much overhang does the roof have?

3. What is the actual length of the common rafters at A?

4. What is the actual length of the hip rafter at B?

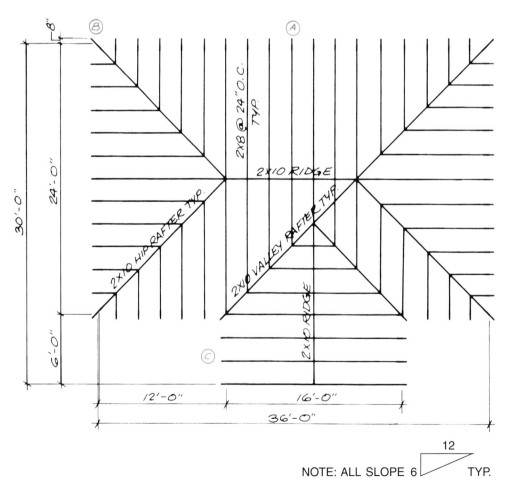

Figure 22-14

5. What is the run of the common rafters at C?

6. What is the actual length of the common rafters at C?

7. What is the length of the short valley rafter?

8. What is the actual length of the shortest hip jack rafter?

9. What is the actual length of the second shortest hip jack rafter?

10. What is the actual length of the shortest valley jack rafter?

11. What is the actual length of the second shortest valley jack rafter?

B. Refer to Lake House Drawings (in the packet) to complete this part of the assignment.

What is the length of the structural steel hip rafter over the dining room?

Notes: • This hip rafter is a steel channel shown on details 5/6 and 6/6, and marked as MC8X8.7

• Remember to allow for the distance from the column centerline and the end of the rafter as dimensioned on the detail drawing.

• This roof has an unusual pitch of 2.96 in 12. This is close enough to use 3 in 12 for calculating rafter lengths.

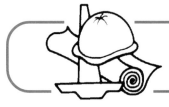

UNIT 23 Cornices

OBJECTIVES

After completing this unit, you will be able to perform the following tasks:

- Describe the cornice construction shown on a set of drawings.
- List the sizes of the individual parts of the cornice shown on a set of drawings.
- Describe the provisions for attic or roof ventilation as shown on a set of drawings.

TYPES OF CORNICES

The *cornice* is the construction at the place where the edge of the roof joins the sidewall of the building. On hip roofs, the cornice is similar on all four sides of the building. On gable and shed roofs, the cornice follows the pitch of the end (*rake*) rafters. The cornice on the ends of a gable or shed roof is sometimes called simply the rake, Figure 23-1. The three main types of cornice are the *box cornice,* the *open cornice,* and the *close cornice.*

Box Cornice

The box cornice boxes the rafter tails. This type of cornice includes a fascia and soffit, Figure 23-2. The *fascia* covers the ends of the rafter tails. The *soffit* covers the underside of the rafter tails. There are three types of box cornices. These types vary in the way the soffit is applied.

Sloping Box Cornice

In the sloping box cornice, the soffit is nailed directly to the bottom edge of the rafter tails. This causes the soffit to have the same slope or pitch as the rafter, Figure 23-3.

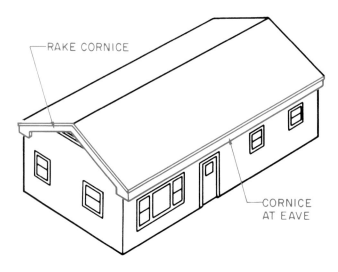

Figure 23-1   The cornice is the construction at the place where the roof and the sidewall meet.

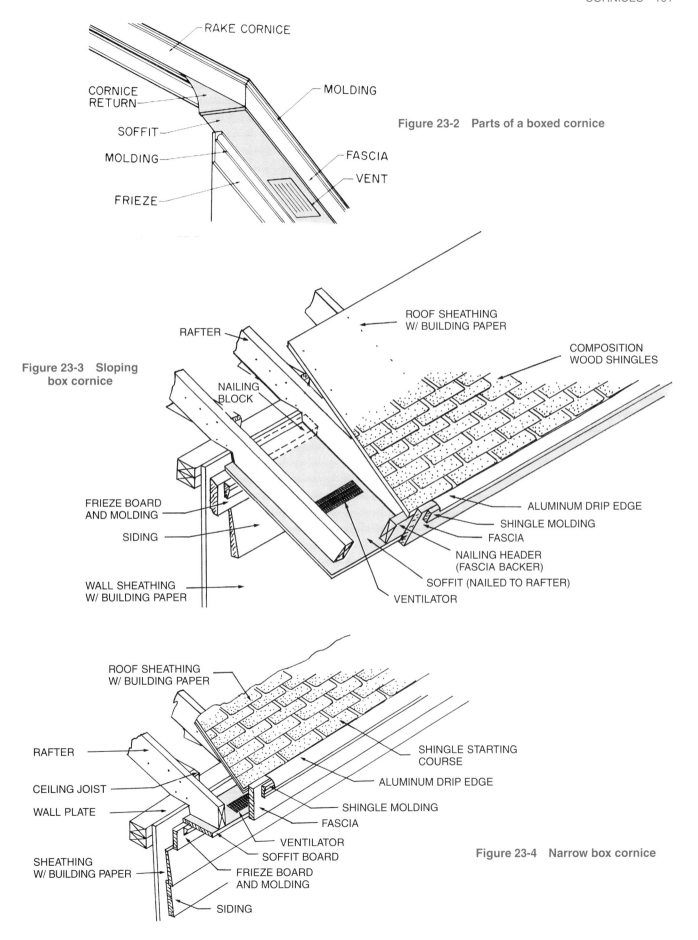

RAKE CORNICE

MOLDING

CORNICE RETURN

SOFFIT

MOLDING

FRIEZE

FASCIA

VENT

Figure 23-2   Parts of a boxed cornice

Figure 23-3   Sloping box cornice

RAFTER

ROOF SHEATHING W/ BUILDING PAPER

COMPOSITION WOOD SHINGLES

NAILING BLOCK

FRIEZE BOARD AND MOLDING

SIDING

WALL SHEATHING W/ BUILDING PAPER

ALUMINUM DRIP EDGE

SHINGLE MOLDING

FASCIA

NAILING HEADER (FASCIA BACKER)

SOFFIT (NAILED TO RAFTER)

VENTILATOR

ROOF SHEATHING W/ BUILDING PAPER

RAFTER

CEILING JOIST

WALL PLATE

SHEATHING W/ BUILDING PAPER

SIDING

SHINGLE STARTING COURSE

ALUMINUM DRIP EDGE

SHINGLE MOLDING

FASCIA

VENTILATOR

SOFFIT BOARD

FRIEZE BOARD AND MOLDING

Figure 23-4   Narrow box cornice

Narrow Box Cornice. In the narrow box cornice, the rafter tails are cut level. The soffit is nailed to this level-cut surface, Figure 23-4.

Wide Box Cornice. In a wide box cornice, the overhang is too wide for a level cut on the rafter tails to hold the full width of the soffit. In conventional wood framing, *lookouts* are installed between the rafter ends and the sidewall. The lookouts provide a nailing surface for the soffit, Figure 23-5. For a metal soffit, special metal channels fastened to the sidewall and back of the fascia hold the soffit, Figure 23-6.

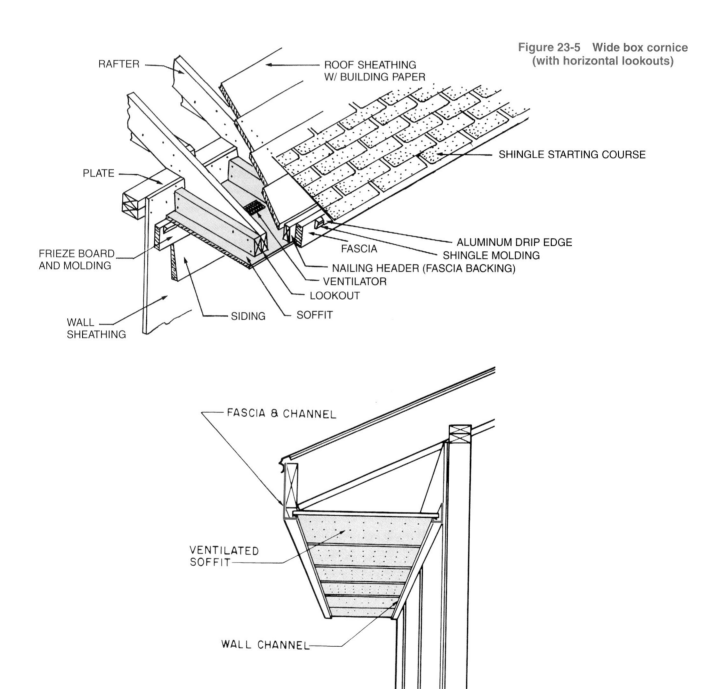

Figure 23-5 Wide box cornice (with horizontal lookouts)

RAFTER

ROOF SHEATHING W/ BUILDING PAPER

SHINGLE STARTING COURSE

PLATE

FRIEZE BOARD AND MOLDING

ALUMINUM DRIP EDGE
SHINGLE MOLDING
FASCIA
NAILING HEADER (FASCIA BACKING)
VENTILATOR
LOOKOUT

WALL SHEATHING

SIDING
SOFFIT

Figure 23-6 Metal and vinyl soffit systems include ventilated soffits and channels.

FASCIA & CHANNEL

VENTILATED SOFFIT

WALL CHANNEL

Open Cornice

In an open cornice, the underside of the rafters is left exposed, Figure 23-7. Blocking is installed between the rafters and above the wall plate to seal the cornice from the weather. An open cornice may or may not include a fascia.

Close Cornice

In a close cornice, the rafters do not overhang beyond the sidewall, Figure 23-8. The interior may be sealed by the sheathing and siding or by a fascia. In either case, there must be some provision for ventilation.

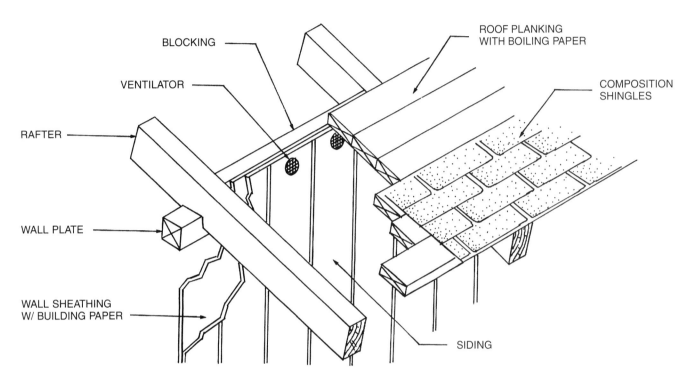

Figure 23-7   Open cornice

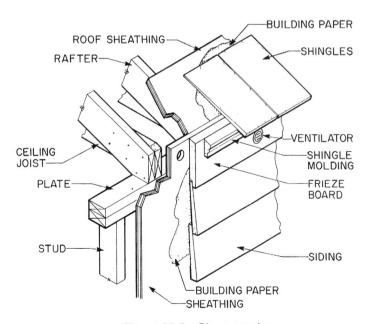

Figure 23-8   Close cornice

CORNICE RETURNS

Any type of construction described for cornices can be used for the rake. When a sloping cornice is used, the fascia and soffit follow the line of the roof up the rake. When a level box cornice us used, a *cornice return* is necessary. This is the construction that joins the level soffit and fascia of the eave with the sloping rake, Figure 23-9.

The style of cornice return is shown on the building elevations, Figure 23-10. Although good architectural drafting practice requires details of all special construction, many architects do not include details of cornice returns. The carpenter is expected to know how to achieve the desired results. Figures 23-11 and 23-12 show the construction of two popular types of cornice returns.

VENTILATION

The cornice usually allows for ventilation of the attic or roof. Attic or roof ventilation is necessary in both hot and cold weather. Without ventilation, the air in the

Courtesy of Richard T. Kreh, Sr.

Figure 23-9   Cornice return

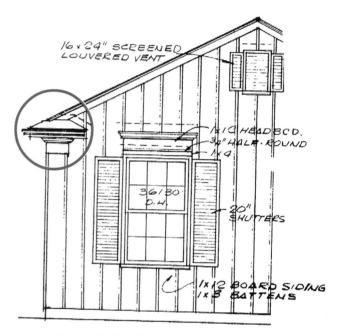

Figure 23-10   The cornice returns can be seen on the building elevations.

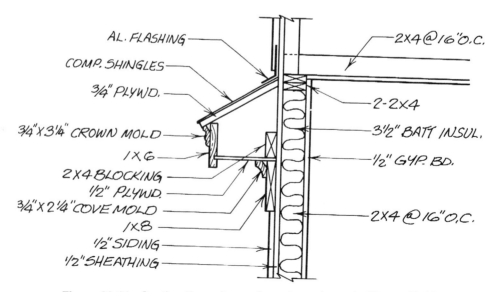

Figure 23-11   Section through cornice return shown in Figure 23-10.

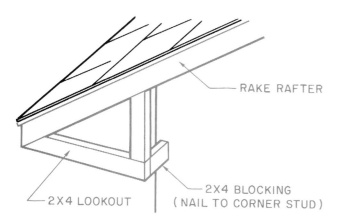

Figure 23-12   Framing for the return shown in Figure 23-9

attic becomes stagnant because it is trapped and unable to circulate.

In hot weather, this stagnant air builds up heat and makes the house warmer. Hot, stagnant air can hold a large amount of moisture. When this moisture-laden air comes in contact with the cooler roof, the moisture condenses. The condensation can reduce the effectiveness of the insulation. Condensation can also cause the wood in the attic to rot.

Attic or roof ventilation also helps prevent ice buildup in cold climates. Without ventilation, the heat from the building melts the snow that falls on the roof. As the melted snow reaches the overhang of the roof, it refreezes. Eventually an ice dam may build up. The ice dam can back up newly melted snow, causing it to seep under the shingles, Figure 23-13.

Ventilation is created by allowing cool air to enter through the cornice and exit at the ridge or through special ventilators, Figure 23-14. A section view of the sidewalls or a special roof and cornice detail shows the construction of the cornice including any ventilators. Notice that each of the soffits in Figures 23-2, 23-4, and 23-5 includes a ventilator.

The heated air can be allowed to exit in one of three ways. Some buildings have a ventilated, metal ridge cap. This may be shown on the building elevations or on a separate detail. When a ventilated ridge is used, an opening is left in the roof decking at the ridge. Metal roof ventilators can be installed in the roof. Ventilators can be installed in the gables. Metal roof ventilators and gable ventilators are usually shown on the building elevations only.

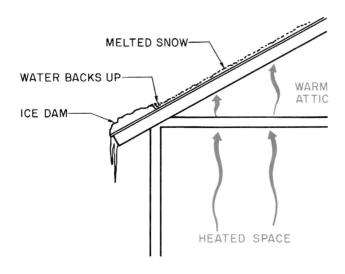

Figure 23-14   Air flow through the attic

Figure 23-13   Ice dam

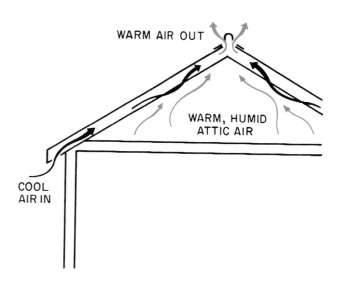

✓ CHECK YOUR PROGRESS ─────────────────────────

Can you perform these tasks?

☐ Describe the construction of a typical sloping box cornice.

☐ Describe the construction of a typical narrow box cornice.

☐ Describe the construction of a typical wide box cornice.

☐ Describe the construction of a typical open cornice.

☐ Describe the construction of a typical close cornice.

☐ Describe the construction of a return for each of the above cornice types.

☐ Trace the flow of ventilating air from the point it enters a cornice to the point it exits the attic.

 ASSIGNMENT ─────────────────────────

Refer to the Two-Unit Apartment and the Lake House drawings (in the packet) to complete the assignment.

1. Which type of cornice does the Apartment have?

2. What material is used for the Apartment cornice?

3. How wide is the Apartment soffit?

4. The Apartment fascia is made of two parts. What are they?

5. What provision does the Apartment cornice have for ventilation?

6. How does attic air exit from the Apartment?

7. Sketch the Lake House cornice and show where air enters for ventilation.

8. There are two ways that air can escape from the Lake House roof. Describe one.

UNIT 24   Windows and Doors

OBJECTIVES

After completing this unit, you will be able to perform the following tasks:

- Interpret information shown on window and door details.

- Find information in window and door manufacturers' catalogs.

WINDOW CONSTRUCTION

Most windows are supplied by manufacturers as completely assembled unit. However, the carpenters who install windows often have to refer to window details for information. Some special installations require knowledge of the construction of the window unit. Also, when a special window is required, the carpenter may build parts of it on the construction site.

Wood Windows

The major types of windows were briefly discussed in Unit 19. All these windows include a frame and sash. The *sash* is the glass and the wood (or metal) that holds the glass. The sash is made of *rails* (horizontal parts) and *stiles* (vertical parts), Figure 24-1. The sash may also include muntins. *Muntins* are small strips that divide the glass into smaller panes. The glass is sometimes called the *lite*.

The window frame is made of the *side jambs*, the *head jamb,* and the *sill*. Stop molding is applied to the inside of the jambs to hold the sash in place. Factory-built windows also come with the exterior casing installed. The *casing* is the molding that goes against the wall around the frame. The interior casing and the apron, if one is included, are applied after the window is installed.

Metal Windows

Many buildings have vinyl or metal windows. Improvements in the design of metal windows have made them competitive with wood windows in both cost and energy efficiency. The most important of these design improvements has been the development of thermal-break windows. *Thermal-break* windows use a combination of air spaces and materials that do not conduct heat easily to separate the exterior from the interior, Figure 24-2.

The basic parts of a metal or vinyl window are similar to those of a wood window. The sash consists of stile, rails, and glazing. The frame is made up of side

Figure 24-1   Parts of a window

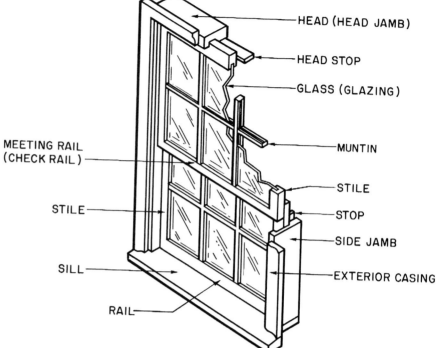

HEAD (HEAD JAMB)

HEAD STOP

GLASS (GLAZING)

MUNTIN

MEETING RAIL
(CHECK RAIL)

STILE

STOP

STILE

SIDE JAMB

SILL

EXTERIOR CASING

RAIL

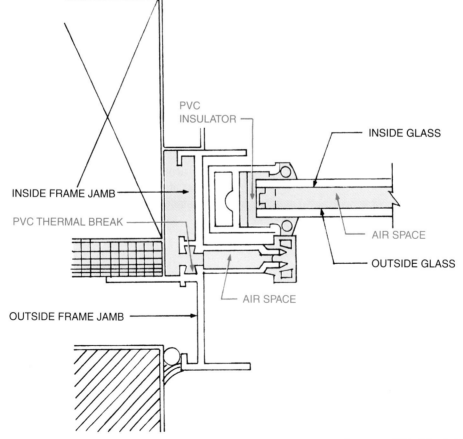

PVC
INSULATOR

INSIDE GLASS

INSIDE FRAME JAMB

AIR SPACE

PVC THERMAL BREAK

OUTSIDE GLASS

AIR SPACE

OUTSIDE FRAME JAMB

Courtesy of Ethyl Capitol Products Corporation

Figure 24-2   **Thermal-break windows use insulating materials and air spaces to separate the interior from the exterior.**

jambs, head jamb, and sill. However, the trim (casing) is not included as part of the window. Often the window frame itself is the only trim used on the exterior. The frame includes a nailing fin for attaching the window to the building framing.

WINDOW DETAILS

All windows include the parts discussed so far. However, to show the smaller parts which vary from one window style to another, architects and manufacturers

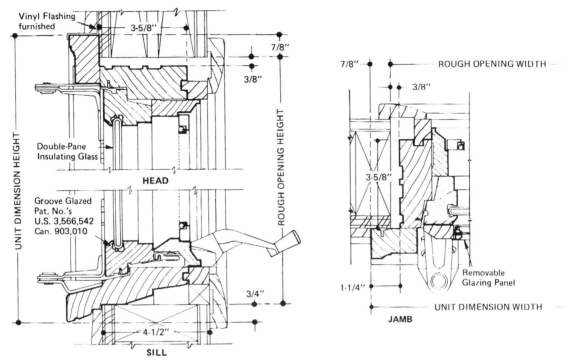

scale: 3" = 1'-0"

Courtesy of Andersen Corporation, Bayport, Minnesota 55003

Figure 24-3   Typical window details drawings.

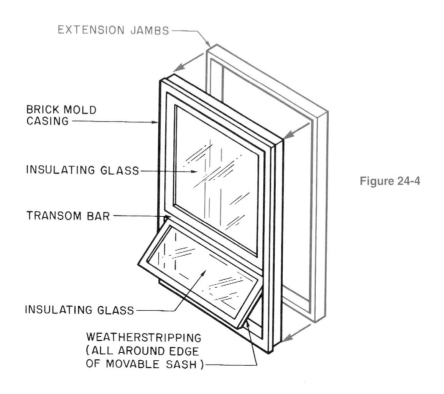

Figure 24-4

use detail drawings. The most common type of window detail is a section, Figure 24-3. All of the parts can be shown in section views of the head, sill, and one side jamb. These sections also usually show the wall framing around the window.

Some of the parts that can be found on window sections are defined here. Find each of the parts on the sections and illustrations in Figures 24-4, 24-5, and 24-6.

- *Weather stripping* is used on windows that open and close. It forms a weather-tight seal around the sash.

- The *transom bar* is the horizontal part of a window frame that separates the upper and lower sash.

- *Meetings rails* or check rails are the rails that meet in the middle of a double-hung window.

- *Insulating glass* is a double layer of glass, creating a dead-air space. The dead air acts as an insulator.

- *Extension jambs* are fastened to standard jambs when the window is installed in a thicker than normal wall.

- A *mullion* is a vertical section of the frame that separates side-by-side sash. If the mullion is formed by butting two windows together, it is called a narrow mullion. If the mullion is built around a stud or other structural support, it is called a support mullion.

Details for metal windows are often drawn as simplified sections of the frame only. Figure 24-7 shows typical details for the thermal-break window in Figure 24-2.

DOOR CONSTRUCTION

Doors include many of the same basic parts as windows, Figure 24-8. A door frame consists of side jambs and a head jamb with stop and casing. Exterior door frames also include a sill. Many doors are made of a framework with panels, Figure 24-9. The parts of panel doors are named similarly to the parts of a window. The vertical parts are *stiles,* and the horizontal parts are *rails.* Doors with glass or louvers are variations of panel doors. The framework is made of rails and stiles, and the glass or louvers

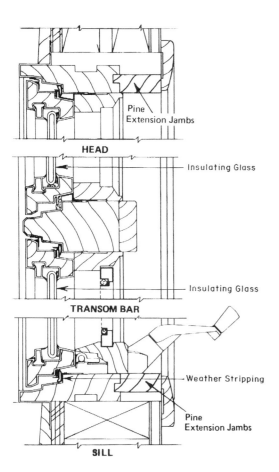

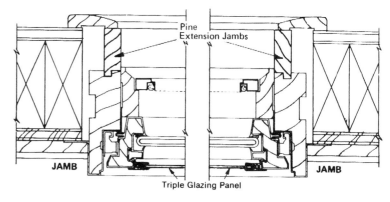

AWNING DETAILS

scale: 3″ = 1′-0″

Courtesy of Andersen Corporation, Bayport, Minnesota 55003

Figure 24-5

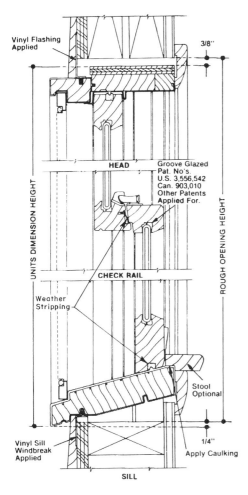

3/8"

HEAD

Groove Glazed
Pat. No's.
U.S. 3,556,542
Can. 903,010
Other Patents
Applied For.

UNITS DIMENSION HEIGHT

ROUGH OPENING HEIGHT

CHECK RAIL

Weather
Stripping

Stool
Optional

Vinyl Sill
Windbreak
Applied

1/4"

Apply Caulking

SILL

INSTALLATION DETAILS
scale: 3" = 1'-0"

Courtesy of Anderson Corporation, Bayport, Minnesota 55003

Figure 24-6

replace the panels. Several manufacturers make molded doors. The most common type of molded door is made of hardboard for interior doors or steel for exterior doors which is manufactured in folds that contour the surface to look like panel doors, Figure 24-10. *Hollow-core doors* consist of an internal frame with flat "skin" applied to each side, Figure 24-11. Some exterior doors have insulation between the two outer steel skins. These insulated doors result in considerable heating and cooling savings.

DOOR DETAILS

Door details are usually less complex than window details. Where security, fire alarm, electronic locks, or special systems must be run in door and window metal frames, the corresponding installers must familiarize themselves with the actual manufacturer details and coordinate their installation with the door and window installation schedule. Carpenters rarely make doors, so all that is needed are simple details of the door frame and its trim, Figure 24-12.

Many doors are sold as prehung units. In these units the frame is assembled, including the trim, and the door is hung in the frame. A section view of the jambs shows how the door is installed. For example, the door detailed in Figure 24-12 is made with two-piece jambs. These split jambs are pulled apart; each side is then slid into the opening for installation. The

Figure 24-7   Typical metal window details

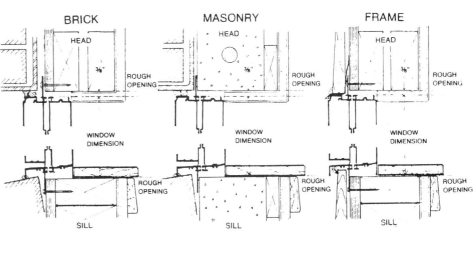

Courtesy of Ethyl Capitol Products Corporation

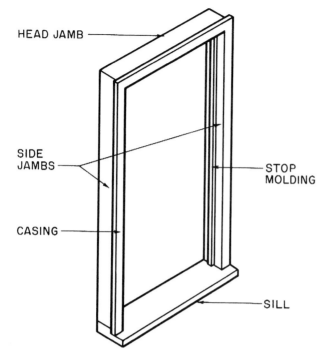

Figure 24-8   Parts of a door frame

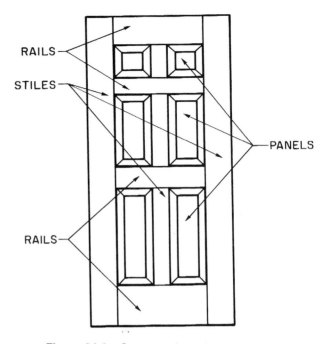

Figure 24-9   Construction of a panel door

Figure 24-10   Molded hardboard door

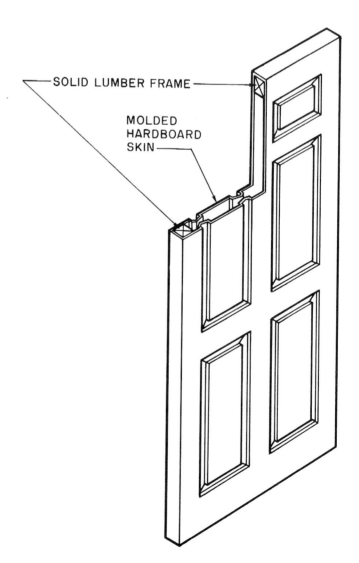

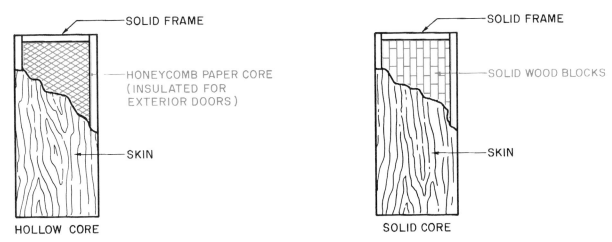

Figure 24-11   Flush doors

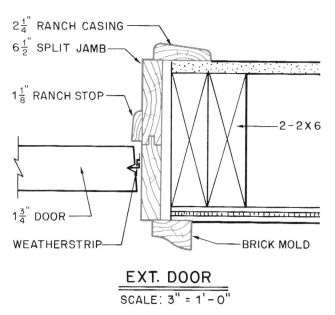

Figure 24-12   Typical jamb detail for an exterior door

stop can be either applied or integral. *Applied stop* is molding which is applied to the jambs with finish nails. *Integral stop* is milled as a part of the jamb when the jamb is manufactured.

READING CATALOGS

It is often necessary to find specific information about windows or doors in the manufacturer's catalog. Usually the catalog has a table of contents listing the types of windows and doors shown. Figure 24-13 shows typical pages reprinted from a manufacturer's catalog. A careful reading of these sample pages will help you to find the information you need in manufacturers' catalogs. For each type of window or door, you will find some or all of the listed information:

- A brief description of the window type and some of the features the manufacturer wants to highlight—a little advertising.

- Installation detail drawings.

- Sizes available—This usually consists of drawings of the various sizes and arrangements, with dimensions for glass size, stud or rough opening, and unit dimensions. (Notice that the nominal size in the catalog sample is written as a four-digit number. The first two digits are the approximate width in feet and inches. The last two digits are the approximate height in feet and inches. For example, a 6030 window is roughly 6 feet wide by 3 feet high.)

- Additional information, such as optional equipment available.

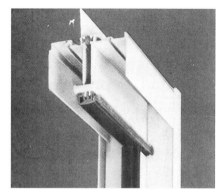

THERMAL-BREAK FRAME SECTIONS—Outside frames at the nailing fins are separated by rigid foamed poly-vinyl chloride, roll pressed to the metal frames. Inside, the frames are separated with extruded poly-vinyl chloride and weather stripped with mylar shielded pile. "Push-pull", panel retaining pin is shown in its position in head section. Pin allows easy removal, and locking into position of the fixed panel (no tools are needed). Integral nailing fins surround the entire frame for easy window installation.

CAPITOL'S

*E-700 ALUMINUM

INSULATED GLASS, THERMAL-BREAK, ROLLING WINDOW

Rated HS-B2-HP (residential)   Rated HS-A2-HP (architectural)

The latest addition to Capitol's fine line of aluminum thermal-break products, the E-700 affords the benefits of thermal-break performance and the convenience of insulated glass . . . frost-free window frames . . . virtual elimination of window glass condensation. In addition to its excellent "energy saving" qualities, the E-700 offers these "sales appealing" features: Window can install with or without sash panels mounted—panels can be installed later from inside the building . . . Panels can be removed and replaced without tools and fixed panel is held in place with a finger operated "push-pull" retaining pin . . . Roll panel latches and automatically locks in closed position. Windows are available in arrangements such as side vents, center vents, sub-, side-, transom lites and fixed/removable lites. All models are available in PPG enamelized and baked-on colors (standard colors at no added cost).

*U.S. Patent No. 4,151,682

HORIZONTAL OR VERTICAL WINDOW MULLING

contact Capitol for complete details.

STACK MULL

SIDE BY SIDE MULL

5" PARTITION MULL

POSITIVE ACTION, SELF-LOCKING

LATCH—Finger-tip ease of operation with this attractive, rugged lock-latch. Latch is located on the interlock stile of the rolling panel and is constructed of Celcon, Ultra-violet stabilized, acetal copolymer M90-08. Latch automatically locks the window when panel is moved into closed position.

Exterior screen of fiberglas mesh, can be inserted or removed from inside the building without removing glass panels. (Optional at added cost)

INTERLOCKING MEETING STILES—The key to the thermal-break feature, at the panel interlocks, is revealed in this photo. Fixed sash interlock stile is at left and rolling sash interlock stile at the right. The stile interlocks are insulated and parted with rigid Celcon separator to maintain thermal-break performance. Insulated glass is cushioned in flexible PVC and full length flexible vinyl flaps seal panels at each interlock.

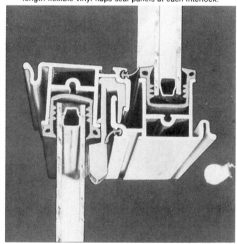

Courtesy of Ethyl Capitol Products Corporation

Figure 24-13   Pages are reprinted from typical manufacturer's catalog.

Fully Effective Thermal Performance Without Breaking (weakening) the Fixed and Rolling Panels!

This detail shows, that we have, without exception, succeeded at each and every part of this design to completely separate the exterior "cold" window surfaces from the interior "warm" window surfaces - both at frame, sash and glass, by the use of enclosed air spaces (as the most effective insulation) and by applied low conductive plastic insulators. And this has been achieved by only thermo-breaking the frame around the center divider and by shielding the unbroken sashes with the divider either from their exposure to the inside or to the outside of the building.

In other words: The fixed outside (cold) sash has been insulated from the interior, and the rolling (warm) sash has been insulated from the exterior.

And we can confirm our theories with a thermal test report by the AAMA authorized Electrical Test Laboratories in New York.

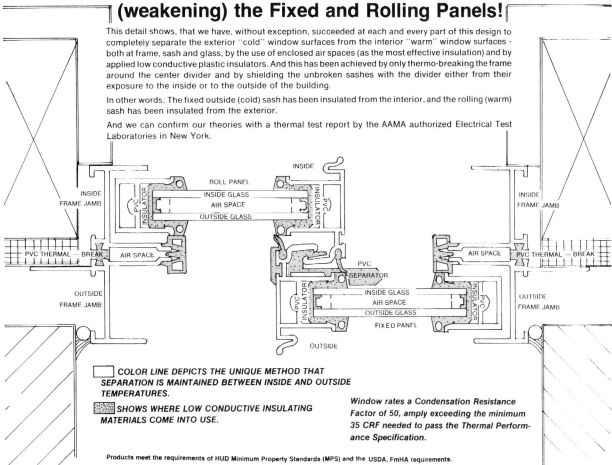

COLOR LINE DEPICTS THE UNIQUE METHOD THAT SEPARATION IS MAINTAINED BETWEEN INSIDE AND OUTSIDE TEMPERATURES.

SHOWS WHERE LOW CONDUCTIVE INSULATING MATERIALS COME INTO USE.

Window rates a Condensation Resistance Factor of 50, amply exceeding the minimum 35 CRF needed to pass the Thermal Performance Specification.

Products meet the requirements of HUD Minimum Property Standards (MPS) and the USDA, FmHA requirements.

SILL AND ROLLING PANEL SECTION—Sloping sill drains water to the outside and is machined to receive one-way weep valves that prevent air infiltration. The thermal-break feature between the inside and outside frame and the weatherstripping are clearly shown here. The roll panel stile shows the Celcon corner section that acts both as a housing for the roller and a guide for the panel as it moves in the sill cavity.

ALUMINUM ROLLERS—Celcon housing is designed to make possible a two position roller height. Roller wheels are of aluminum and ride on a rigid vinyl track for smooth panel operation. Notice that panel frames do not butt at corners. They telescope to achieve a rigid corner and eliminate gaps at the corner joints.

FIXED PANEL REMOVAL—Retaining pins at the frame head and sill lock the fixed window panels in place. Fixed panel cannot be removed when roll panel is in closed position. Retaining pins can be moved without the use of tools to easily remove the fixed panel.

To provide a weather tight seal at the machined areas at the top and bottom of the interlocks, "finger-operated" PVC sealer blocks are used.

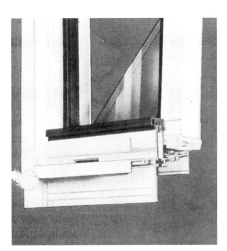
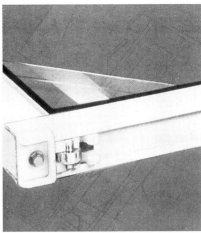
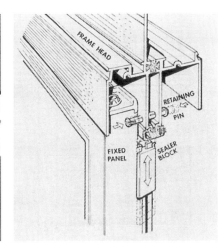

Figure 24-13 (continued)

SPECIFICATIONS

GENERAL: Windows shown are identified as series E-700, rated HS-B2-HP (residential) and series E-705, rated HS-A2-HP (Architectural), as manufactured by Capitol Products Corporation. All horizontally rolling windows and window arrangements (side-vents, center-vents, sub-, side-, transom lites and fixed removable lites) shall be thermally insulated aluminum windows. AAMA requires that the largest size of any particular model we offer for sale be tested. That size has been tested and meets AAMA's specifications. No other size of the model will be tested and no representation as to air infiltration is made except as to the model tested. Windows furnished for job will duplicate the test window in all mechanical parts and details within standard commercial tolerances as required under ANSI/AAMA 302.9-77.

OPERATION: Window frames can be installed without sashes. Fixed and rolling sashes shall be loaded into frames from the building interior and shall be removable without the use of any tools for cleaning, maintenance or re-glazing. The rolling sash shall automatically lock when in closed position. Insect screens shall be installed or removed from the building interior without removing a sash or the use of tools, and shall be reversible for handed rolling sash operation.

MATERIALS: All frame members, rails and stiles shall be fabricated from extruded aluminum alloy 6063-T5 with nominal wall thicknesses of .062″ for the Series E700 windows, and with nominal wall thicknesses of .062″ for all members except the sill, which shall have a nominal wall thickness of .078″, for the Series E705 windows.

Thermal-break frame separators shall be of extruded rigid and rigid-foamed poly-vinyl chloride. Glazing gaskets shall be 70 durometer extruded poly-vinyl chloride. Weatherstripping shall be 50 durometer extruded poly-vinyl chloride.

WEATHERSTRIPPING: Weatherstripping at the frame perimeters and the interlocks shall be extruded 50 durometer poly-vinyl chloride. Adjustable, dual-durometer PVC end seals shall be provided at all interlock ends at the point of contact on the frame separators.

Weather stripping at head and sill and at rolling sash to be mylar shielded pile.

Fasteners shall be stainless steel. Hardware shall be aluminum and Celcon, Ultra-violet stabilized, Acetal Copolymer M90-08. Sash bumpers shall be adhesive backed poly-vinyl chloride.

CONSTRUCTION: Frame members shall be thermally insulated by a crimped insulator and a snap-on type dual-durometer separator which functions as insulator and weatherstrip for the exterior fixed sash and for the rolling sash. Jambs shall be machined at both ends to receive telescopingly the head and sill. The joint shall be made watertight by applying a die-cut gasket and fastening with two screws per joint. Sills shall be machined to receive one-way valves and fixed sash retaining pins. Window thermal design to be such that no thermal break of sash aluminum components is required. All sash stiles shall be tubular aluminum shapes for torsional rigidity and shall be machined at both ends to telescope head and sill. Jambs shall be fastened with one screw per joint together with top rail guides, fixed sash bottom rail setting block and rolling sash roller housing.

Top and bottom rails shall be machined at both ends to receive stiles and shall have weep slots for draining the bottom rail glazing cavity. Interlock stiles shall be designed to provide a pocket for the insect screen frame and shall be machined to receive the spring loaded lock and lock keeper.

HARDWARE: Exterior fixed sashes shall have injection molded rail guides and sash lock keeper. Interior rolling sashes shall have injection molded two-position roller housings, aluminum wheels and injection molded, spring loaded lock at interlock stile. The exterior fixed sash shall be locked in place by a push-pull retaining pin at head and sill. Sills shall have PVC one-way weep valves. Jambs shall have adhesive backed bumpers at mid-height.

FINISH: All aluminum frame and sash members shall have a finish that provides a smooth uniform appearance. Electrostatically applied paint over alodine base provided in standard colors at no added cost. Extruded PVC and exposed injection molded parts shall be of charcoal grey color.

GLASS: Glass shall be ½″ sealed double glazing with standard glass thickness as required by AAMA.

SCREENS: Frames shall be fabricated of a roll formed aluminum rail with corners accurately mitred and fitted to a hairline joint. Screen cloth shall be fiberglas. Frames shall have 2 springloaded plungers on one side rail and a finger pull on the opposite side rail. Screens and finish on screen frames shall be available at additional cost.

INSTALLATION: (by others) Frames should be installed straight plumb and level without springing or twisting, and securely fastened in place in accordance with recommendations or details. Mastic or caulking compounds must be applied, before installation, between fin and adjacent construction to provide weather-tight installation and to maintain the integrity of the energy saving features of the product. Loading of sashes and final adjustments shall be made by the installers to assure proper sash operation and window performance.

PERFORMANCE LEVEL SPECIFICATIONS	Air Infiltration 25mph wind		Water Resistance		Operating Force		Uniform Load Deflection		
	measured	allowed	measured	allowed	measured	allowed	measured	allowed	
	cfm/ft		psf no flow		pounds		inches		
B2	.37	.375	2.86	2.86	10	15	.197	.331	
A2	.37	.375	4.34	3.34	20	20	.200	.351	

AAMA requires that the largest size of any particular model we offer for sale be tested. That size has been tested and meets AAMA's specifications. No other size of the model will be tested and no representation as to air infiltration is made except as to the model tested.

TWO LITE SLIDER

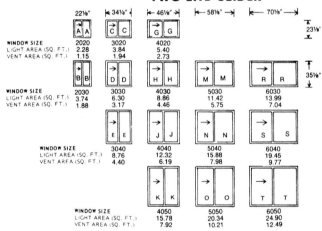

SIZE CHART
E-700 ROLLING WINDOWS
(interior dimensions shown in inches)

SIDE VENT

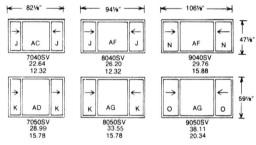

CENTER VENT

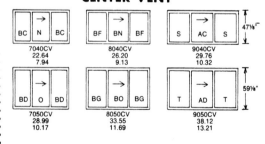

GLASS SIZES (in inches)

A	10″ x 20″	R	34″ x 32″
B	10″ x 32″	S	34″ x 44″
C	16″ x 20″	T	34″ x 56″
D	16″ x 32″		
E	16″ x 44″	AC	36″ x 44″
		AD	36″ x 56″
G	22″ x 20″	AF	48″ x 44″
H	22″ x 32″	AG	48″ x 56″
J	22″ x 44″	BC	26″ x 44″
K	22″ x 56″	BD	26″ x 56″
M	28″ x 32″	BF	30″ x 44″
N	28″ x 44″	BG	30″ x 56″
O	28″ x 56″	BN	32″ x 44″
		BO	32″ x 56″

SCREEN SIZES (in inches)

WIDTHS	HEIGHTS
for 2-Lite	for All Styles
20=10″	20=20¾″
30=16″	30=32¾″
40=22″	40=44¾″
50=28″	50=56¾″
60=34″	
for Side Vent	
70=22″	
80=22″	
90=28″	
for Center Vent	
70=27⁹⁄₁₆″	
80=31⁵⁄₁₆″	
90=35⁵⁄₁₆″	

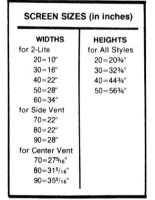

Ethyl Capitol Products   CAPITOL PRODUCTS CORPORATION

Mechanicsburg, Pennsylvania 17055 (717) 766-7661

NATIONAL SASH & DOOR JOBBERS ASSOCIATION ASSOCIATE MEMBER

CPN95558 7/80 ©

Figure 24-13   (continued)

CAPITOL PRODUCTS CORP.
Mechanicsburg, PA 17055

page D 1 —10/78
SUPERCEDES—2/77 CPN95002

E·700 DETAILS

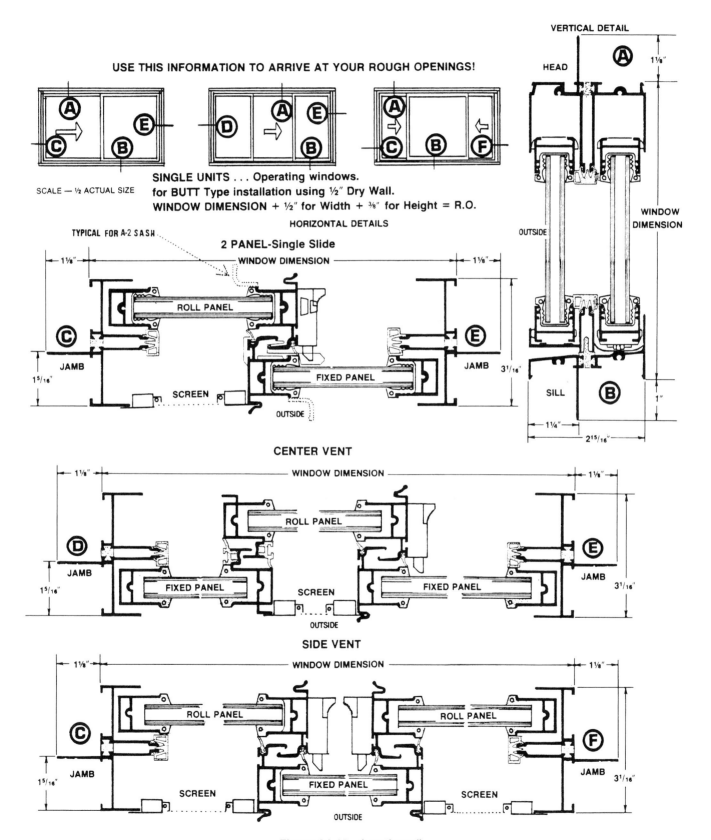

USE THIS INFORMATION TO ARRIVE AT YOUR ROUGH OPENINGS!

SCALE — ½ ACTUAL SIZE

SINGLE UNITS . . . Operating windows.
for BUTT Type installation using ½" Dry Wall.
WINDOW DIMENSION + ½" for Width + ⅜" for Height = R.O.

HORIZONTAL DETAILS

2 PANEL-Single Slide

CENTER VENT

SIDE VENT

Figure 24-13 (continued)

Ethyl **CP** CAPITOL PRODUCTS CORP
Mechanicsburg, PA 17055

page D 2 —10/78
SUPERCEDES—New CPN95002

E-700 DETAILS

TYPICAL BUTT TYPE INSTALLATION DETAILS

ROUGH OPENING WIDTH = WINDOW DIMENSION + ½"   ROUGH OPENING HEIGHT = WINDOW DIMENSION + ⅜"

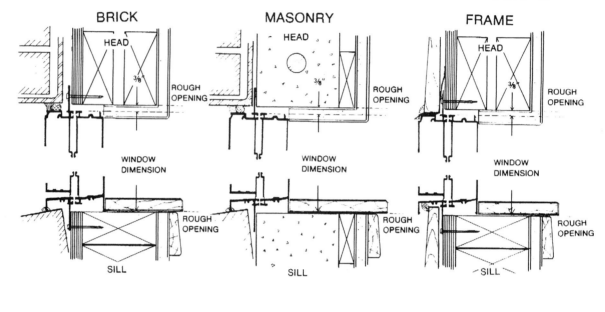

SCALE — 3" = 1'

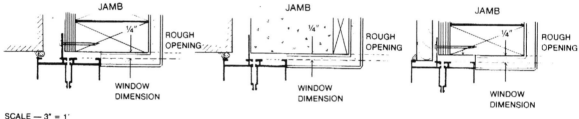

IMPORTANT—Mastic or caulking compounds must be applied before installation, between window fin and adjacent construction to provide weather-tight installation and maintain the integrity of the products energy saving features.

THE CAPITOL MULLING AND FIXED LITE WINDOW WALL SYSTEM

Fin and finless windows can be combined in fixed lite and operating window configurations. Either Side-by-Side or Stack Mulled. Fixed Lites can be mixed and matched with side vent, center vent and two-lite rolling windows to achieve window wall arrangements.

Windows can be ordered with fins already removed from "mulled side" of window, or fins can easily be removed in the field for mulling.

Fixed Lite System features Removable Inserts. Glass is enclosed in its own sash frame which can be removed from the main window frames.

Figure 24-13   (continued)

✓ CHECK YOUR PROGRESS ───────────────────────────

Can you perform these tasks?

☐ Identify the following parts of a window: sash, rails, stiles, muntins, glazing, side jambs, head jambs, sill, stop, exterior casing, and interior casing.

☐ Identify each part of a window on a window detail.

☐ Explain the basic features of the following door types: panel, flush, hollow core, molded, and solid core.

☐ Explain what is meant by applied stop and integral stop.

☐ Find the rough opening, unit size, and glass size for windows in a manufacturer's catalog.

ASSIGNMENT ───────────────────────────

Refer to the Lake House drawings when necessary to complete the assignment.

1. Name the lettered parts (a through f) in Figure 24-14.

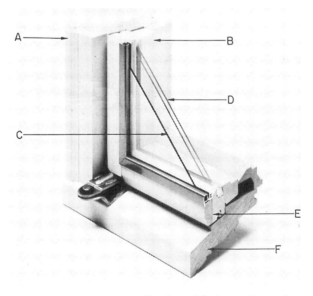

*Courtesy of Andersen Corporation,
Bayport, Minnesota 55003*

Figure 24-14

2. What is the nominal size of the window in the south end of the Lake House dining room?

3. In the catalog sample shown in Figure 24-13, what is the width and height of the rough opening for a #4030 window?

4. In the catalog sample, what are the rough opening dimensions for a window with a 28-inch by 44-inch glass size?

5.   In the catalog sample in Figure 24-13, what is the glass size of the window in bedroom #2 of the Lake House?

6.   In the catalog sample, what is the R.O. for the window in bedroom #2 of the Lake House?

7.   Is the exterior door in the Lake House kitchen to be prehung or site hung?

8.   What type and size are the doors in the Lake House playroom closet?

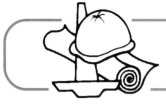

UNIT 25 Exterior Wall Coverings

OBJECTIVES

After completing this unit, you will be able to perform the following tasks:

- Describe the exterior wall covering planned for all parts of a building.
- Explain how flashing, drip caps, and other devices are used to shed water.
- Describe the treatment to be used at corners and edges of the exterior wall covering.

WOOD SIDING

Wood is a popular siding material because it is available in a variety of patterns, it is easy to work with, and it is durable. Wood siding includes horizontal boards, vertical boards, shingles, plywood, and hardboard.

Boards can be cut into a variety of shapes for use as horizontal siding, Figure 25-1. These boards are nailed to the wall surface starting at the bottom and working toward the top. With wood siding, a starting strip of wood furring is nailed to the bottom of the wall. This starting strip holds the bottom edge of the first piece of siding away from the wall. Hardboard siding often has an insert on the back. This insert replaced the starting strip, Figure 25-2. Each board covers the top edge of the one below. The amount of each board left exposed to the weather is called the *exposure* of the siding, Figure 25-3.

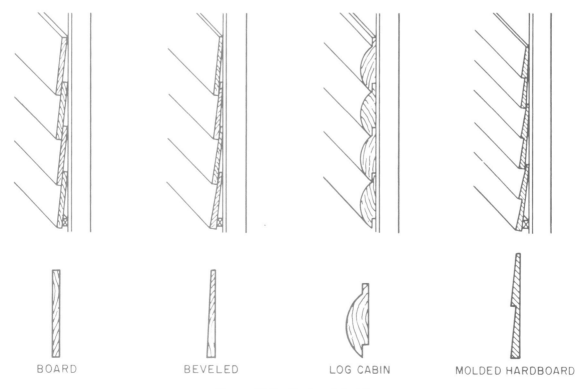

BOARD   BEVELED   LOG CABIN   MOLDED HARDBOARD

Figure 25-1   Horizontal siding

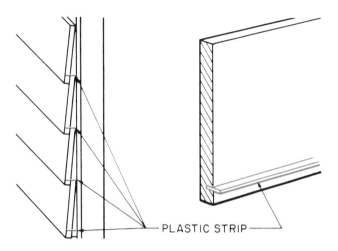

—PLASTIC STRIP—

Figure 25-2   Many manufacturers of hardboard siding include a plastic insert which helps align the pieces.

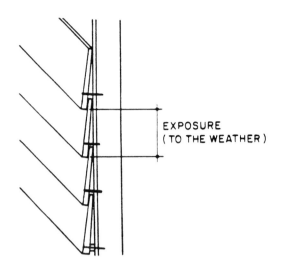

EXPOSURE (TO THE WEATHER)

Figure 25-3   The exposure is the amount of the siding exposed to the weather.

Vertical siding patterns can be created by boards, plywood, or hardboard, Figure 25-4. These materials are applied directly to the wall with no special starting strips. However, where horizontal joints are necessary, they should be lapped with rabbet joints or *Z* flashing should be applied. The flashing can be concealed with battens, Figure 25-5.

Shingles take longer to apply, but make an excellent siding material. In place of a starting strip of furring, the bottom *course* (row) of shingles is doubled.

Drip Caps and Flashing

It is important to prevent water from getting behind the siding where it can cause dry rot. Where a horizontal surface meets the siding, water is apt to collect. Flashing is used to prevent this water from running behind the siding. Aluminum is the most common flash-

ing material. The aluminum flashing is nailed to the wall before the siding is applied. The lower edge of the flashing extends over the horizontal surface far enough to prevent the water from running behind the siding. Areas to be flashed are noted on building elevations. The flashing is shown on the detail drawings or building elevations, Figure 25-6.

The heads of windows and doors may form a small horizontal surface. Wood drip cap molding can be used in these places to shed water, Figure 25-7. Drip caps are shown on details and elevations.

Corner and Edge Treatment

Regardless of the kind of siding used, the edges must be covered to prevent water from soaking into the end grain or running behind the siding. Around windows, vents, doors, and other wall openings; the trim

BOARD & BATTEN   REVERSE BOARD & BATTEN   TEXTURE 1-11 PLYWOOD

Figure 25-4   Three popular vertical siding patterns

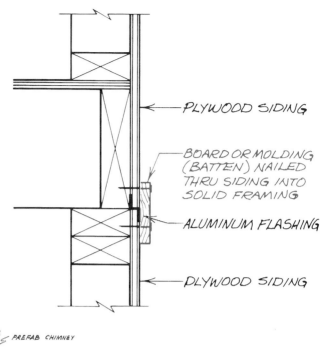

Figure 25-5   Horizontal hardboard siding should be covered with battens.

PLYWOOD SIDING

BOARD OR MOLDING (BATTEN) NAILED THRU SIDING INTO SOLID FRAMING

ALUMINUM FLASHING

PLYWOOD SIDING

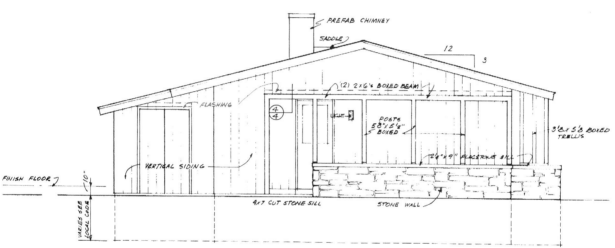

LEFT SIDE ELEVATION

PREFAB CHIMNEY
SADDLE
(2) 2×6's BOXED BEAM
POSTS 5⅝"×5½" BOXED
3'⅜×5'⅜ BOXED TRELLIS
2'4"×9" FLAGSTONE SILL
FLASHING
VERTICAL SIDING
FINISH FLOOR
VARIES SEE LOCAL CODE
4×7 CUT STONE SILL
STONE WALL
12 3

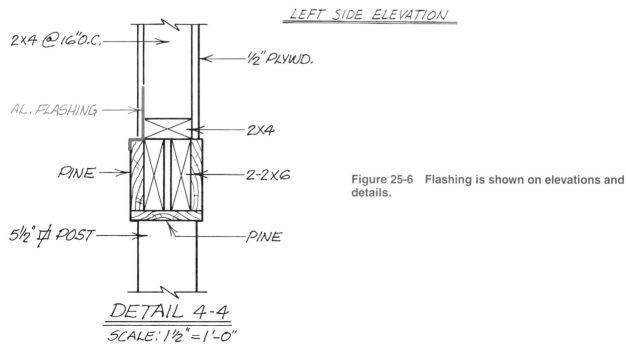

2×4 @ 16"O.C.
½" PLYWD.
AL. FLASHING
2×4
PINE
2-2×6
5½" ☐ POST
PINE

DETAIL 4-4
SCALE: 1½"=1'-0"

Figure 25-6   Flashing is shown on elevations and details.

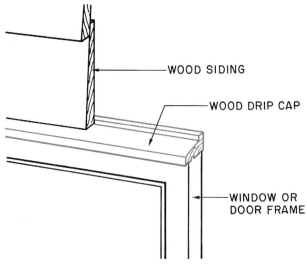

Figure 25-7   Drip cap

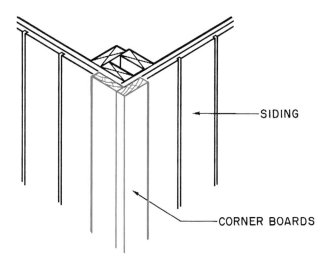

Figure 25-8   Corner boards can be used with most types of siding

Figure 25-9   Metal corners

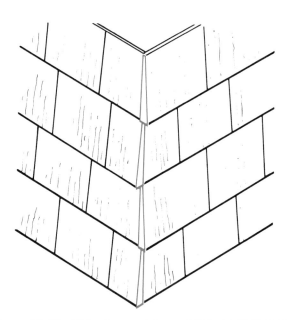

Figure 25-10   Shingles are trimmed to form their own corners.

around the opening usually covers the end grain of the siding. In inside corner, a strip of wood can be used to form a corner bead for the siding. Outside corners in plywood, hardboard sheets, or vertical boards are usually built with corner boards, Figure 25-8. With horizontal siding, outside corners can be built with corner boards or metal corner caps, Figure 25-9. Shingles are trimmed to form their own outside corner, Figure 25-10, or they are butted against corner boards.

The corner treatment to be used can usually be seen on the building elevations. If no corner treatment is shown on the elevations, look for a special detail of a typical corner.

METAL AND PLASTIC SIDING

Several manufacturers produce aluminum, steel, and vinyl siding and trim. The most common type is made to look like horizontal beveled wood siding. A variety of trim pieces is available for any type of application, Figure 25-11.

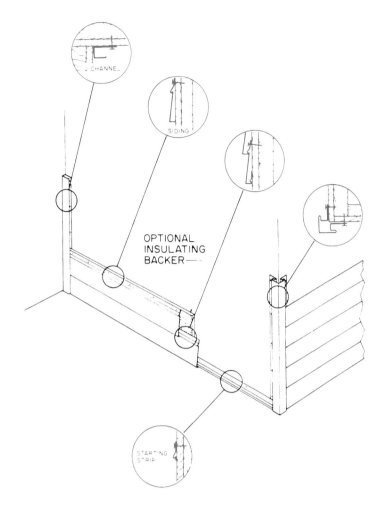

J CHANNEL

SIDING

OPTIONAL
INSULATING
BACKER—

STARTING
STRIP

Figure 25-11   Accessories for metal and vinyl siding

STUCCO

Stucco is a plaster made with portland cement. The wall sheathing is covered with waterproof building paper. Next the lath (usually wire netting) is stapled to the wall, Figure 25-12. Finally, the stucco is troweled on—a rough scratch coat first, Figure 25-13, then a brown coat to build up to the approximate thickness, and finally a finish coat. Outside corners and edges are formed with galvanized metal beads nailed to the wall before the scratch coat is applied.

MASONRY VENEER

Masonry veneer is usually either brick or natural stone. It is called veneer because it is a thin layer of masonry over some kind of structural wall. The structure may be wood framing, metal framing, or concrete blocks. The masonry has to rest on a solid foundation. Normally the building foundation is built with a four-inch ledge to support the masonry veneer. This ledge can be seen in a typical wall section, Figure 25-14.

Interesting patterns can be created by using half bricks and bricks in different positions. Some of the most common bond patterns, as they are called, are shown in Figure 25-15. If no pattern is indicated, the bricks are normally laid in running bond.

Above and below windows, above doors, and in other special areas, bricks may be laid in varying positions, Figure 25-16. If bricks are to be laid in any but the stretcher position, they will be shown on the detail drawings, Figure 25-17. The details for openings in masonry walls will also indicate a lintel at the top of the opening. The lintel is usually angle iron. It carries the weight of the masonry above the opening.

GENERAL

The electrical installer needs to be familiar with the rough-in-requirements for all outside lighting and devices and any special requirements that may be needed with the specified exterior wall finish. The required backboard or rough-in frame and box must be properly installed to prevent any finished wall damage.

Figure 25-12   Wall prepared for stucco

Figure 25-13   Scratch coat applied

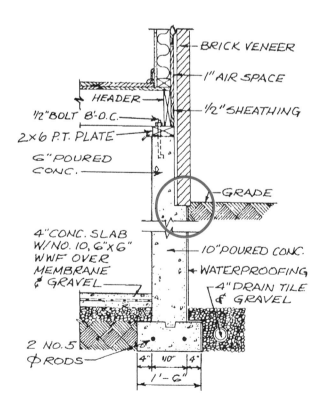

Figure 25-14   Typical foundation section with ledge for masonry veneer.

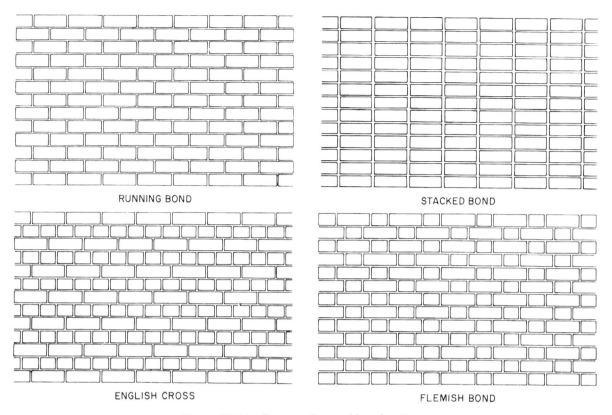

Figure 25-15   Frequently used bond patterns

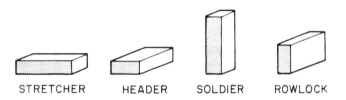

Figure 25-16   Brick positions

Figure 25-17   Notice the rowlock and
soldier bricks on the window detail

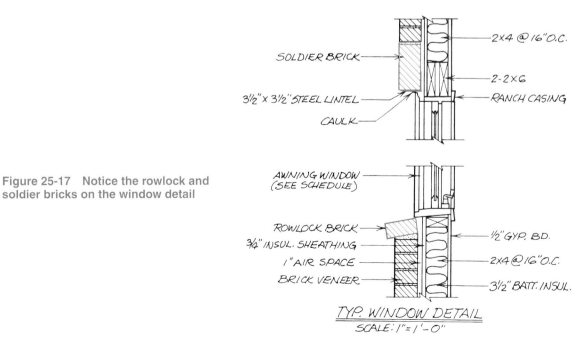

✓ CHECK YOUR PROGRESS ———————————————————————————————

Can you perform these tasks?

- [ ] Describe the exterior wall covering on all parts of a building.
- [ ] List the locations of flashing on all parts of a building.
- [ ] List the dimensions of all drip caps to be installed on a building.
- [ ] Describe the support for masonry veneer.
- [ ] List locations where bricks are to be used in stretcher, header, soldier, and rowlock positions.

ASSIGNMENT ———————————————————————————————

Refer to the Lake House drawings (in the packet) to complete the assignment.

1. What material is used for the Lake House siding?

2. How are the outside corners of the Lake House siding finished?

3. In some places, the Lake House siding extends below the wood framing onto the foundation. How is the siding fastened to the foundation?

4. What prevents water from running under the siding at the heads of the Lake House windows?

5. Detail 1/7 of the Lake House shows aluminum screen nailed behind the top edge of the siding. What is the purpose of the opening covered by this screen?

6. Describe one use of aluminum flashing under the siding on the Lake House.

UNIT 26   Decks

OBJECTIVES

After completing this unit, you will be able to perform the following tasks:

- Explain how a deck is to be supported.

- Describe how a deck is to be anchored to the house.

- Locate the necessary information to build handrails on decks.

Wood decks are used to extend the living area of a house to the outdoors. A deck may be a single-level platform, or it may be a complex structure with several levels and shapes. However, nearly all wood decks are made of wood planks laid over joists or beams. Figure 26-1. The planks are laid with a small space between them, so rainwater does not collect on the deck.

The same construction methods are used for decks and porches as for other parts of the house. The parts of deck construction that require special attention or that were not covered earlier in the text are discussed here.

SUPPORT

The deck must be supported by stable earth. The support must also extend below the frostline in cold climates. The most common method of support is by wood posts which rest on concrete floorings, Figure 26-2. All

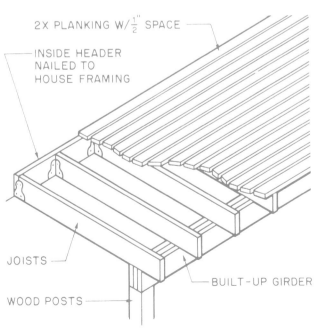

Figure 26-1   Typical deck construction

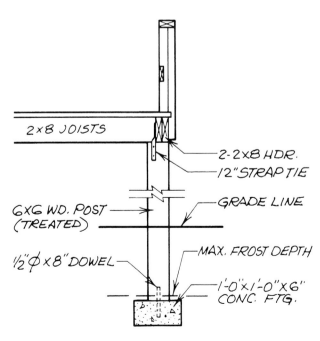

Figure 26-2   Foundation for a typical wood deck

wood used in the construction of a deck should be treated with a chemical to prevent decay. For example, Deck Detail 3/6 for the Lake House includes a note that all wood is to be CCA treated. CCA is the abbreviation for *chromated copper arsenate*—a common wood preservative.

Decks are usually included on the floor plans for the house. The floor plans show the overall dimensions of the decks and the locations of posts, piers, or other support, Figure 26-3. If the decks are complex, they may be shown on a separate plan or detail. On the Lake House, the decks are shown on the floor framing plan.

Typically the posts or piers support a beam or girder, which in turn supports joists. The beam may be solid wood or built-up wood. The joists can be butted against or rested on top of the beam in any of the ways discussed for floor framing.

ANCHORING THE DECK

The most common ways of anchoring the deck to the building are shown in Figure 26-4. If the deck is at the same level as the house floor, the deck joists can be cantilevered from the floor joists. In this case, blocking is required between the joists. If through bolts are used, the holes are drilled when the deck is built. If the deck is secured with anchor bolts in a concrete foundation, the anchor bolts must be positioned when the foundation is poured.

The anchor bolts or through bolts hold a joist header. The deck joists either rest on top of the header or butt against it. If the joists butt against the header, they are supported by joist hangers or a ledger strip.

RAILINGS

Most decks have a railing because they are several feet from the ground. Although metal railings are available in ready-to-install form, the architectural style of most wood decks calls for a carpenter-built railing. The simplest type of railing is made of uprights and two or three horizontal rails. The uprights are bolted to the deck frame and the rails are bolted, screwed, or nailed to the uprights. The style of the railing and the hardware involved are usually indicated on a detail drawing, Figure 26-5.

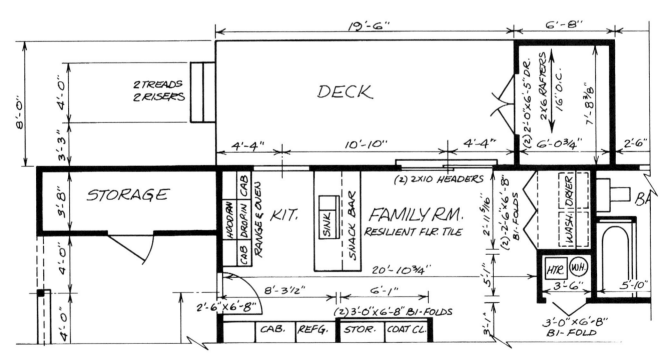

Figure 26-3 Deck shown on floor plan

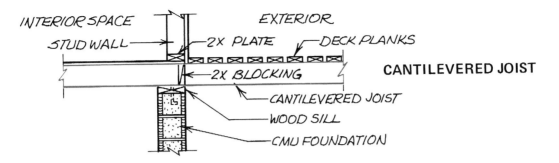

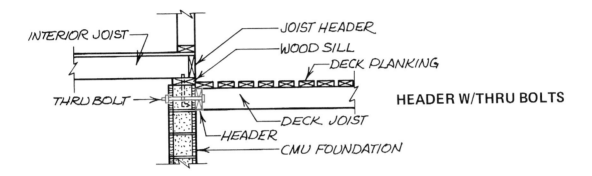

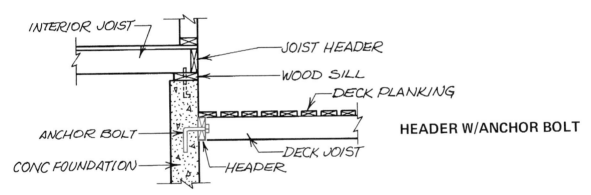

Figure 26-4   Three methods of anchoring the deck to the house

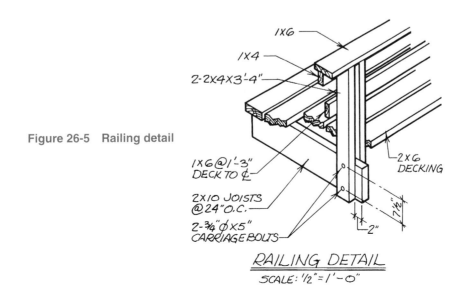

Figure 26-5   Railing detail

✓ CHECK YOUR PROGRESS

Can you perform these tasks?

☐ List the dimensions for the size and location of posts or columns to support a deck or porch.

☐ Describe the framing for a deck.

☐ Explain how a deck is to be tied to the building.

☐ Describe the construction of railings for a deck.

☐ List the sizes of all parts of a deck.

ASSIGNMENT

Refer to the Lake House drawings (in the packet) to complete the assignment.

1. What supports the south edge of the decks located outside the Lake House living and dining rooms?

2. How far from the outside of the house foundation is the centerline of these supports?

3. How far apart are these supports?

4. How many anchor bolts are required to fasten both of these decks to the Lake House foundation?

5. What is the purpose of the aluminum flashing shown on Deck Detail 3/6?

6. What material is used for the railings on the Lake House south decks?

7. How many lineal feet of horizontal rails are there on these decks?

8. What is the total rise from the lower deck to the higher deck? Which deck is higher?

9. What supports the west edge of the deck between the Lake House kitchen and the garage?

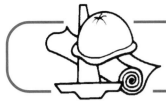

UNIT 27   Finishing Site Work

OBJECTIVES

After completing this unit, you will be able to perform the following tasks:

- Describe retaining walls, planter, and other constructed landscape features shown on a set of drawings.

- Find the dimensions of paved areas.

- Identify new plantings and other finished landscaping shown on a site plan or landscape plan.

As the exterior of the building is being finished or soon after it is finished, the masons, carpenters, and landscapers begin the finished landscape work. Any constructed features (called *site appurtenances*) are completed first. Then trees and shrubs are planted. Finally, lawns are planted.

RETAINING WALLS

Retaining walls are used where sudden changes in elevation are required, Figure 27-1. The retaining wall retains, or holds back, the earth. Where the height of the retaining wall is several feet, the earth may put

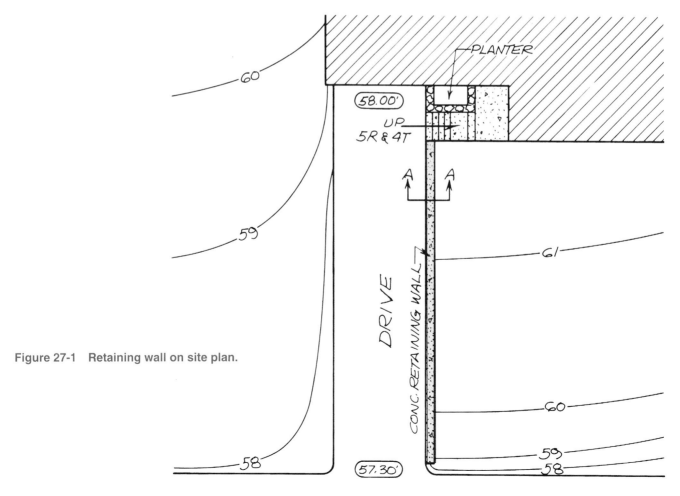

Figure 27-1   Retaining wall on site plan.

considerable stress on the wall. Therefore, it is important to build the wall according to the plans of the designer. In some cases, a section through the wall may be included to show the thickness of the wall, its foundation, and any reinforcing steel, Figure 27-2. For low retaining walls, the site plan may be the only drawing included.

A low retaining wall is sometimes built around the base of a tree when the finished grade is higher than the natural grade. This retaining wall forms a *well* around the tree, allowing the roots of the tree to "breathe." An example of a tree that will require a well can be seen in Figure 27-3, taken from the Lake House site plan. The 24" oak is at an elevation of approximately 333 feet, but the finished grade at this point is 336 feet. Therefore, a well 3 feet deep is required.

PLANTERS

Planters are sometimes included in the construction of retaining walls or attached to the building. In these cases, the information needed to build the planter is included with the information for the building or retaining wall, Figure 27-4. The planter is built right along with the house or retaining wall. If a planter that is separate from other construction is included, it is usually shown with dimension on the site plan. A special section may be included with the details and sections to show how the planter is constructed, Figure 27-5.

The planter should be lined with a waterproof membrane, such as polyethylene (common plastic sheeting), or coated with asphalt waterproofing. This keeps the acids and salts in the soil from seeping through the planter and staining it. The planter should also include some way for water to escape. This can be through the bottom or through weep holes. *Weep holes* are openings just above ground level. In cold climates, the planter may be lined with compressible plastic foam. This allows the earth in the planter to expand as it freezes, without cracking the planter. If the planter is to have landscape lighting, automatic watering, etc., additional waterproofing may be required where these utilities penetrate the waterproof membrane.

PAVED AREAS

Paved areas on housing sites are drives, walks, and patios. Drives and walks are usually described most fully in the specifications for the project. However, the site plan includes dimensions and necessary grading information for paved areas, Figure 27-3. These dimensions are usually quite straight forward and easy to understand.

Patios are similar to drives and walks in that they are flat areas of paving with easy-to-follow dimensions. They may differ from drives and walks by having different paving materials, such as slate, brick, and flagstone, for example. Patios may also be made of a concrete slab with different surface material.

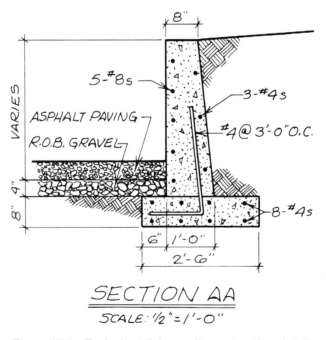

SECTION AA
SCALE: 1/2" = 1'-0"

Figure 27-2   Typical retaining wall construction detail

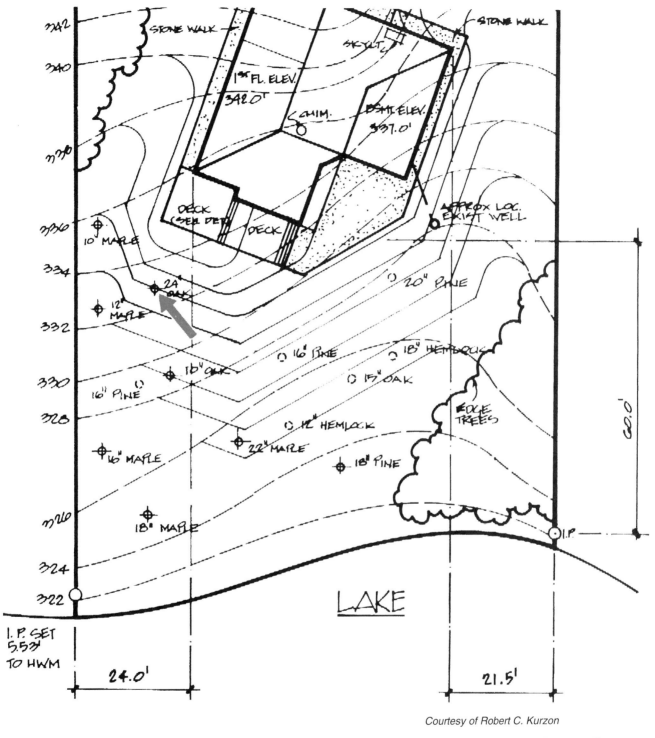

Courtesy of Robert C. Kurzon

Figure 27-3 Typical site plan. The 24-inch oak tree near the SW corner of the house requires a well.

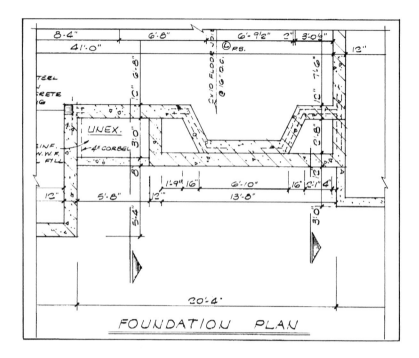

FOUNDATION PLAN

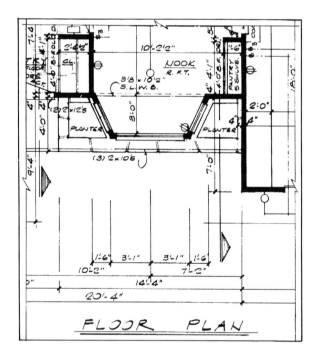

FLOOR   PLAN

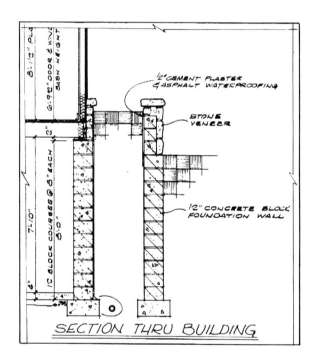

SECTION THRU BUILDING

Figure 27-4   Because this planter is a part of the foundation, it is included on the normal drawings for the house.

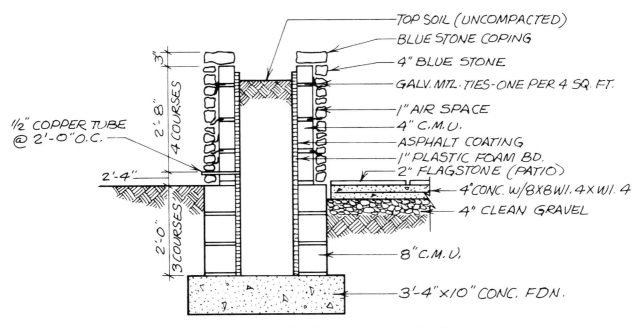

TOP SOIL (UNCOMPACTED)
BLUE STONE COPING
4" BLUE STONE
GALV. MTL. TIES-ONE PER 4 SQ. FT.
1" AIR SPACE
4" C.M.U.
ASPHALT COATING
1" PLASTIC FOAM BD.
2" FLAGSTONE (PATIO)
4" CONC. W/8X8 W1.4 X W1.4
4" CLEAN GRAVEL
8" C.M.U.
3'-4" X10" CONC. FDN.

1/2" COPPER TUBE @ 2'-0" O.C.

3"
2'-8"
4 COURSES
2'-4"
2'-0"
3 COURSES

Figure 27-5   Section through planter and patio

PLANTINGS

Plantings include three types: grass or lawns, shrubs, and trees. On some projects, such items are planted by the owner. When the builder/contractor does the landscaping, the trees and shrubs are planted first; then, the lawns are planted. Some or all of the trees included in the landscape design may have been left when the site was cleared. Any new trees to be planted are shown with a symbol and an identifying note, as shown in Figure 27-3. There is no widely accepted stan-dard for the symbols used to represent trees and shrubs. Most architects and drafters use symbols that represent deciduous (hardwood) trees, coniferous (ev-ergreen) trees, palms, and low shrubs, Figure 27-6. The trees and shrubs may also be listed in a schedule of plantings, Figure 27-7.

Grass is planted by seeding or sodding. Although a note on the site plan may indicate seeded or sodded areas, more detailed information is usually given on the schedule of plantings or in the specification.

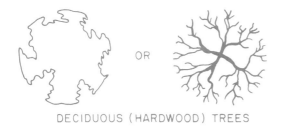

OR

DECIDUOUS (HARDWOOD) TREES

CONIFEROUS (NEEDLE LIKE) TREES

PALM TREES

LOW SHRUBS

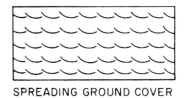

SPREADING GROUND COVER

Figure 27-6   Typical symbols for plantings

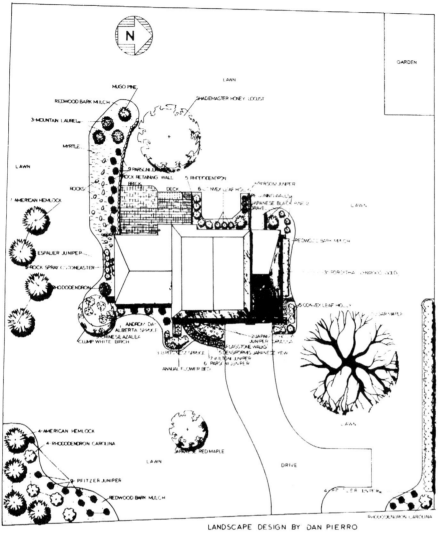

LANDSCAPE PLANTING PLAN FOR:
MR. & MRS. C. BURTON
KENNEDY ROAD
COBLESKILL, N.Y.

LANDSCAPE DESIGN BY DAN PIERRO

P L A N T   L I S T

BOTANICAL NAME	COMMON NAME	NUMBER USED	BOTANICAL NAME	COMMON NAME	NUMBER USED
	~TREES~			~TREES~	
ACER PALMATUM	JAPANESE RED MAPLE	1	JUNIPERUS CHINENSIS TORULOSA	TORULOSA JUNIPER	1
ACER SACCHARUM	SUGAR MAPLE	1	JUNIPERUS CHINENSIS WILTONI	WILTONI JUNIPER	7
BETULA PAPYRIFERA	CLUMP WHITE BIRCH	1	KALMIA LATIFOLIA	MOUNTAIN LAUREL	3
GLEDITSIA TRIACANTHOS INERMIS	SHADEMASTER HONEY LOCUST	1	PICEA ABIES NIDIFORMIS	BIRDS NEST SPRUCE	3
	~SHRUBS~		PICEA GLAUCA ALBERTIANA	ALBERTA SPRUCE	1
AZALEA MOLLIS	CHINESE AZALEA	5	PIERIS JAPONICA	ANDROMEDA	1
COTONEASTER HORIZONTALIS	ROCK SPRAY COTONEASTER	5	PINUS MUGO MUGHUS	MUGO PINE	1
FORSYTHIA INTERMEDIA	FORSYTHIA LYNWOOD GOLD	31	PINUS THUNBERGI	JAPANESE BLACK PINE	2
ILEX CRENATA	JAPANESE HOLLY	2	RHODODENDRON CAROLINA	RHODODENDRON CAROLINA	11
ILEX CRENATA CONVEXA BULLATA	CONVEX LEAF HOLLY	11	TAXUS CUSPIDATA DENSIFORMIS	DENSIFORMIS JAPANESE YEW	5
JUNIPERUS CHINENSIS ESPALIER	ESPALIER JUNIPER	1	TSUGA CANADENSIS	AMERICAN HEMLOCK	8
JUNIPERUS CHINENSIS PARSONI	PARSONI JUNIPER	18		~GROUND COVERS~	
JUNIPERUS CHINENSIS PFITZERIANA	PFITZER JUNIPER	13	VINCA MINOR	MYRTLE	1200

Figure 27-7   A landscape plan complete with plant list

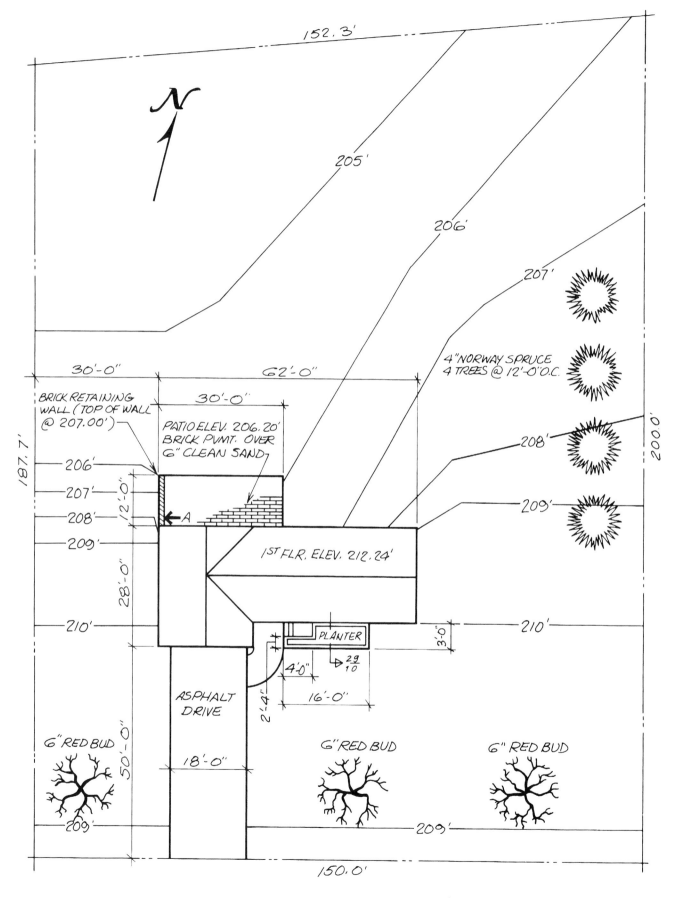

Figure 27-8   Site plan for assignment questions

✓ CHECK YOUR PROGRESS

Can you perform these tasks?

- ☐ List the dimensions of a retaining wall.
- ☐ Describe the footing for a retaining wall.
- ☐ Describe provisions for drainage through a retaining wall.
- ☐ List the dimensions of a planter.
- ☐ Explain how the inside of a planter is to be waterproofed.
- ☐ List the dimensions of all paved areas.
- ☐ Describe the pitch of all paved areas.
- ☐ Name the types of plantings shown on a site or landscape plan.

ASSIGNMENT

Refer to Figures 27-8 and 27-9 to complete the assignment.

1.   What is the height of the retaining wall at A?

2.   How long is the retaining wall?

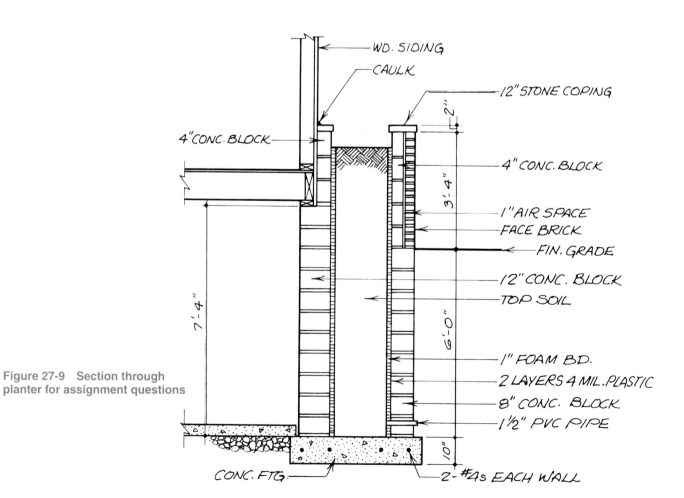

Figure 27-9   Section through planter for assignment questions

3. Of what material is the retaining wall constructed?

4. What is the width and length of the patio?

5. What materials are used in the construction of the patio?

6. Describe the weep holes in the planter.

7. How is the planter treated to prevent acids and salts from staining its surface?

8. How many deciduous trees are to be planted?

9. What is the area of the driveway?

10. Assuming that the driveway is 4 inches thick, how many cubic yards of asphalt does it require? (See Math Review 22.)

UNIT 28   Fireplaces

After completing this unit, you will be able to perform the following tasks:

- Describe the foundation, firebox, throat, and chimney of a fireplace using information from a set of construction drawings.

- Explain the finish of the exposed parts of the fireplace, using information from a set of construction drawings.

BASIC CONSTRUCTION AND THEORY OF OPERATION

A fireplace can be divided into four major parts or zones: foundation, firebox, throat area, and chimney, Figure 28-1. Each of these zones has a definite function. To understand the construction details, it is necessary to know how these zones work.

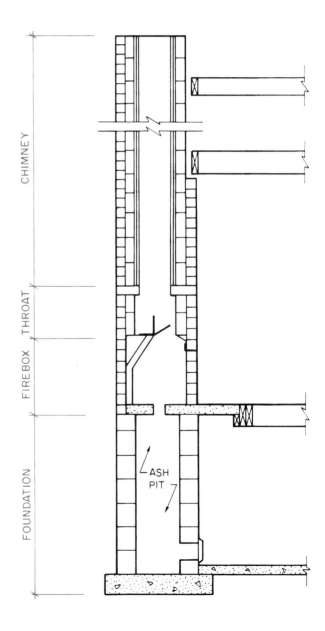

Figure 28-1   Four zones of a fireplace

Foundation

The fireplace foundation serves the same purpose as the foundation of the house—it supports the upper parts and spreads the weight over an area of stable earth. The foundation consists of a footing and walls capable of supporting the necessary weight, Figure 28-2. The fireplace foundation sometimes houses an ash pit. The *ash pit* is a reservoir to hold ashes that are dropped through an *ash dump* (a small door) in the floor of the fireplace, Figure 28-3. When the foundation includes an ash pit, a *cleanout door* is installed near the bottom of the ash pit, Figure 28-4.

Firebox

The firebox is the area where the fire is built. In all-masonry fireplaces, the firebox is constructed of two layers of masonry, Figure 28-2. Each layer is called a *wythe.* The floor of the firebox consists of firebricks laid over a concrete base. The concrete base may extend beyond the face of the firebox to support the hearth. The *hearth,* which may be tile, stone, brick, or slate, forms a noncombustible floor area in front of the fireplace. The outer walls of the firebox are most often common brick. The inner walls are of firebrick to withstand the heat of the fire. The back wall of the firebox slopes in to direct the smoke and gas into the throat area. The masonry over the fireplace opening is supported by a steel lintel. The lintel is long enough so that four inches of it can rest on the masonry at each end.

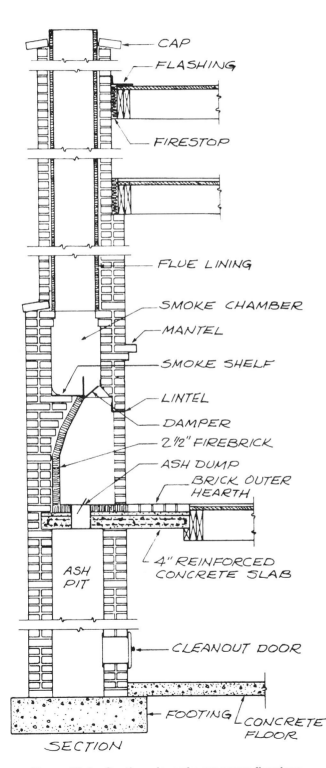

Figure 28-2   Section view of a masonry fireplace

Courtesy of Majestic Company

Figure 28-3   Ash dump

Courtesy of Majestic Company

Figure 28-4   Cleanout door

Throat

The throat of the fireplace is the area where the firebox narrows into the chimney. Modern wood-burning fireplaces are built with a metal damper in the throat, Figure 28-5. The *damper* is a door which can be closed

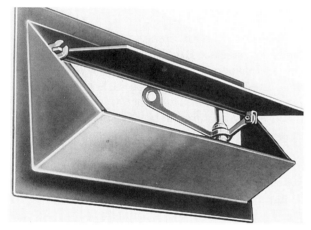

Courtesy of Majestic Company

Figure 28-5   Damper

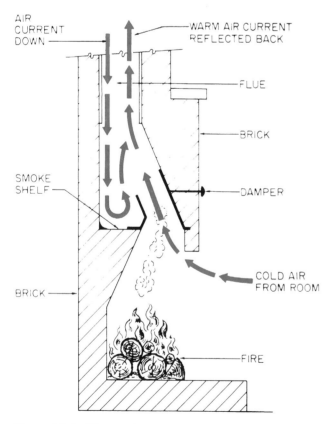

Figure 28-6   The smoke shelf turns the incoming air back up the chimney.

to prevent heat from escaping up the chimney when the fireplace is not in use. The damper is placed on top of the firebox with one-inch clearance on all four sides. This clearance allows the metal damper to expand as it gets hot.

The flat area behind the damper (above the sloped back of the firebox) is called the *smoke shelf.* The smoke shelf is especially important for the proper operation of the fireplace. The cold air coming down the chimney hits the smoke shelf and turns back up with the rising hot gas and smoke from the fire box. This helps carry the smoke and gas up the chimney, Figure 28-6. If the smoke shelf is not built properly and kept clean, the falling cold air can force the smoke and gas back into the firebox.

Chimney

The chimney carries the smoke and hot gas from the throat to above the house. The top of the chimney must be high enough above the roof, trees, and other nearby obstructions to insure that the air flows evenly across its top. As a general rule the chimney should extend 2 feet through the roof and 2 feet above anything within 10 feet. However, the dimensions on the drawings should always be followed.

To insure fire safety and a smooth inner surface, masonry chimneys are lined with a clay flue. This flue is installed in 2-foot sections as the chimney is built. A 1-inch air space is allowed between the flue lining and the chimney masonry, Figure 28-7.

In recent years insulated, metal chimneys have become quite popular. These chimneys are lightweight; they do not require massive foundation for their support. It also takes less time to install them. Because the outer wall of a metal chimney remains cool, it can be enclosed in wood, Figure 28-8. The chimney sections are slipped together and fastened with sheet metal screws; then, the chimney is framed with wood and covered with the specified siding and trim. A chimney enclosure of this type is called a *chase.*

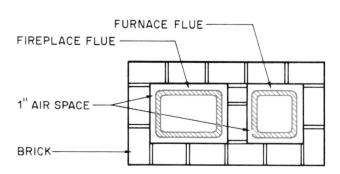

Figure 28-7   Plan view of a two-flue chimney

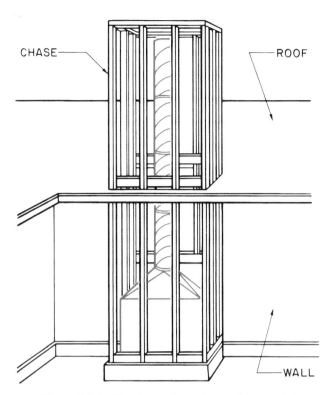

CHASE

ROOF

WALL

Figure 28-8   The framed enclosure for a metal chimney is called a chase.

PREFABRICATED METAL FIREPLACES

Constructing masonry fireplaces is time consuming. Their great weight requires massive foundations. Engineered, metal fireboxes have been developed which can be installed in very little time and require only modest foundations, Figure 28-9. The prefabricated units are available from several manufacturers, and in a variety of styles. However, they are all similar in that they have double walls and a complete throat with the damper in place. Most also have a firebrick floor.

The double wall improves the heating ability of the metal fireplace. Cool air enters the space between the walls through openings near the bottom. The air absorbs hear from the fire and, because warm air naturally rises, it exits through openings near the top of the unit, Figure 28-10. The outside surfaces of the prefabricated unit are cooled by the circulating air; the unit can be enclosed in wood if recommended by the manufacturer. For a more traditional appearance, the exposed face of the fireplace can be covered with masonry veneer.

Courtesy of Majestic Company

Figure 28-9   Prefabricated metal fireplace

FIREPLACE DRAWINGS

Much of the information about a fireplace is shown on the building plans and elevations. The foundation is

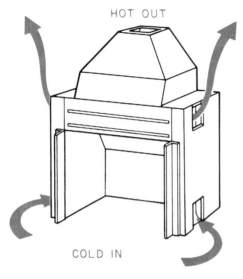

HOT OUT

COLD IN

Figure 28-10 In a heat-circulating fireplace, cold air enters the double wall at the bottom and warm air exits at the top.

normally included on the foundation plan for the building. The dimensions and notes show the location of the fireplace foundation, its size, and the size and type of material to be used. The floor plan of the house shows where the fireplace is located and its overall dimensions. The building elevations show the chimney.

More detailed information about the fireplace is shown on the fireplace details, which usually include a cross section, Figure 28-11. The following information is often included on a section view of the fireplace:

- Dimensions of the firebox

- Materials used inside the firebox (firebrick)

- Materials used for the outside of the firebox and chimney

- Ash dump, if any is included

- Lintel over the opening

- Dimensions of the hearth

- Mantle, if any is included

- Dimension from smoke shelf to flue

- Size of the flue

- Materials for the chimney

An elevation of the fireplace shows the exterior finish of the fireplace, Figure 28-12. Only those features that could not be adequately described on the section view are called out on the elevation. However, this view shows the exterior finish—the mantle and the trim—better than the section view does.

Figure 28-11 A section view shows the construction of the firebox and throat area.

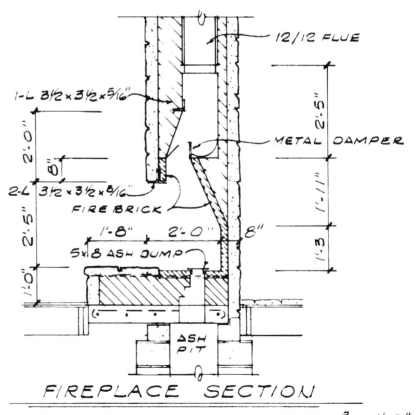

1-L 3½ x 3½ x 5/16"

2'-0"

8"

2-L 3½ x 3½ x 5/16"

2.5"

1'-0"

FIRE BRICK

1'-8"   2'-0"   8"

5x8 ASH DUMP

ASH PIT

12/12 FLUE

2'-5"

METAL DAMPER

1'-11"

.3

FIREPLACE SECTION

SCALE: ⅜ = 1'-0"

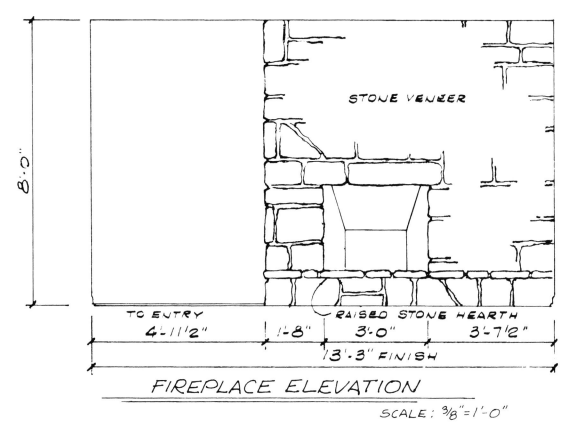

FIREPLACE ELEVATION

SCALE: $3/8" = 1'-0"$

Courtesy of Home Planners, Inc.

Figure 28-12   An elevation view is a good guide to the finished appearance of the fireplace.

✓ CHECK YOUR PROGRESS

Can you perform these tasks?

☐ Identify the firebox, smoke shelf, damper, flue, and hearth on a fireplace section.

☐ List the dimensions of the outside of a fireplace, firebox opening, inside of the firebox, hearth, and height of the chimney for a fireplace.

☐ Describe the construction, including dimensions, for the enclosure of a prefabricated fireplace.

☐ Describe the finish and trim for a fireplace.

ASSIGNMENT

Refer to the Lake House drawings (in the packet) to complete the assignment.

1.   Does the Lake House have an all-masonry fireplace or a prefabricated metal fireplace?

2.   How wide is the opening of the firebox?

3.   How high is the opening of the firebox?

4. What is the opening next to the fireplace?

5. Determine the overall width and length of the fireplace including the hearth.

6. Of what material is the hearth constructed?

7. What is used for a lintel over the firebox opening? (Include dimensions.)

8. Briefly describe the foundation of the fireplace.

9. How far above the highest point on the roof is the top of the chimney?

10. What is the total height from the playroom floor to the top of the chimney?

11. What is the overall height of the brickwork involved in the fireplace construction?

12. The top of the fireplace is covered with plastic laminate on $3/4$-inch plywood. How much clearance is there between that plywood and the chimney?

UNIT 29   Stairs

After completing this unit, you will be able to perform the following tasks:

- Identify the parts of stairs.
- Calculate tread size and riser size.

STAIR PARTS

In order to discuss the layout and construction of stairs you need to know the names of the parts of a set of stairs. The main parts of a stair are defined here and also shown in Figure 29-1.

- *Stringers,* sometimes called *strings,* are the main support members. The assembly made up of the stringers and vertical supports is called a stair *carriage.*

- *Treads* are supported by the stringers. The treads are the surfaces one steps on.

- *Risers* are the vertical boards between the treads.

- A *landing* is a platform in the middle of the stairs. Landings are used in stairs that change directions or in very long flights of stairs.

- The *run* of the stairs is the horizontal distance covered by the stairs.

Figure 29-1   Basic stair parts

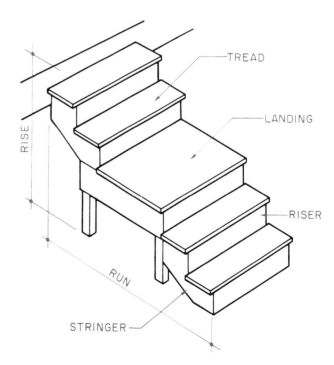

- The *rise* of the stairs is the total vertical dimension of the stairs.

The trim parts and balustrade are shown in Figure 29-2.

- The *nosing* is the portion of the tread that projects beyond the riser.

- The underside of the nosing may be trimmed with molding called *stair cove*.

- A *handrail* is usually required on any stairs that are not completely enclosed.

- *Balusters* are the vertical pieces that support the handrail at each step.

- The *newel post* is a heavier vertical support used at the bottom of the stair.

- The balusters, newel post, and handrail together are called a *balustrade*.

TYPES OF STAIRS

Stairs can be built with open stringers or housed stringers. *Open stringers* are cut in a saw tooth pattern to form a surface for fastening the treads, Figure 29-3. *Housed stringers* are routed out, so the treads fit between them, Figure 29-4. Stairs are also called open or closed depending on whether the space between the treads is enclosed with risers. Both the kind of stringers

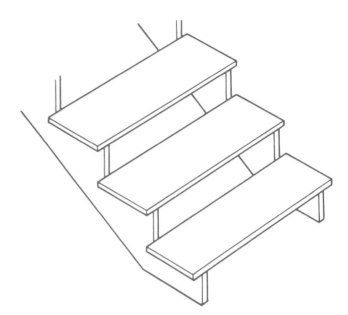

Figure 29-3   Open stringers and open risers

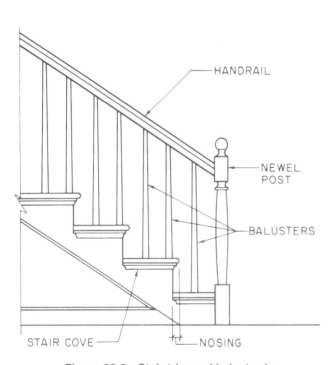

Figure 29-2   Stair trim and balustrade

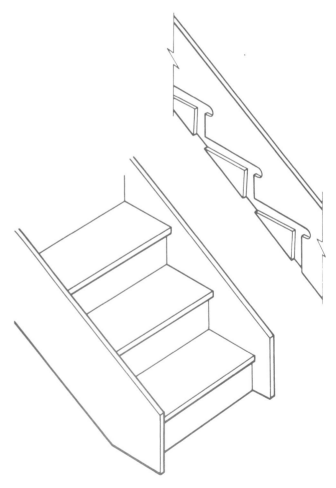

Figure 29-4   Housed stringers and closed risers

and the risers are shown on a section through the stairs. If the stringers are housed, the top of the far stringer shows on the section view. When stair lighting is specified to be installed in the stair riser, the electrical installer must coordinate with the carpenters to avoid a possible conflict with the center or mid-span stringer.

Stairs are also named according to their layout as seen in a plan view, Figure 29-5. Straight stairs are the simplest design. They may or may not include a landing. L-shaped and U-shaped stairs are used where space does not permit a straight run. L-shaped and U-shaped stairs have a landing at the change in directions. The carriages for L-shaped and U-shaped stairs include vertical supports under the landing, Figure 29-6. Winder stairs use tapered treads to change directions. Winder stairs are not permitted by some building codes. Spiral stairs require the least floor space. They are made up of winder treads, usually supported by a center column.

Temporary stairs used during construction only are usually of cruder construction. They are often made of 2 × 10 stringers with wooden cleats to support the treads. Service stairs in unoccupied spaces, such as a cellar, are often made of 2 × 12 open strings with 2 × 10 treads. Stairs for exterior decks are also sometimes built with 2 × 12 open strings, Figure 29-7. However, the carpenter should remember that these stairs are important to the overall appearance of the deck.

CALCULATING RISERS AND TREADS

Nearly all stair details include a notation showing the number of treads with their width and the number of risers with their height, Figure 29-6. Occasionally only the total run, total rise, number of treads, and number of risers is given on the drawings, Figure 29-7. Notice that there is one more riser than there are treads. The builder must calculate the size of the treads and risers. The steepness of the stairs depends on the relationship between tread width and riser height. If the treads are wide and the risers low, the stairs are gradual. If the treads are narrow and the risers high, the stairs are steep. To climb stairs with wide treads and high risers requires an uncomfortably long stride. To climb stairs with narrow treads and low risers requires an uncomfortably short stride. Building codes specify the maximum height of risers. For example, the Uniform Building Code requires that risers are no more than 8 inches high. Also, to ensure that the relationship between risers is comfortable, one of the following three rules should be followed:

1. The sum of one riser and one tread should be between 17 inches and 18 inches.

2. The sum of two risers and one tread should be between 24 inches and 25 inches.

3. Multiplying the width of one tread by the height of one riser should give a product of 70 inches to 75 inches.

To build the stairs in Figure 29-7 within these rules, the risers must be $7\frac{1}{2}$ inches high and the treads must be $9\frac{3}{4}$ inches wide. The height of the risers can be found by dividing the total rise by the number of risers. (3'-1$\frac{1}{2}$" or 37$\frac{1}{2}$ ÷ 5 = 7$\frac{1}{2}$") (See Math Review 10.) The width of the treads is found by dividing the total run by the number of treads. (3'-3" or 39" ÷ 4 = 9$\frac{3}{4}$") Now check the tread and riser dimensions against the three rules just listed:

1. $7\frac{1}{2}$ inches (riser) plus $9\frac{3}{4}$ inches (tread) equals $17\frac{1}{4}$ inches.

2. $7\frac{1}{2}$ inches plus $7\frac{1}{2}$ inches (2 risers) plus $9\frac{3}{4}$ inches (1 tread) equals $24\frac{3}{4}$ inches.

3. $9\frac{3}{4}$ inches (tread) multiplied by $7\frac{1}{2}$ inches (riser) equals $73\frac{1}{8}$ inches.

To calculate the size of the treads and risers in stairs with landings, treat each part as a separate stair. However, the treads and risers should be the same size in each part.

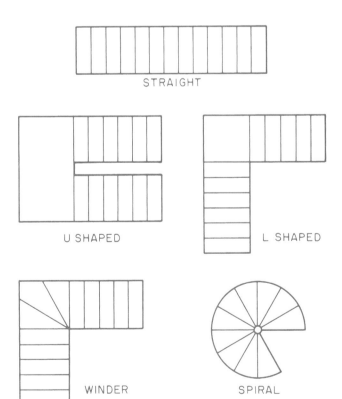

STRAIGHT

U SHAPED

L SHAPED

WINDER

SPIRAL

Figure 29-5 Stair layouts

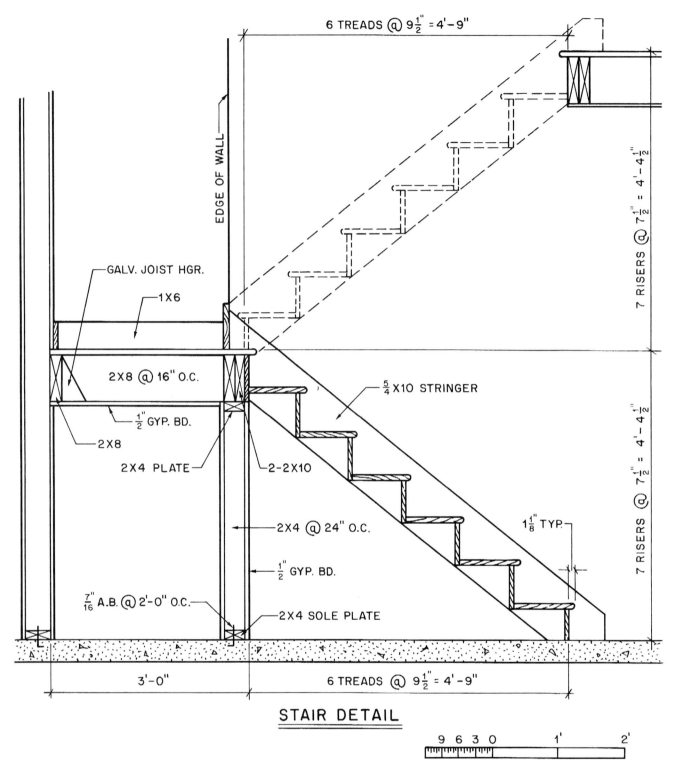

6 TREADS @ 9½" = 4'-9"

EDGE OF WALL

GALV. JOIST HGR.

1X6

2X8 @ 16" O.C.

½" GYP. BD.

2X8

2X4 PLATE

5/4 X10 STRINGER

2-2X10

2X4 @ 24" O.C.

½" GYP. BD.

7/16" A.B. @ 2'-0" O.C.

2X4 SOLE PLATE

1⅛" TYP.

7 RISERS @ 7½" = 4'-4½"

7 RISERS @ 7½" = 4'-4½"

3'-0"

6 TREADS @ 9½" = 4'-9"

STAIR DETAIL

9 6 3 0   1'   2'

Figure 29-6   Stair detail with 2 × 4 studs to support the landing

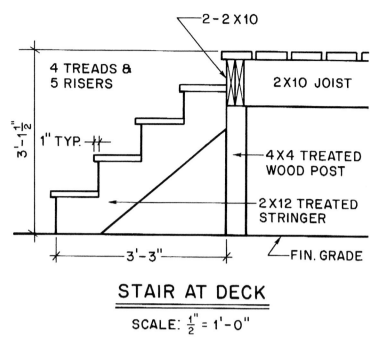

STAIR AT DECK

SCALE: $\frac{1}{2}$" = 1'-0"

Figure 29-7   Typical detail for deck stairs

✓ CHECK YOUR PROGRESS ─────────────────────────

Can you perform these tasks?

☐ Identify the following stair parts on drawings: stringers, treads, risers, landings, nosing, stair cover molding, newel post, handrail, and baluster.

☐ Calculate desired tread width and riser heights from the total run and rise.

☐ List the sizes of all of the parts of a stair.

ASSIGNMENT ──────────────────────────────

Refer to the Lake House drawings (in the packet) to complete the assignment.

Questions 1 through 6 refer to the stairs between the south decks of the Lake House.

1.   What is the total run of the stairs between the decks?

2.   What is the width of each tread?

3.   What is the length of each tread?

4.   What is the rise of each step in these stairs?

5.   How many stringers are used under these stairs?

6.   What size material is used for the stringers?

Questions 7 through 13 refer to the stairs from the kitchen to the bedroom level in the Lake House.

7.   How many risers are there and how high is each riser?

8.   How many treads are there and how wide (front to back including nosing) is each tread?

9.   What is the total rise of the stairs?

10. Is this stair built with open or housed stringers?

11. The railing at this stair extends the length of the kitchen hall. If vertical railing supports are spaced at 16" O.C., how many vertical supports are used?

12. How long is each vertical railing support?

13. What material is used for the horizontal rails?

UNIT 30   Insulation and Room Finishing

OBJECTIVES

After completing this unit, you will be able to perform the following tasks:

- Identify the insulation to be used in walls, floors, and ceilings.
- Identify the wall, ceiling, and floor covering material to be used.
- List all of the kinds of interior molding to be used.

INSULATION

Thermal insulation is any material that is used to resist the flow of heat. In very warm climate, thermal insulation is used to resist the flow of heat from the outside to the inside. In cold climates, thermal insulation is used to resist the flow of heat from inside to the outside. Insulating material is rated according to its ability to resist the flow of heat. The measure of this resistance is the *R value* of the material. The higher the *R* value, the better the material insulates. Typical *R* values for sidewall or attic insulation range from R-3 to R-38, Figure 30-1.

Thermal insulation is generally available in four forms. *Foamed-in-place* materials are synthetic compounds that are sprayed onto a surface, and then produce an insulating foam by a chemical reaction, Figure 30-2. *Rigid boards* are plastic foams which have been produced in board form at a factory. Another common type of rigid insulation is made of fiberglass which is

TYPE	TYPICAL THICKNESSES AND R VALUES	COMMENTS
Fiberglass blankets & batts	3/4", R-3 2 1/2", R-8 3 1/2", R-11 6", R-19 9", R-30 12", R-38	Flexible blanket-like material. Available with or without vapor barrier on one side.
Fiberglass blowing wool	R value depends on depth of coverage, but is sligtly less than fiberglass blankets.	Other types of blowing or pouring insulation include cellulose and mineral wool.
Rigid fiberglass board	1", R-4.4	Material is similar to fiberglass blankets, but with rigid binder to create rigid boards. Usually faced with aluminum foil.
Rigid urethane foamed board	1/2", R-3.6 3/4", R-5.4 1", R-7.2 1 1/2", R-10.8 2", R-14.6	Plastic that has been cured in a foamed state to introduce bubbles of air. This creates a rigid board which is usually faced with aluminum foil.
Foamed-in-place urethane	R value depends on thickness depth, but is approximately the same as rigid urethane boards.	Other plastic materials may also be foamed in place.

Figure 30-1   Common types of thermal insulation

Courtesy of The Upjohn Company

Figure 30-2   Foamed-in-place insulation

Courtesy of Owens-Corning Fiberglas Corporation

Figure 30-3   Rigid insulation

manufactured in rigid form instead of as flexible blankets, Figure 30-3. Rigid boards are frequently used for foundation insulation, under concrete slabs, and for sheathing. *Blanket insulation* is in the form of flexible rolls or batts usually made of fiberglass wool, Figure 30-4. Loose *pouring insulation* can be any of a variety of materials that can be poured into place and has good insulating property. A common pouring insulation is loose fiberglass, Figure 30-5.

The insulation is shown in the building sections. If the insulation is to be installed between studs, joists, or rafters; it will be sized accordingly. For example, if batts are to be used in a 2 × 4 wall, the insulation will be indicated as 3½ inches thick—the width of a 2 × 4. Where insulation is used in ventilated spaces, there should be room for the necessary air circulation, Figure 30-6.

Where the insulation includes a vapor barrier, such as Kraft paper, foil, or polyethylene, the vapor barrier is installed on the heated side of the wall. This prevents the moisture in the warm air from passing through the wall and condensing on the cold side of the wall. Such condensation can reduce the R value of the insulation and cause painted surfaces to blister and peel.

Courtesy of Owens-Corning Fiberglas Corporation

Figure 30-4   Blanket insulation

Courtesy of Owens-Corning Fiberglas Corporation

Figure 30-5   Blowing wool

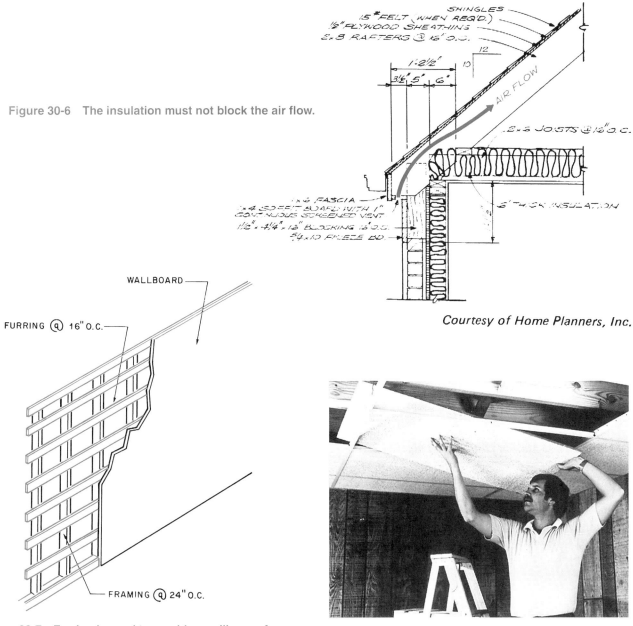

Figure 30-6 The insulation must not block the air flow.

Courtesy of Home Planners, Inc.

Figure 30-7 Furring is used to provide a nailing surface when framing is spaced too wide or over masonry and concrete.

Courtesy of Celotex Corporation

Figure 30-8 Suspended ceiling

WALL AND CEILING COVERING

By far the most widely used wall surface material is *gypsum wallboard*. The most common thicknesses are $3/8$ inch, $1/2$ inch, and $5/8$ inch. If gypsum board (sometimes called *sheet rock*) is to be used over framing that is spaced more than 16 inches O.C. or over masonry, the designer may call for furring. *Furring* consists of narrow strips of wood, usually spaced at 16 or 12 inches, O.C., to which the wall covering is fastened, Figure 30-7. The furring, if any is to be used, and the thickness of the gypsum wallboard are indicated on the wall sections.

The electrical installer should review the room finish requirements for any specified finish that would require special electrical device box extension rings (plaster rings) other than the standard extension rings for the corresponding standard gypsum wallboard thicknesses previously listed.

Ceramic tile also is frequently used on bathroom walls. Ceramic tile may be installed over a base of water-resistant gypsum board, plaster, or plywood.

Gypsum wallboard and suspended ceilings are the most common ceiling treatments. Suspended ceilings consist of panels or ceiling tiles supported in a lightweight metal framework. The metal framework is

ROOM	FLOOR	WALLS	CEILING
KITCHEN	QUARRY TILE	GYP. BD. w/WALL PAPER	12"x12" TILE
DINING ROOM	OAK PARQUET	GYP. BD.	GYP. BD.
LIVING ROOM	CARPET/PART. BD.	GYP. BD.	GYP. BD.
FAMILY ROOM	CARPET/PART. BD.	H.D. BD. PANEL/GYP. BD.	GYP. BD.
BEDROOM #1	CARPET/PART. BD.	GYP. BD.	GYP. BD.
BEDROOM #2	CARPET/PART. BD.	GYP. BD.	GYP. BD.
BEDROOM #3	CARPET/PART. BD.	GYP. BD.	GYP. BD.
BATH #1	CERAMIC TILE	CERAMIC TILE/GYP. BD.	GYP. BD.
BATH #2	CERAMIC TILE	CERAMIC TILE/GYP. BD.	GYP. BD.
CLOSETS	CARPET/PART BD.	GYP. BD.	GYP. BD.
FOYER	SLATE	GYP. BD. w/WALL PAPER	GYP. BD.

Figure 30-9   Room finish schedule

suspended several inches below the ceiling framing on steel wires, Figure 30-8.

Building sections or wall sections usually include a typical wall and ceiling. This is representative of what is planned for most of the walls and ceilings in the house. However, there may be some exceptions, such as the bathrooms, or kitchen where water-resistant wallboard is required. Somewhere within the contract documents, you should find a complete list of all room finishes. This may be on one of the drawings or it may be written into the specifications. Figure 30-9 shows a room finish schedule that might be included on the drawings. It is common for things like finish color to be left for the owner to choose.

FINISHED FLOORS

A list of possible floor materials would be very long. However, most of them fall into one of the following categories: wood, carpet, ceramic, masonry, and resilient materials such as vinyl tile. The finished floor covering is easily found in a schedule of room finishes, but the underlayment for each category is different. *Underlayment* is any material that is used to prepare the subfloor to receive the finished floor.

Architectural drafters seem to differ in how they indicate what underlayment is to be used and how the finished floor is to be installed. Some do not include any underlayment on the drawings, but rely on the builder's knowledge of good construction practices. This is a dan-

gerous practice. If you find this situation, the architect should be asked for a clarification. Sometimes when the drawings do not describe the underlayment, the specifications do include it. Some drafters indicate the underlayment on the floor plans. Other drafters include details and section of each area with different types of finished floors, so the underlayment can be shown there.

INTERIOR MOLDING

Molding is used to decorate surfaces, protect the edges and corners of surfaces, and to conceal joints or seams, between surfaces. Molding may be made of wood, plastic, or metal and is available in many shapes and styles. The shapes of commonly used wood molding have been standardized. Each shape is identified by a number, Figure 30-10.

Most interior molding is shown on detail drawings. The following are some of the most common uses for interior molding:

- Window casing
- Window stool
- Door casing
- Base (bottom of wall)
- Cove (top of wall)
- Chair rail (middle of wall)
- Trim around fireplace mantle
- Trim around built-in cabinets

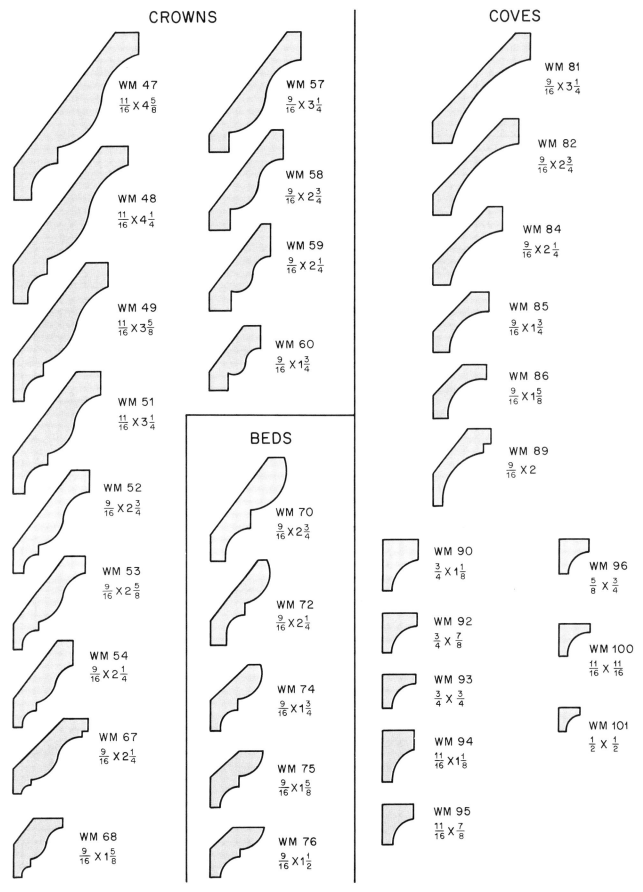

Figure 30-10   Common shapes of wood molding

QUARTER ROUNDS

WM 103
$1\frac{1}{16} \times 1\frac{1}{16}$

WM 104
$\frac{11}{16} \times 1\frac{3}{8}$

WM 105
$\frac{3}{4} \times \frac{3}{4}$

WM 106
$\frac{11}{16} \times \frac{11}{16}$

WM 107
$\frac{5}{8} \times \frac{5}{8}$

WM 108
$\frac{1}{2} \times \frac{1}{2}$

WM 109
$\frac{3}{8} \times \frac{3}{8}$

WM 110
$\frac{1}{4} \times \frac{1}{4}$

HALF ROUNDS

WM 120
$\frac{1}{2} \times 1$

WM 122
$\frac{3}{8} \times \frac{11}{16}$

WM 123
$\frac{5}{16} \times \frac{5}{8}$

WM 124
$\frac{1}{4} \times \frac{1}{2}$

FLAT ASTRAGALS

WM 133
$\frac{11}{16} \times 1\frac{3}{4}$

WM 134
$\frac{11}{16} \times 1\frac{3}{8}$

WM 135
$\frac{7}{16} \times \frac{3}{4}$

BASE SHOES

WM 126
$\frac{1}{2} \times \frac{3}{4}$

WM 129
$\frac{7}{16} \times \frac{11}{16}$

WM 127
$\frac{7}{16} \times \frac{3}{4}$

WM 131
$\frac{1}{2} \times \frac{3}{4}$

SHELF EDGE/ SCREEN MOULD

WM 137
$\frac{3}{8} \times \frac{3}{4}$

WM 141
$\frac{1}{4} \times \frac{5}{8}$

WM 138
$\frac{5}{16} \times \frac{5}{8}$

WM 142
$\frac{1}{4} \times \frac{3}{4}$

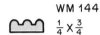
WM 140
$\frac{1}{4} \times \frac{3}{4}$

WM 144
$\frac{1}{4} \times \frac{3}{4}$

GLASS BEADS

WM 147
$\frac{1}{2} \times \frac{9}{16}$

WM 148
$\frac{3}{8} \times \frac{3}{8}$

BASE CAPS

WM 163
$\frac{11}{16} \times 1\frac{3}{8}$

WM 167
$\frac{11}{16} \times 1\frac{1}{8}$

WM 164
$\frac{11}{16} \times 1\frac{1}{8}$

WM 172
$\frac{5}{8} \times \frac{3}{4}$

WM 166
$\frac{11}{16} \times 1\frac{1}{4}$

BRICK MOULD

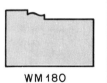
WM 175
$1\frac{1}{16} \times 2$

WM 180
$1\frac{1}{4} \times 2$

WM 176
$1\frac{1}{16} \times 1\frac{3}{4}$

DRIP CAPS

WM 187
$1\frac{11}{16} \times 2$

WM 196
$\frac{11}{16} \times 1\frac{3}{4}$

WM 188
$1\frac{1}{16} \times 1\frac{5}{8}$

WM 197
$\frac{11}{16} \times 1\frac{5}{8}$

CORNER GUARDS

WM 199
1×1

WM 200
$\frac{3}{4} \times \frac{3}{4}$

WM 201
$1\frac{5}{16} \times 1\frac{5}{16}$

WM 204
$1\frac{5}{16} \times 1\frac{5}{16}$

WM 202
$1\frac{1}{8} \times 1\frac{1}{8}$

WM 205
$1\frac{1}{8} \times 1\frac{1}{8}$

WM 203
$\frac{3}{4} \times \frac{3}{4}$

WM 206
$\frac{3}{4} \times \frac{3}{4}$

BATTENS

WM 224
$\frac{9}{16} \times 2\frac{1}{4}$

WM 229
$\frac{11}{16} \times 1\frac{5}{8}$

ROUNDS

WM 232 $1\frac{5}{8}$

WM 233 $1\frac{5}{16}$

WM 234 $1\frac{1}{16}$

SQUARES

WM 236 $1\frac{5}{8} \times 1\frac{5}{8}$

WM 237 $1\frac{5}{16} \times 1\frac{5}{16}$

WM 238 $1\frac{1}{16} \times 1\frac{1}{16}$

WM 239 $\frac{3}{4} \times \frac{3}{4}$

Figure 30-10 (continued)

SHINGLE PANEL MOULDINGS

WM 207
$\frac{11}{16} \times 2\frac{1}{2}$

WM 209
$\frac{11}{16} \times 2$

WM 210
$\frac{11}{16} \times 1\frac{5}{8}$

WM 212
$\frac{11}{16} \times 2\frac{1}{2}$

WM 213
$\frac{9}{16} \times 2$

WM 217
$\frac{11}{16} \times 1\frac{3}{4}$

WM 218
$\frac{11}{16} \times 1\frac{1}{2}$

HAND RAIL

WM 230
$1\frac{1}{2} \times 1\frac{11}{16}$

WM 231
$1\frac{1}{2} \times 1\frac{11}{16}$

WM 240
$1\frac{1}{4} \times 2\frac{1}{4}$

PICTURE MOULDINGS

WM 273
$\frac{11}{16} \times 1\frac{3}{4}$

WM 276
$\frac{11}{16} \times 1\frac{3}{4}$

SCREEN/S4S STOCK

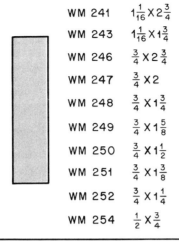

WM 241	$1\frac{1}{16} \times 2\frac{3}{4}$
WM 243	$1\frac{1}{16} \times 1\frac{3}{4}$
WM 246	$\frac{3}{4} \times 2\frac{3}{4}$
WM 247	$\frac{3}{4} \times 2$
WM 248	$\frac{3}{4} \times 1\frac{3}{4}$
WM 249	$\frac{3}{4} \times 1\frac{5}{8}$
WM 250	$\frac{3}{4} \times 1\frac{1}{2}$
WM 251	$\frac{3}{4} \times 1\frac{3}{8}$
WM 252	$\frac{3}{4} \times 1\frac{1}{4}$
WM 254	$\frac{1}{2} \times \frac{3}{4}$

LATTICE

WM 265	$\frac{9}{32} \times 1\frac{3}{4}$
WM 266	$\frac{9}{32} \times 1\frac{5}{8}$
WM 267	$\frac{9}{32} \times 1\frac{3}{8}$
WM 268	$\frac{9}{32} \times 1\frac{1}{8}$

BACK BANDS

WM 280
$\frac{11}{16} \times 1\frac{1}{16}$

WM 281
$\frac{11}{16} \times 1\frac{1}{8}$

WAINSCOT/PLY CAP MOULDINGS

WM 290
$\frac{11}{16} \times 1\frac{3}{8}$

WM 292
$\frac{9}{16} \times 1\frac{1}{8}$

WM 294
$\frac{11}{16} \times 1\frac{1}{8}$

WM 295
$\frac{1}{2} \times 1\frac{1}{4}$

WM 296
$\frac{3}{4} \times \frac{3}{4}$

CHAIR RAILS

WM 297
$\frac{11}{16} \times 3$

WM 298
$\frac{11}{16} \times 2\frac{1}{2}$

WM 300
$1\frac{1}{16} \times 3$

WM 303
$\frac{9}{16} \times 2\frac{1}{2}$

WM 304
$\frac{1}{2} \times 2\frac{1}{4}$

WM 390
$\frac{11}{16} \times 2\frac{5}{8}$

FLAT STOOLS

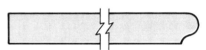

WM 1021 $\frac{11}{16} \times$ WIDTH SPECIFIED

T-ASTRAGALS

WM 1300
$1\frac{1}{4} \times 2\frac{1}{4}$

WM 1305
$1\frac{1}{4} \times 2$

WM 1310
$1\frac{1}{4} \times 2\frac{1}{4}$

WM 1315
$1\frac{1}{4} \times 2$

Figure 30-10 (continued)

CASING

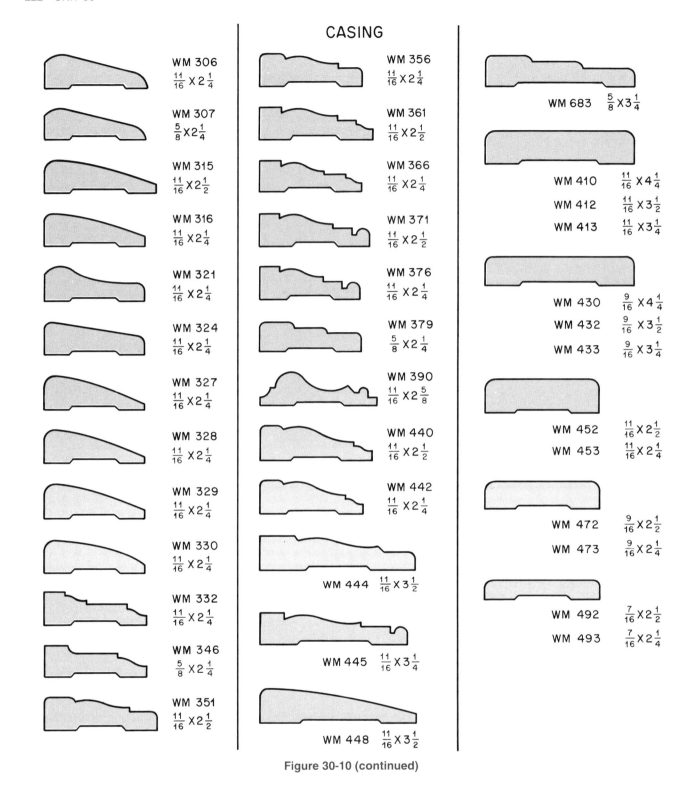

WM 306 $\frac{11}{16} \times 2\frac{1}{4}$

WM 307 $\frac{5}{8} \times 2\frac{1}{4}$

WM 315 $\frac{11}{16} \times 2\frac{1}{2}$

WM 316 $\frac{11}{16} \times 2\frac{1}{4}$

WM 321 $\frac{11}{16} \times 2\frac{1}{4}$

WM 324 $\frac{11}{16} \times 2\frac{1}{4}$

WM 327 $\frac{11}{16} \times 2\frac{1}{4}$

WM 328 $\frac{11}{16} \times 2\frac{1}{4}$

WM 329 $\frac{11}{16} \times 2\frac{1}{4}$

WM 330 $\frac{11}{16} \times 2\frac{1}{4}$

WM 332 $\frac{11}{16} \times 2\frac{1}{4}$

WM 346 $\frac{5}{8} \times 2\frac{1}{4}$

WM 351 $\frac{11}{16} \times 2\frac{1}{2}$

WM 356 $\frac{11}{16} \times 2\frac{1}{4}$

WM 361 $\frac{11}{16} \times 2\frac{1}{2}$

WM 366 $\frac{11}{16} \times 2\frac{1}{4}$

WM 371 $\frac{11}{16} \times 2\frac{1}{2}$

WM 376 $\frac{11}{16} \times 2\frac{1}{4}$

WM 379 $\frac{5}{8} \times 2\frac{1}{4}$

WM 390 $\frac{11}{16} \times 2\frac{5}{8}$

WM 440 $\frac{11}{16} \times 2\frac{1}{2}$

WM 442 $\frac{11}{16} \times 2\frac{1}{4}$

WM 444 $\frac{11}{16} \times 3\frac{1}{2}$

WM 445 $\frac{11}{16} \times 3\frac{1}{4}$

WM 448 $\frac{11}{16} \times 3\frac{1}{2}$

WM 683 $\frac{5}{8} \times 3\frac{1}{4}$

WM 410 $\frac{11}{16} \times 4\frac{1}{4}$

WM 412 $\frac{11}{16} \times 3\frac{1}{2}$

WM 413 $\frac{11}{16} \times 3\frac{1}{4}$

WM 430 $\frac{9}{16} \times 4\frac{1}{4}$

WM 432 $\frac{9}{16} \times 3\frac{1}{2}$

WM 433 $\frac{9}{16} \times 3\frac{1}{4}$

WM 452 $\frac{11}{16} \times 2\frac{1}{2}$

WM 453 $\frac{11}{16} \times 2\frac{1}{4}$

WM 472 $\frac{9}{16} \times 2\frac{1}{2}$

WM 473 $\frac{9}{16} \times 2\frac{1}{4}$

WM 492 $\frac{7}{16} \times 2\frac{1}{2}$

WM 493 $\frac{7}{16} \times 2\frac{1}{4}$

Figure 30-10 (continued)

BASE MOULDINGS

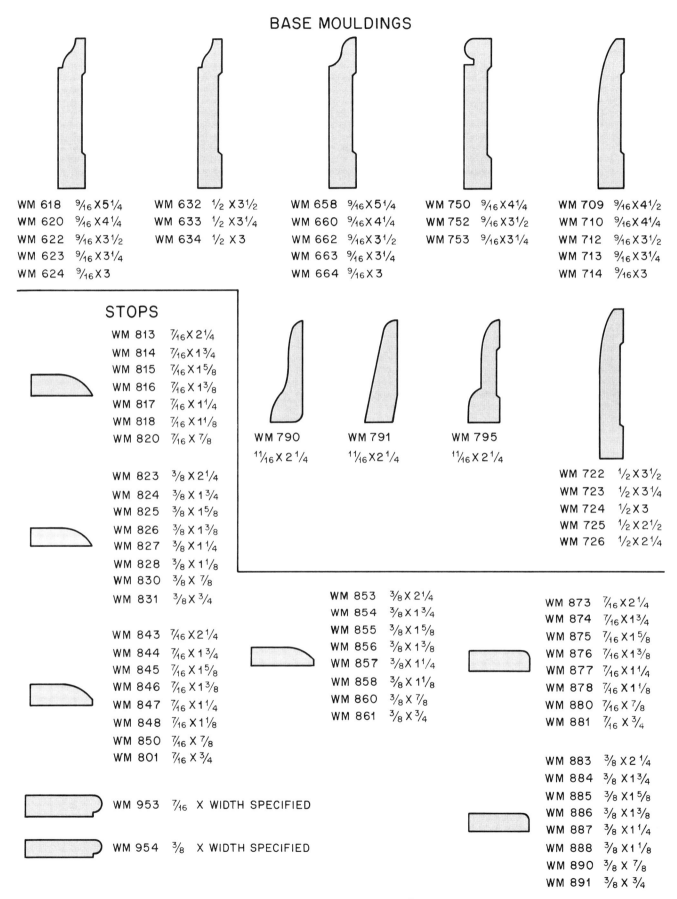

WM 618 $^9/_{16}$ X $5^1/_4$	WM 632 $^1/_2$ X $3^1/_2$	WM 658 $^9/_{16}$ X $5^1/_4$	WM 750 $^9/_{16}$ X $4^1/_4$	WM 709 $^9/_{16}$ X $4^1/_2$
WM 620 $^9/_{16}$ X $4^1/_4$	WM 633 $^1/_2$ X $3^1/_4$	WM 660 $^9/_{16}$ X $4^1/_4$	WM 752 $^9/_{16}$ X $3^1/_2$	WM 710 $^9/_{16}$ X $4^1/_4$
WM 622 $^9/_{16}$ X $3^1/_2$	WM 634 $^1/_2$ X 3	WM 662 $^9/_{16}$ X $3^1/_2$	WM 753 $^9/_{16}$ X $3^1/_4$	WM 712 $^9/_{16}$ X $3^1/_2$
WM 623 $^9/_{16}$ X $3^1/_4$		WM 663 $^9/_{16}$ X $3^1/_4$		WM 713 $^9/_{16}$ X $3^1/_4$
WM 624 $^9/_{16}$ X 3		WM 664 $^9/_{16}$ X 3		WM 714 $^9/_{16}$ X 3

STOPS

WM 813 $^7/_{16}$ X $2^1/_4$
WM 814 $^7/_{16}$ X $1^3/_4$
WM 815 $^7/_{16}$ X $1^5/_8$
WM 816 $^7/_{16}$ X $1^3/_8$
WM 817 $^7/_{16}$ X $1^1/_4$
WM 818 $^7/_{16}$ X $1^1/_8$
WM 820 $^7/_{16}$ X $^7/_8$

WM 823 $^3/_8$ X $2^1/_4$
WM 824 $^3/_8$ X $1^3/_4$
WM 825 $^3/_8$ X $1^5/_8$
WM 826 $^3/_8$ X $1^3/_8$
WM 827 $^3/_8$ X $1^1/_4$
WM 828 $^3/_8$ X $1^1/_8$
WM 830 $^3/_8$ X $^7/_8$
WM 831 $^3/_8$ X $^3/_4$

WM 843 $^7/_{16}$ X $2^1/_4$
WM 844 $^7/_{16}$ X $1^3/_4$
WM 845 $^7/_{16}$ X $1^5/_8$
WM 846 $^7/_{16}$ X $1^3/_8$
WM 847 $^7/_{16}$ X $1^1/_4$
WM 848 $^7/_{16}$ X $1^1/_8$
WM 850 $^7/_{16}$ X $^7/_8$
WM 801 $^7/_{16}$ X $^3/_4$

WM 953 $^7/_{16}$ X WIDTH SPECIFIED

WM 954 $^3/_8$ X WIDTH SPECIFIED

WM 790
$^{11}/_{16}$ X $2^1/_4$

WM 791
$^{11}/_{16}$ X $2^1/_4$

WM 795
$^{11}/_{16}$ X $2^1/_4$

WM 722 $^1/_2$ X $3^1/_2$
WM 723 $^1/_2$ X $3^1/_4$
WM 724 $^1/_2$ X 3
WM 725 $^1/_2$ X $2^1/_2$
WM 726 $^1/_2$ X $2^1/_4$

WM 853 $^3/_8$ X $2^1/_4$
WM 854 $^3/_8$ X $1^3/_4$
WM 855 $^3/_8$ X $1^5/_8$
WM 856 $^3/_8$ X $1^3/_8$
WM 857 $^3/_8$ X $1^1/_4$
WM 858 $^3/_8$ X $1^1/_8$
WM 860 $^3/_8$ X $^7/_8$
WM 861 $^3/_8$ X $^3/_4$

WM 873 $^7/_{16}$ X $2^1/_4$
WM 874 $^7/_{16}$ X $1^3/_4$
WM 875 $^7/_{16}$ X $1^5/_8$
WM 876 $^7/_{16}$ X $1^3/_8$
WM 877 $^7/_{16}$ X $1^1/_4$
WM 878 $^7/_{16}$ X $1^1/_8$
WM 880 $^7/_{16}$ X $^7/_8$
WM 881 $^7/_{16}$ X $^3/_4$

WM 883 $^3/_8$ X $2^1/_4$
WM 884 $^3/_8$ X $1^3/_4$
WM 885 $^3/_8$ X $1^5/_8$
WM 886 $^3/_8$ X $1^3/_8$
WM 887 $^3/_8$ X $1^1/_4$
WM 888 $^3/_8$ X $1^1/_8$
WM 890 $^3/_8$ X $^7/_8$
WM 891 $^3/_8$ X $^3/_4$

Figure 30-10 (continued)

WM 903	$\frac{7}{16}$ X 2$\frac{1}{4}$	WM 913	$\frac{3}{8}$ X 2$\frac{1}{4}$	WM 933	$\frac{7}{16}$ X 2$\frac{1}{4}$	WM 943	$\frac{3}{8}$ X 2$\frac{1}{4}$
WM 904	$\frac{7}{16}$ X 1$\frac{3}{4}$	WM 914	$\frac{3}{8}$ X 1$\frac{3}{4}$	WM 934	$\frac{7}{16}$ X 1$\frac{3}{4}$	WM 944	$\frac{3}{8}$ X 1$\frac{3}{4}$
WM 905	$\frac{7}{16}$ X 1$\frac{5}{8}$	WM 915	$\frac{3}{8}$ X 1$\frac{5}{8}$	WM 935	$\frac{7}{16}$ X 1$\frac{5}{8}$	WM 945	$\frac{3}{8}$ X 1$\frac{5}{8}$
WM 906	$\frac{7}{16}$ X 1$\frac{3}{8}$	WM 916	$\frac{3}{8}$ X 1$\frac{3}{8}$	WM 936	$\frac{7}{16}$ X 1$\frac{3}{8}$	WM 946	$\frac{3}{8}$ X 1$\frac{3}{8}$
WM 907	$\frac{7}{16}$ X 1$\frac{1}{4}$	WM 917	$\frac{3}{8}$ X 1$\frac{1}{4}$	WM 937	$\frac{7}{16}$ X 1$\frac{1}{4}$	WM 947	$\frac{3}{8}$ X 1$\frac{1}{4}$
WM 908	$\frac{7}{16}$ X 1$\frac{1}{8}$	WM 918	$\frac{3}{8}$ X 1$\frac{1}{8}$	WM 938	$\frac{7}{16}$ X 1$\frac{1}{8}$	WM 948	$\frac{3}{8}$ X 1$\frac{1}{8}$
WM 910	$\frac{7}{16}$ X $\frac{7}{8}$	WM 920	$\frac{3}{8}$ X $\frac{7}{8}$	WM 940	$\frac{7}{16}$ X $\frac{7}{8}$	WM 950	$\frac{3}{8}$ X $\frac{7}{8}$
WM 911	$\frac{7}{16}$ X $\frac{3}{4}$	WM 921	$\frac{3}{8}$ X $\frac{3}{4}$	WM 941	$\frac{7}{16}$ X $\frac{3}{4}$	WM 951	$\frac{3}{8}$ X $\frac{3}{4}$

PANEL STRIPS MULLION CASINGS

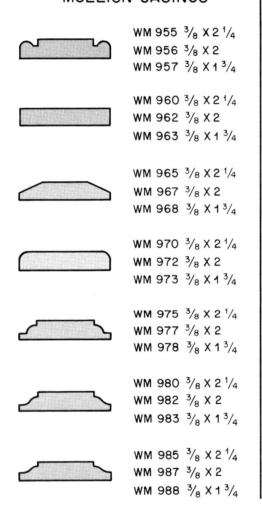

WM 955	$\frac{3}{8}$ X 2$\frac{1}{4}$
WM 956	$\frac{3}{8}$ X 2
WM 957	$\frac{3}{8}$ X 1$\frac{3}{4}$
WM 960	$\frac{3}{8}$ X 2$\frac{1}{4}$
WM 962	$\frac{3}{8}$ X 2
WM 963	$\frac{3}{8}$ X 1$\frac{3}{4}$
WM 965	$\frac{3}{8}$ X 2$\frac{1}{4}$
WM 967	$\frac{3}{8}$ X 2
WM 968	$\frac{3}{8}$ X 1$\frac{3}{4}$
WM 970	$\frac{3}{8}$ X 2$\frac{1}{4}$
WM 972	$\frac{3}{8}$ X 2
WM 973	$\frac{3}{8}$ X 1$\frac{3}{4}$
WM 975	$\frac{3}{8}$ X 2$\frac{1}{4}$
WM 977	$\frac{3}{8}$ X 2
WM 978	$\frac{3}{8}$ X 1$\frac{3}{4}$
WM 980	$\frac{3}{8}$ X 2$\frac{1}{4}$
WM 982	$\frac{3}{8}$ X 2
WM 983	$\frac{3}{8}$ X 1$\frac{3}{4}$
WM 985	$\frac{3}{8}$ X 2$\frac{1}{4}$
WM 987	$\frac{3}{8}$ X 2
WM 988	$\frac{3}{8}$ X 1$\frac{3}{4}$

RABBETED STOOLS

SPECIFY WIDTH OF RABBET AND DEGREE OF BEVEL

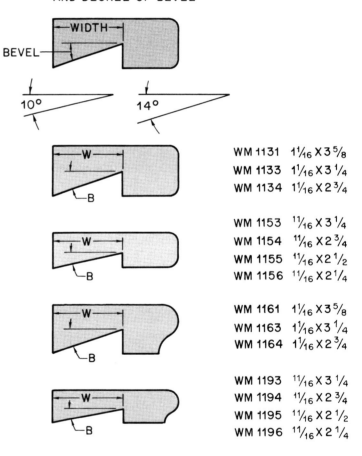

WM 1131	1$\frac{1}{16}$ X 3$\frac{5}{8}$
WM 1133	1$\frac{1}{16}$ X 3$\frac{1}{4}$
WM 1134	1$\frac{1}{16}$ X 2$\frac{3}{4}$
WM 1153	$\frac{11}{16}$ X 3$\frac{1}{4}$
WM 1154	$\frac{11}{16}$ X 2$\frac{3}{4}$
WM 1155	$\frac{11}{16}$ X 2$\frac{1}{2}$
WM 1156	$\frac{11}{16}$ X 2$\frac{1}{4}$
WM 1161	1$\frac{1}{16}$ X 3$\frac{5}{8}$
WM 1163	1$\frac{1}{16}$ X 3$\frac{1}{4}$
WM 1164	1$\frac{1}{16}$ X 2$\frac{3}{4}$
WM 1193	$\frac{11}{16}$ X 3$\frac{1}{4}$
WM 1194	$\frac{11}{16}$ X 2$\frac{3}{4}$
WM 1195	$\frac{11}{16}$ X 2$\frac{1}{2}$
WM 1196	$\frac{11}{16}$ X 2$\frac{1}{4}$

Figure 30-10 (continued)

✓ CHECK YOUR PROGRESS ──────────────────────────────

Can you perform these tasks?

☐ Describe the kinds of insulation to be used in walls, around foundations, under slabs, in floors, in attics, and as sheathing.

☐ List the kind and thickness of material used for the interior surface of walls, ceilings, and floors.

☐ Identify areas where furring is to be used and the spacing of the furring.

☐ List all of the types of molding to be used for interior trim.

ASSIGNMENT ──

Refer to the Lake House drawings (in the packet) to complete the assignment.

1. What size or rating and kind of insulation is to be used in each of the following locations?

   a. Framed exterior walls

   b. Roof

   c. Under heat sink

   d. Masonry walls of playroom

2. What type molding is to be used as casing around interior doors?

3. What type of molding is used at the bottom of interior walls and partitions?

4. What kind of trim is used to cover the lower edges of exposed box beams?

5. Describe the wall finish in the playroom including:

   a. What the wall finish material is fastened to.

   b. The kind of material used for wall finish.

6. What material is used for ceiling finish in the playroom?

7. What material is used for underlayment on typical framed floors?

8. What is the finished floor material at the heat sink?

9. What covers the interior faces of wood box beams?

10. What is the finished wall material in the bedrooms?

UNIT 31 Cabinets

OBJECTIVES

After completing this unit, you will be able to perform the following tasks:

- List the sizes and types of cabinets shown on a set of drawings.

- Identify cabinet types and dimensions in manufacturers' literature.

SHOWING CABINETS ON DRAWINGS

Cabinets are typically found in the kitchen, bathroom, and laundry room. Most cabinets are supplied by a manufacturer, so the carpenter's job is only to install them. When custom-built cabinets are to be included, they are assembled in a cabinet shop and delivered to the site.

The layout of the cabinets can be determined by reading the floor plan and cabinet elevations together. The floor plan normally includes reference marks indicating each cabinet elevation, Figure 31-1. The floor plan shows the location of major appliances and may include some overall layout dimensions.

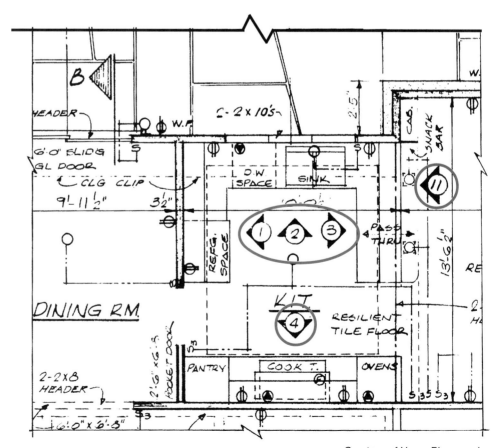

Figure 31-1 Kitchen floor plan with key to cabinet elevations

Courtesy of Home Planners, Inc.

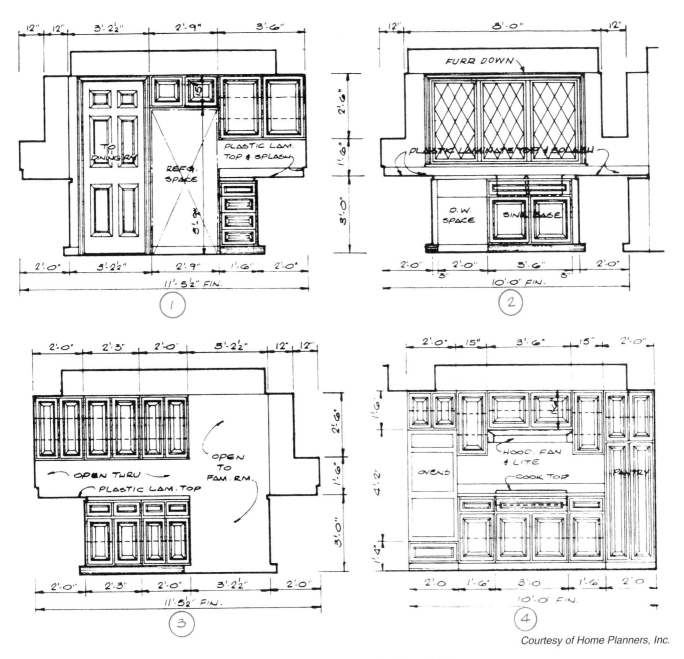

Courtesy of Home Planners, Inc.

Figure 31-2   Elevations for figure 31-1

When the project has separate electrical drawings, the electrical installer must coordinate these designated outlet locations between the electrical and architectural drawings. There may be electrical outlets shown on the architectural drawings and not shown on the electrical drawings. These variances must be resolved early in the rough-in stage of construction.

More complete cabinet information is usually shown on the cabinet elevations, Figure 31-2. Cabinet elevations are drawn for each direction from which cabinets can be viewed. These elevations show the types of cabinets, their sizes, and their arrangement. Cabinet types and sizes are recognized by a combination

of drawing representation and commonly used letter/number designations. Some drawings rely heavily on standard dimensioning and pictorial representation. This is true of the case shown in Figure 31-2. Other drawings include a letter/number designation for each cabinet, as shown in Figure 31-3. The letter part of the designation represents the type of cabinet. Architects and drafters vary in how they use letter designations, but they are usually easy to understand after a moments' study. Some typical letter designations are shown in Figure 31-4.

The numbers in the cabinet designation represent the dimensions of the cabinet. Base cabinets are a standard height (usually 3'-0" including the countertop) and a standard front-to-back depth (usually 2'-0") Therefore, base cabinet designations only include two digits to represent width in inches. Wall cabinets are a

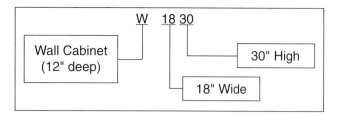

standard front-to-back depth (usually 1'-0"), but the width and height vary. Therefore, wall cabinet designations include four digits—two for width and two for height. The first and second digits usually indicate width and the third and fourth digits usually indicate height. The following example explains a typical cabinet designation.

In addition to cabinets, most cabinet manufacturers provide a variety of accessories:

- Shelves are finished to match the cabinets, so they can be used in open areas.

- Valances are prefinished decorative pieces to use between wall cabinets (over a sink, for example).

- Filler pieces are prefinished boards used to enclose narrow spaces between cabinets.

- A variety of molding may be used for trim.

READING MANUFACTURERS' LITERATURE

Cabinet manufacturers publish catalogs and specifications describing their cabinets. The numbering systems used in this literature are similar to those used on construction drawings. There may be slight differences between manufacturers, but these are usually easy to see. Figure 31-5 is reprinted from one manufacturer's literature. A few tips to reading this manufacturer's literature are listed:

- Cabinet heights and standard depths are dimensioned on the left side of the page.

- Referring to these dimensions, you can see that the first and second digits of wall cabinet numbers indicate their height. Therefore, the third and fourth digits indicate width.

- Standard base cabinets with one drawer above and doors below are designated by three digits only. The first digit (4) indicates a standard base cabinet. The second and third digits indicate width.

- Base cabinets with other than one drawer above a door are designated by a letter, but the last two digits still indicate width.

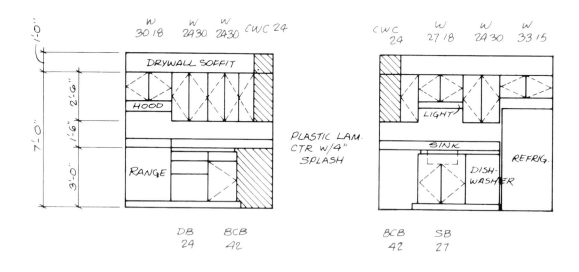

KITCHEN ELEVATIONS

Courtesy of Robert C. Kurzon

Figure 31-3 Letter/number designations for cabinets

LETTER	CABINET TYPE
W	WALL CABINET
WC	CORNER WALL CABINET
B	BASE CABINET
D OR DB	DRAWER BASE CABINET
BC	BASE CORNER CABINET
RC	REVOLVING CORNER CABINET
SF	SINK FRONT (ALSO FOR COOKTOP)
SB	SINK BASE (ALSO FOR COOKTOP)
U	UTILITY OR BROOM CLOSET
OV	OVEN CABINET

Figure 31-4   Key to typical cabinet designations

✓ CHECK YOUR PROGRESS

Can you perform these tasks?

- ☐ Orient cabinet elevations to floor plans.
- ☐ Describe cabinet types and sizes indicated by typical letter/number designations.
- ☐ List all of the cabinets shown on the drawings for a house.

ASSIGNMENT

Refer to Lake House drawings (in the packet) to complete the assignment.

Using the catalog pages shown in Figure 31-5 on the following pages, list all of the cabinets for the Lake House kitchen.

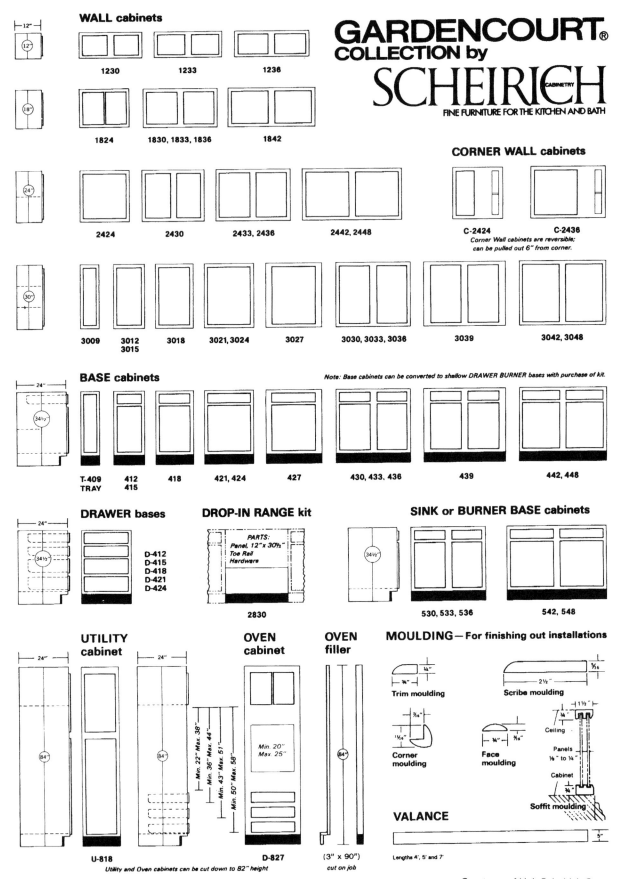

Figure 31-5   Cabinet catalog

After three years of intensive research and development the H. J. Scheirich Company is proud to present the new Gardencourt Collection. We wished to make the handsomest cabinets possible—and the toughest . . . cabinets so well styled they would fit into any interior, yet so well built they would pass the most rigorous test of all—everyday use in a busy kitchen.

We discarded all the old woodworking concepts and created a wholly new kind of cabinetry of exceptional structural rigidity and dimensional stability. The basic material for all parts is made of wood fibers bonded solidly together to provide extraordinary strength and resistance to variations in temperature and humidity. Framework, panels, shelves and doors—all are made of the bonded material. Then they are totally sheathed in Vinyl with exterior surfaces in pecan grain finish overlayed with a clear Vinyl for added protection. These are truly "carefree" cabinets. Produced in a modern air-conditioned factory employing highly sophisticated machinery, Gardencourt maintains the well known Scheirich standard of excellence.

All Scheirich cabinets are delivered complete with mounting hardware and detailed instructions for installation. Single door cabinets can hinge left or right. Shipped with doors attached, they can be reversed very simply by moving one hinge screw for each hinge into pre-drilled holes on the opposite inside front of the cabinet frame.

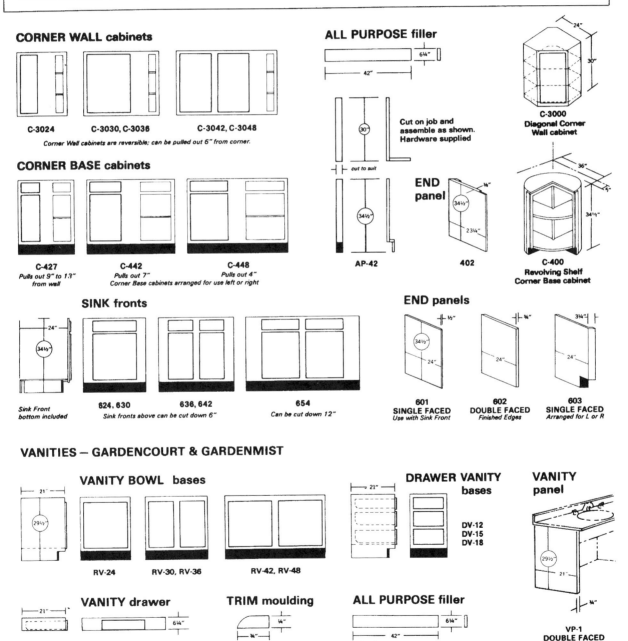

Figure 31-5 Cabinet catalog (continued)

PART II   Test

A. Refer to the Lake House drawings to determine each of the following dimensions.

1. Northwest corner of the site to the high water mark at the west boundary
2. Septic tank to the nearest property line
3. Width of the walk on the west side of the house
4. Width of the south end of the drive (in front of garage)
5. Depth of earth fill over septic drainfield
6. Elevation at the butt of the most southerly maple tree to remain
7. Outside of garage foundation (width × length)
8. Outside of garage footing (width × length)
9. North to south spacing of square steel columns
10. Length of northwest square steel column
11. Length of haunch under the slab between the north steel columns
12. East-west dimension inside the lower level bathroom at the widest end
13. Closet in bedroom #1 (width × length)
14. Inside of bedroom #1 (width × length)
15. Deck between the kitchen and the garage (width × length)
16. Length of the joists in the loft
17. Height of foundation wall at the overhead garage door
18. Elevation at the bottom of the concrete piers for the kitchen deck
19. Difference in elevations at the bottom of the south garage footing and the bottom of the nearest house footing
20. Elevation at the bottom of the deepest excavation
21. Width of the cabinet over the refrigerator
22. Width of the cabinet closest to the living room stairs
23. Total thickness of a typical exterior frame wall (allow ½ inch for siding)
24. Lineal feet of soldier-course bricks in fireplace
25. Lineal feet of 2 × 10 lumber in eave of corrugated roof
26. Finish floor to top surface of the free-standing closet/cabinet unit in the kitchen (include surface material)
27. Length of studs in wall separating the playroom from the crawl space under the kitchen
28. Length of the steel pins which anchor the foundation walls to the footings
29. Length of C8 × 11.5 structural steel beam
30. Outside surface of foundation under the dining room to the centerline of the deck footings
31. Door in south end of garage (width × height × thickness)
32. Length of 2 × 4 studs under living room bench
33. Width of the treads in the stairs from the kitchen to the bedroom level
34. Rough opening for the door from the hall into bedroom #1 (width × height)
35. Thickness of the concrete at haunches

B. Describe the material at each of the following locations in the Lake House. Include such considerations as the kind of material, nominal size of masonry units, and nominal size of lumber.

1. Tile field pipe
2. Reinforcement in concrete slabs
3. Rungs in ladder to loft
4. Hip rafter in northwest corner of house
5. Roof insulation

6. Roof deck
7. Exterior wall studs
8. Rafter headers at skylight
9. Purlins under corrugated roof
10. Finished surface on west wall of playroom
11. Reinforcement in footings
12. Anchor bolts in wood sill
13. Wood sill
14. Vapor barrier under concrete slab
15. Floor underlayment in bedrooms
16. Floor joists in kitchen
17. Floor joists in bedrooms
18. Stair stringers between decks
19. Railing around south decks
20. Top course of foundation at heat sink
21. Expansion joint at edge of heat sink
22. Rafters above loft
23. Posts supporting box beams
24. Girder under west wall of bedroom #1
25. Base plates on posts supporting box beams
26. Girder under kitchen deck joists
27. Finished surface of living room bench
28. Housed stringer on north side of stair to bedrooms
29. Stair treads
30. Door casings (interior)
31. Door casings (exterior)
32. Window casings (interior)
33. Glazing (light) in window on south side of bedroom #1
34. Planks on kitchen deck
35. Foundation at west side of kitchen deck

C. Answer each of these questions about the Lake House.

1. How many cubic yards of concrete are required for the concrete slab at elevation 337.00 feet?
2. What is the length of the concrete piers between the kitchen and the garage?
3. How many 4' × 8' sheets of plywood are needed to build the box beams?
4. How long are the rafters in the corrugated roof? (Refer to the illustrated rafter table.)
5. How long is the 2 × 12 hip rafter? (Refer to the illustrated rafter table.)

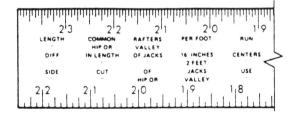
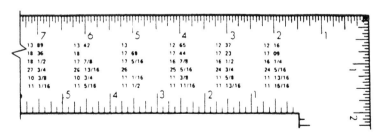

Part III

MULTIFAMILY CONSTRUCTION

Part III provides an opportunity for you to extend your ability to read construction drawings. Once you have mastered the contents of Parts I and II, you should be able to read and thoroughly understand most residential construction drawings. Part III helps you apply the skills developed earlier to other types of construction and to the work of the mechanical and electrical trades.

The Town House drawings were selected for reference in Part III because they represent quality construction in a geographic region where many construction practices are different from those in the rest of North America.

The complete set of drawings for the Town House is far too large to include in the packet that accompanies this textbook. Therefore, selected sheets or portions of sheets are used here. To use the available space most efficiently, some of the drawings have been reorganized and combined—with parts of two sheets printed on one. The original sheet and drawing numbers have been retained; therefore, all references printed on the drawings are still applicable.

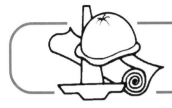

UNIT 32 *Orienting the Drawings*

OBJECTIVES

After completing this unit, you will be able to perform the following tasks:

- Locate a particular building or plan within a large development.

- Explain the relationships between drawings for construction projects where several plans are to be adjoined in one building.

- Visualize a building design by reading the drawings.

IDENTIFYING BUILDINGS AND PLANS

Multifamily dwellings are often built in large developments—with many similar buildings in a single development. Some developments are completed in phases. One phase is completely constructed and begins earning income for the developer before the next stage is started. Figure 32-1 shows the site plan for Hidden Valley, the development for which the Town House was designed. Hidden Valley is to be developed in four phases. Each phase is outlined with a heavy broken line on the site plan.

The first phase of Hidden Valley includes eleven buildings. Each building is labeled as to building type and the parts that make up the building, Figure 32-2. Building Type II is made up of four separate units, each with its own floor plan. Each unit is built like a separate building joined to the next, Figure 32-3. Each plan is identified by a letter or letter and numeral. (The term *plan* is used to refer to a particular arrangement of rooms or design. This should not be confused with the use of *plan* to refer to a type of drawing.)

The plans in building Type II are A, B1, B2, and A R. Each Plan type is described on separate drawings. Drawings 1 through 5 are for Plan A. Drawings 6 through 10 are for Plan B.

One technique that is used to create similar, yet different, plans is to reverse them. A reversed plan is created by building the plan as though seen in a mirror. The reversed plan has the same features and the same dimensions, but their arrangement is reversed, Figure 32-4. In building Type II of the Hidden Valley development, the south end is a reversal of Plan A. This is designated on the site plan and on Figure 32-2 as A R. The letter *R* indicates a reverse plan.

ORGANIZATION OF THE DRAWINGS

The drawings for a large project must be systematically organized, so information can be found easily. Architects follow similar pattern in organizing their drawings. The general order of drawing sheets is similar to that of specifications: site plans are first; foundation plans and floor plans, next; building elevations, third;

Courtesy of berkus-group architects

Figure 32-1   Site plan

then, structural and architectural details; mechanical, next to last; and electrical, last. The cover sheet usually includes an index or table of contents for the drawing set, Figure 32-5.

The index for the Hidden Valley drawing does not indicate mechanical or electrical drawings. In the western part of the United States, it is common practice not to prepare separate drawings for mechanical or electrical work on town houses. The essential information for these trades is included on the drawings for the other trades.

VISUALIZING THE PLAN

The first step in becoming familiar with any plan should be to mentally walk through the plan. This technique was described earlier in Unit 18 for the Lake House. Refer to the Hidden Valley drawings as you mentally walk through Plan A of the Town House.

Enter through the overhead garage door on the east side of the lowest level. The garage is an open area with stairs leading up to the main floor. Next to the stairs there is an area of dropped ceiling beneath

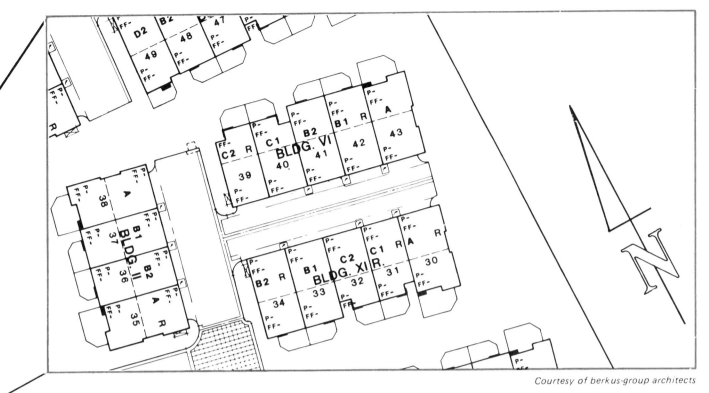

Figure 32-2 This is a section of the site plan at the size it was drawn.

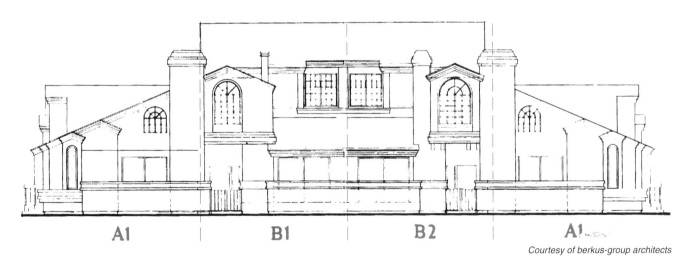

Figure 32-3 Building type II includes four housing units.

the stairs on the main floor. A note on Drawing 2 refers to Section B.

This note introduces a new consideration for anyone reading construction drawings. Although the architect reviews the drawings carefully, there is always a possibility that errors will appear as one did in this instance. The dropped ceiling beneath the stairs is actually shown on Section E.

The drawings for earlier units of this textbook were reviewed and corrected more thoroughly than can be reasonably expected for actual construction drawings. The drawings for the Town House are printed exactly as they were prepared for the construction job. A few errors may remain on these drawings. Any errors that do remain give you valuable experience in detecting and dealing with error.

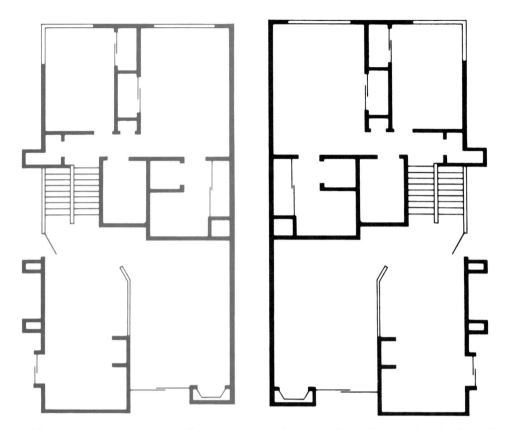

Figure 32-4   A reversed plan is similar to what would be seen by looking at the plan in a mirror.

At the top of the garage stairs, there is an entry area with an exterior door. A boxed number $\boxed{33}$ refers to the plan notes printed on Drawing 2. This note indicates a stub wall dividing the entry from the dining room. Boxed note references are used throughout the Town House drawings. Reading each related note as it is encountered will help you to better understand the plan.

As you enter the dining room and kitchen to the west of the entry area, the floor material changes. The kitchen ceiling is dropped to 7'-6" as dimensioned on Section A-A, Drawing 4. The areas with dropped ceiling above are shown by light cross hatching on the floor plan. The kitchen has base cabinets on three walls and a peninsula of cabinets on three walls and a peninsula of cabinets separating the kitchen and dining room. Opposite the peninsula of base cabinets, there is a 4'-8" alcove with cabinets, Figure 32-6.

Visually walk back through the dining room along a 36" high guard rail, plan note $\boxed{13}$, to a 2-riser stair down to the living room, Figure 32-7. At the (front) west end of the large, 18'-4" × 13'-4" living room there is a fireplace, Figure 32-8. To the left (north) of the fireplace is a 7'-0" × 8'-0" sliding glass door that opens onto a small courtyard. A note on the floor plan indicates a *36 MOJ RADIUS ABOVE.* The legend of symbols and abbreviations on sheet O shows that MOJ means *mea-*

sure on the job. Looking at the Front Elevation, Drawing 5, you can see that this note refers to a window with a 36-inch radius at the top.

From the entry area, another stair leads up to a hall serving the bedrooms and bath, Figure 32-9. Across the hall, a pair of 2'-6" × 6'-8" hollow-core doors open into bedroom #2. By the stairs, a door opens into bathroom #2. This is a small bathroom with a tub, water closet, and lavatory. At the south end of the hall, you enter the master bedroom. There is an 8'-0" × 5'-0" aluminum sliding window at the east end of the master bedroom. On the north wall near the hall door, there is a wardrobe closet with a shelf and pole (S&P). At the west end of the room there is a master bath. The master bath has a 42" × 60" tub, a vanity (called pullman here), a large wardrobe closet, and a toilet room.

To thoroughly understand the plan, you should read all of the plan notes on Drawing A-2 and locate the features to which they refer. Then locate each of the details, indicated by a circle. The numerals above the horizontal line in these circles indicate the detail being referenced. The letter and numeral below the horizontal line indicate on which drawing the detail appears. As you find these plan notes and details, refer to the framing plans, building sections, and elevations to thoroughly understand each.

table of contents

		COVER SHEET AND INDEX
a		SITE PLAN
b		GENERAL NOTES
c		GENERAL NOTES
1	**PLAN A**	FOUNDATION PLAN
2		FLOOR PLAN
3		FRAMING PLAN
4		SECTIONS
5		ELEVATIONS
6	**PLAN B**	FOUNDATION PLAN
7		FLOOR PLAN
8		FRAMING PLAN
9		SECTIONS
10		ELEVATION (1)
11		ELEVATION (2)
12	**PLAN C**	FOUNDATION PLAN
13		FLOOR PLAN
14		FRAMING PLAN
15		SECTIONS
16		ELEVATIONS (1)
17		ELEVATIONS (2)
18	**PLAN D**	FOUNDATION PLAN
19		FLOOR PLAN
20		FRAMING PLAN
21		SECTIONS
22		ELEVATIONS (1)
23		ELEVATION (2)
24	PLAN A-B	INTERIOR ELEVATIONS -PLAN A & B
25	PLAN C-D	INTERIOR ELEVATIONS -PLAN C & D
26		BUILDING TYPE I
27		BUILDING TYPE II
28		BUILDING TYPE III
29		BUILDING TYPE IV
30		BUILDING TYPE V
31		BUILDING TYPE VI
32		BUILDING TYPE VII
33		BUILDING TYPE VII
34		BUILDING TYPE VIII
35		BUILDING TYPE VIII
36		BUILDING TYPE IX
37		BUILDING TYPE IX
38		BUILDING TYPE X
39		BUILDING TYPE X
40		BUILDING TYPE XI
41		BUILDING TYPE XII
42		METER ENCLOSURE
43		REC. CENTER FOUND FLR PLAN
44		REC CENTER SECTION ELEV
D1		FOUNDATION DETAILS
D2		FRAMING DETAILS
D3		FRAMING DETAILS
D4		FRAMING DETAILS

Courtesy of berkus-group architects

Figure 32-5   Table of contents for Hidden Valley construction drawings

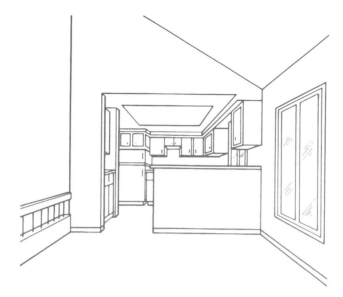

Figure 32-6   Town House kitchen as seen from entry

Figure 32-7   Hidden Valley stairs

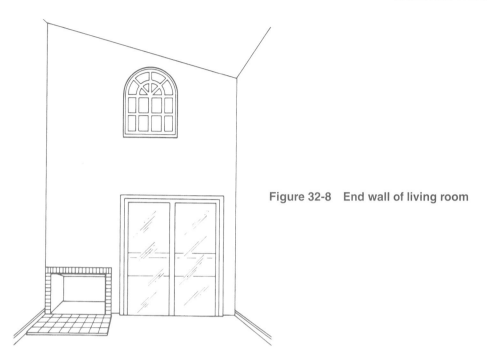

Figure 32-8   End wall of living room

Figure 32-9   View from Hidden Valley living room

✓ CHECK YOUR PROGRESS

Can you perform these tasks?

☐ Locate a particular building on the site plan for a large development.

☐ Locate a particular plan within a multi-plan building.

☐ Locate the drawings for a particular plan type within a large drawing set consisting of several buildings and plan types.

 ASSIGNMENT

Refer to the Town House drawings (including those in the packet and Figures 32-1 and 32-2) to complete the assignment.

1. How many buildings are included in phase one?

2. What plans are included in building Type VI?

3. In phase one, Building II, on which side (compass direction) is the courtyard?

4. How thick is the concrete slab for the garage floor in Plan B?

5. What size are the floor joists under the dining room in Plan B?

6. In Plan B, what supports the kitchen floor joists under the back wall of the kitchen?

7. On which drawing would you find elevations of the kitchen cabinets for Plan A? For Plan B?

8. On which drawing would you find details of concrete piers under bearing posts for girders?

9. What is the height of the handrail at the dining room/living room stairs in Plan B?

10. For each of the major rooms of Plan B listed below, indicate the overall dimensions. Do not include closets, stairs, or minor irregularities. Allow for the thickness of all walls. Walls are dimensioned to the face of the framing.

    a. Living room   d. Deck   f. Bedroom #2
    b. Dining room   e. Library   g. Master bedroom
    c. Kitchen

11. What is the tread width and riser height for the stairs between the living room and dining room in Plan B?

12. What important feature of the living room is beside the entry in Plan B?

13. How long are the studs in the partition between the master bedroom and bedroom #2 in Plan B?

14. How high above finished grade is the top of the privacy fence in front of Plan B?

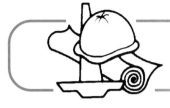

UNIT 33   Regional Variations

OBJECTIVES

After completing this unit, you will be able to describe important variations in the construction methods used for the buildings in the drawing packet with this textbook.

The Two-Unit Apartment and the Lake House were designed and built in the Eastern states. The Town House was designed and built in the West. Each of these buildings is quite different from the others in design and construction methods. Some of the differences are the result of differences in the availability of materials; some, the result of different building codes and structural requirements. You should be familiar with the applicable local or regional building codes and their related amendments for the specific code year being used locally for field inspection. This is necessary for every construction project. There are many areas that do not keep the adopted building codes and the related amendments updated to the most recent national standards. You could be *cited* for a building code violation if you install to a more current code or standard than has been adopted in that area or region. Other differences are the result of factors such as life-style or heating and air-conditioning requirements. Some of the characteristics of buildings in the same regions and of the same basic types as the Two-Unit Apartment, Lake House, and Town House are presented in the following paragraphs.

THE TWO-UNIT APARTMENT

The Apartment building discussed in the first part of this textbook was designed to be built in upstate New York. Some of the important points that the architect considered in designing the Apartment are listed:

- Very cold winters
- Increasing scarcity of wood in this region
- Ready availability in this region of metal building products
- Type of property: low-investment, income-earning property for the owner

Climate Considerations

In this region, temperatures of 20° below zero (Fahrenheit) are common. This drives the frost deep into the earth. Therefore, foundations must be more than three feet deep. The foundation and perimeter of the concrete slab are insulated to prevent excessive heat loss to the frozen earth. Also, all exterior walls and ceilings are heavily insulated. Notice that even the sheathing is rated according to its *R* value.

Use of Metals

In the part of the country where the Two-Unit Apartment was built, metal and plastic are used extensively to replace wood as a building material. These man-made materials are less expensive than wood, they resist the damaging effects of the severe weather, and they can be installed quickly. For these reasons, aluminum siding and aluminum cornices were chosen for the Apartment. Plywood siding is used at the corners. The vertical pattern of the plywood breaks up the long, straight lines of the horizontal aluminum siding. The plywood siding also serves as corner bracing to prevent wracking of the walls. The plywood corners alone would not be sufficient to prevent wracking in some regions. Upstate New York does not have the threat of earthquakes or extreme winds; therefore, the plywood corners are adequate.

Cost Considerations

The Apartment is designed as a low-cost project. The design is simple, and the drawings can be completed in relatively little space. Most homes in the Northeast have basements, but the Apartment is designed without a basement to conserve costs. The rectangular shape of the Apartment also is inexpensive. Generally, the fewer corners a building has and the fewer floor levels it has, the less expensive it is to build.

THE LAKE HOUSE

The Lake House is designed as a year-round vacation home on a lake in Virginia. Although the winters are not as extreme as those in upstate New York, the temperature is frequently well below freezing. The thermal insulation methods used in the Apartment are also used in the Lake House. In addition, the Lake House receives a lot of its heat from the sun.

Cost was less of a restriction in designing the Lake House. Therefore, the Lake House has many corners and many levels. For this reason, more drawings are needed to completely describe the Lake House.

For one- and two-family homes the electrical, plumbing, and HVAC contractors usually prepare any drawings needed for their work separately. The architectural drawings include basic information, such as the locations of major appliances, but separate electrical, plumbing, or HVAC drawings are not included.

THE TOWN HOUSE

The Town House is quite different from the Two-Unit Apartment or the Lake House because it was designed in southern California and it is part of a large project for condominiums and high rent housing. In designing buildings for this region, the architect must consider the threat of earthquakes and high winds. Also, the warmer temperatures affect the foundation design and insulation requirements.

Foundations

In the Southwest where frost is not a concern but earthquakes are, foundations are usually shallow and poured in one piece, Figure 33-1. A one-piece foundation is called a *monolithic* foundation. The foundation design varies depending on what it supports and where it is within the building.

For example, find the detail callout for the foundation between the house and garage in Plan A (near the middle of Drawing 1). The callout is $\frac{2b}{d1}$. Now find the callout for the detail of the exterior foundation of the garage (in the top left corner of Drawing 1). The callout is $\frac{4}{d1}$. Now find Details 2 and 4 on Drawing d-1 and notice the differences. The foundation at the exterior walls has a high ledge to receive the thickened edge of the garage slab.

The foundation details include letters in place of actual dimensions in many places. Frequent notes direct you to 10/d4. Sheet d4 is not included in the drawing package with this book, but Detail 10 from that

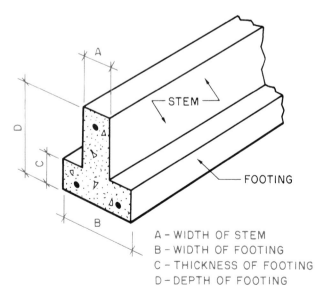

A – WIDTH OF STEM
B – WIDTH OF FOOTING
C – THICKNESS OF FOOTING
D – DEPTH OF FOOTING

Figure 33-1   Dimensions of monolothic foundation

FOUNDATION RECOMMENDATIONS

FOOTINGS				SLAB IN LIVING AREAS		SLAB IN GARAGE AREA	
EXTERIOR		INTERIOR					
DEPTH	REINFORCEMENT	DEPTH	REINFORCEMENT	REINFORCEMENT	MOISTURE	REINFORCEMENT	MOISTURE
E	F	G	F	N/A	N/A	H	X

LEGEND FOR FOUNDATION RECOMMENDATIONS
PHASE 1

G — 12 INCHES BELOW LOWEST ADJACENT GRADE.

E — 18 INCHES BELOW LOWEST ADJACENT GRADE.

F — ONE NO. 4 REBAR AT TOP AND ONE AT BOTTOM.

H — SLAB SHOULD BE DESIGNED AS FLOATING MEMBER AND SHOULD BE SEPARATED FROM PERIMETER FOOTINGS BY $\frac{1}{2}$ INCH CONSTRUCTION FELT OR EQUIVALENT.

X — NO SPECIAL REQUIREMENT.

NOTE

1.   MINIMUM FOOTING WIDTH SHOULD BE 12 INCHES.

2.   EXTERIOR AND INTERIOR FOOTINGS FOR TWO-STORY STRUCTURES SHOULD HAVE A MINIMUM EMBEDMENT OF 18 INCHES BELOW LOWEST ADJACENT GRADE.

3.   INTERIOR SLABS SHOULD BE STRUCTURALLY TIED TO PERIMETER FOOTINGS IN LIVING AREAS.

4.   PROVIDE FOOTING AT GARAGE OPENING.

U.B.C. FOOTING REQUIREMENTS
(THESE ARE THE MINIMUM REQUIREMENTS. SEE SOILS REPORT FOR MORE STRINGENT CONDITIONS ONLY.)

	A WIDTH OF STEM	B WIDTH OF FTG.	C THICK-NESS OF FTG.	D DEPTH OF FTG.
A — 1 — STORY	6"	12"	6"	12"
B — 2 — STORY	8"	15"	7"	18"
C — 3 — STORY	10"	18"	8"	24"

Courtesy of berkus-group architects

Figure 33-2   Foundation recommendations

CB COLUMN BASES

Model No.	W	L	Material Stirrups	Bolts	Uplift Design Loads
CB44	3 9/16"	3 5/8"	3/16"x2"	(2) 5/8"	5030
CB46	3 9/16"	5 1/2"	3/16"x2"	(2) 5/8"	5030
CB48	3 9/16"	7 1/2"	3/16"x2"	(2) 5/8"	5030
CB5	5 1/4"	Specify	3/16"x3"	(2) 5/8"	5030
CB66	5 1/2"	5 1/2"	3/16"x3"	(2) 5/8"	5030
CB68	5 1/2"	7 1/2"	3/16"x3"	(2) 5/8"	5030
CB610	5 1/2"	9 1/2"	3/16"x3"	(2) 5/8"	5030
CB612	5 1/2"	11 1/2"	3/16"x3"	(2) 5/8"	5030
CB7	6 7/8"	Specify	1/4"x3"	(2) 3/4"	7230
CB88	7 1/2"	7 1/2"	1/4"x3"	(2) 3/4"	7230
CB810	7 1/2"	9 1/2"	1/4"x3"	(2) 3/4"	7230
CB812	7 1/2"	11 1/2"	1/4"x3"	(2) 3/4"	7230
CB9	8 7/8"	Specify	1/4"x3"	(2) 3/4"	7230
CB1010	9 1/2"	9 1/2"	1/4"x3"	(2) 3/4"	7230
CB1012	9 1/2"	11 1/2"	1/4"x3"	(2) 3/4"	7230
CB1212	11 1/2"	11 1/2"	1/4"x3"	(2) 3/4"	7230

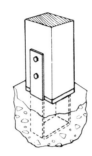

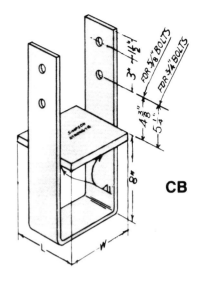

HD HOLDOWNS

Model No.	Bolt Attachment			Average Test Ultimate	Design Load Value* When Installed on Stud Thickness of				
	Concrete Dia.	Concrete Min.† Embedment	Stud		1 1/2"	2	2 1/2"	3	3 1/2"
HD 2	5/8"	9"	(2)-5/8"MB	13,200	2450	2520	2520	2520	2520
HD 5	3/4"	11"	(2)-3/4"MB	19,000	3375	3610	3610	3610	3610
HD 6	1"	14"	(3)-3/4"MB	18,600	5060	5410	5410	5410	5410
HD 7	1"	14"	(3)-7/8"MB	28,600	6350	6500	6500	6500	6500
HD 7	1 1/8"	15"	(3)-7/8"MB	28,600	6350	7100	7500	7500	7500
HD 2N	5/8"	9"	(2)-5/8"MB	8,800	2450	2520	2520	2520	2520
HD 5N	3/4"	11"	(2)-3/4"MB	11,600	3375	3610	3610	3610	3610
HD 7N	1"	14"	(2)-1"MB	20,300	4800	5640	6480	6500	6500

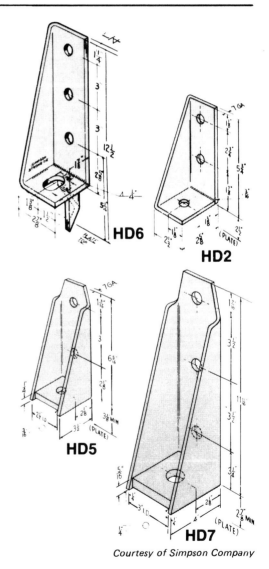

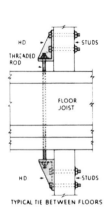

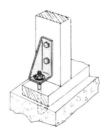

Courtesy of Simpson Company

Figure 33-3   Ties and anchors for Town House

HL ANGLES

Model No.	DIMENSIONS			Bolts (total)	Gussets	BOLT LOAD VALUES	
	Mat.	W_1 & W_2	L			Parallel to Grain	Perpend. to Grain
HL33	3/16″	3¼″	2½″	2-⅝″ M.B.	None	1255	725
HL35	3/16″	3¼″	5″	4-⅝″ M.B.	None	2510	1450
HL35G	3/16″	3¼″	5″	4-⅝″ M.B.	One	2510	1450
HL37	3/16″	3¼″	7½″	6-⅝″ M.B.	None	3765	2175
HL37G	3/16″	3¼″	7½″	6-⅝″ M.B.	Two	3765	2175
HL53	3/16″	5¾″	2½″	4-⅝″ M.B.	None	2510	1450
HL55	3/16″	5¾″	5″	8-⅝″ M.B.	None	5025	2900
HL55G	3/16″	5¾″	5″	8-⅝″ M.B.	One	5025	2900
HL57	3/16″	5¾″	7½″	12-⅝″ M.B.	None	7535	4250
HL57G	3/16″	5¾″	7½″	12-⅝″ M.B.	Two	7535	4250
HL43	¼″	4¼″	3″	2-¾″ M.B.	None	1805	970
HL46	¼″	4¼″	6″	4-¾″ M.B.	None	3610	1940
HL46G	¼″	4¼″	6″	4-¾″ M.B.	One	3610	1940
HL49	¼″	4¼″	9″	6-¾″ M.B.	None	5435	2910
HL49G	¼″	4¼″	9″	6-¾″ M.B.	Two	5435	2910
HL73	¼″	7¼″	3″	4-¾″ M.B.	None	3610	1940
HL76	¼″	7¼″	6″	8-¾″ M.B.	None	7225	3880
HL76G	¼″	7¼″	6″	8-¾″ M.B.	One	7225	3880
HL79	¼″	7¼″	9″	12-¾″ M.B.	None	10875	5820
HL79G	¼″	7¼″	9″	12-¾″ M.B.	Two	10875	5820

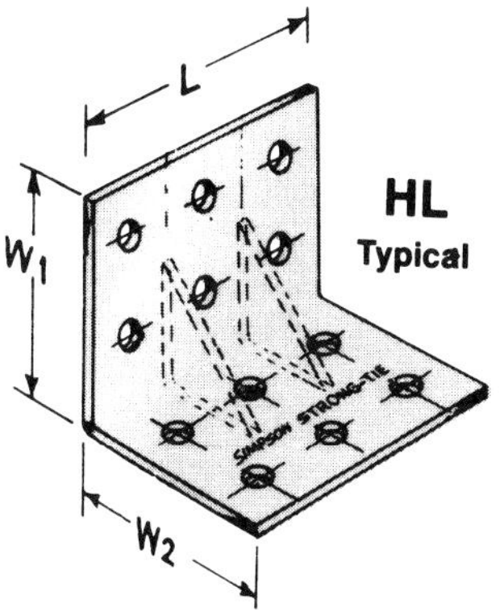

STC/DTC ROOF TRUSS CLIPS

Model No.	DIMENSIONS		MATERIAL Ga.	NAILING	
	Plate Base	Vertical Leg		Base	Slot
STC	1¼″ x 1¾″	1¼″ x 2¾″	18 Ga. Galv.	2-8d	1-8d
STCT	1¼″ x 1¾″	1¼″ x 4¼″	18 Ga. Galv.	2-8d	1-8d
DTC	2½″ x 1¾″	2½″ x 2¾″	18 Ga. Galv.	4-8d	2-8d

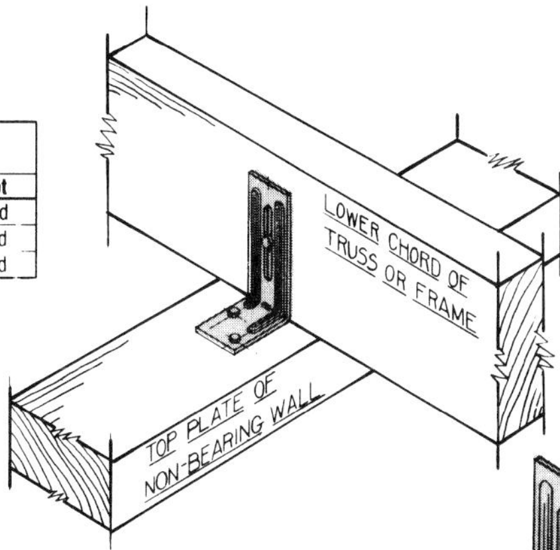

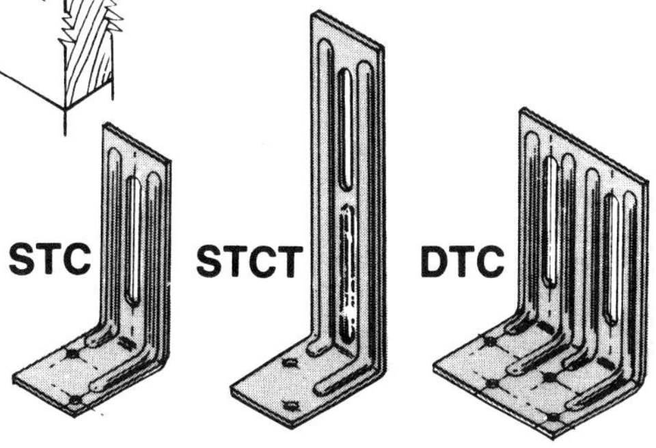

PA PURLIN ANCHORS

Model No.	Length	Connectors To Purlins	DESIGN LOADS	
			Normal	Max.
PA18	18½″	(12)-16d	1600	2130
PA23	23¾″	(18)-16d	2410	3200
PA28	29″	(24)-16d	3140	4170

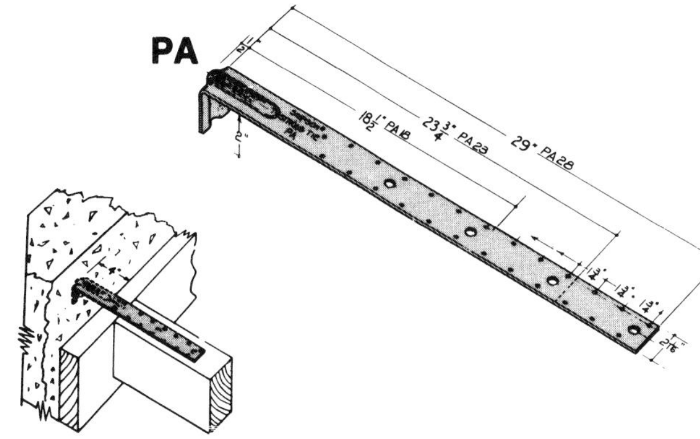

Figure 33-3 (continued)

A35 FRAMING ANCHORS

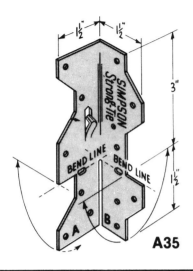

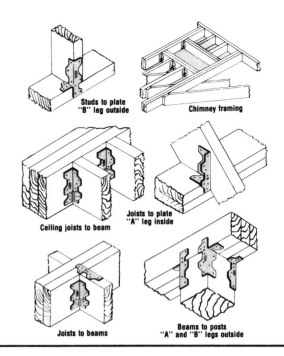

ST STRAP TIES

Model No.	DIMENSIONS			FASTENERS		DESIGN LOADS		
						Nails	BOLTS	
	Material	Width	Length	Nails	Bolts	Nails	Single Shear	Double Shear
ST292	20 ga. galv.	2¹/₁₆″	9⁵/₁₆″	12-16d	—	805	—	—
ST2122	20 ga. galv.	2¹/₁₆″	12¹³/₁₆″	16-16d	—	1170	—	—
ST2115	20 ga. galv.	¾″	16⁵/₁₆″	10-16d	—	670	—	—
ST2215	20 ga. galv.	2¹/₁₆″	16⁵/₁₆″	20-16d	—	1340	—	—
ST6215	16 ga. galv.	2¹/₁₆″	16⁵/₁₆″	20-16d	—	1340	—	—
ST6224	16 ga. galv.	2¹/₁₆″	23⁵/₁₆″	28-16d	—	1875	—	—
ST6236	16 ga. galv.	2¹/₁₆″	33¹³/₁₆″	40-16d	—	2580	—	—
ST9	16 ga. galv.	1¼″	9″	8-16d	—	535	—	—
ST12	16 ga. galv.	1¼″	11⅝″	10-16d	—	670	—	—
ST18	16 ga. galv.	1¼″	17¾″	14-16d	—	935	—	—
ST22	16 ga. galv.	1¼″	21⅝″	18-16d	—	1205	—	—

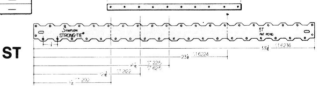

ST

SA STRAP CONNECTORS

Model No.	Strap Section	DIMENSIONS ALTERNATE CONNECTIONS				DESIGN LOADS*	
		L₁	L₂	Bolts ea. side	Nails ea. side	Bolts only	Nails only
SA34	7 ga x 2¹/₁₆″	34″	9″	2-¾″	—	2860	—
SA45	7 ga x 2¹/₁₆″	45″	19½″	2-¾″	—	2860	—
SA36	12 ga x 2¹/₁₆″	36″	9″	2-½″	11-16d	1630	1475
SA47	12 ga x 2¹/₁₆″	47″	19½″	2-½″	11-16d	1630	1475
SAL36¹	12 ga x 2¹/₁₆″	36″	9″	2-½″	11-16d	3260	2950
SAL47¹	12 ga x 2¹/₁₆″	47″	19½″	2-½″	11-16d	3260	2950

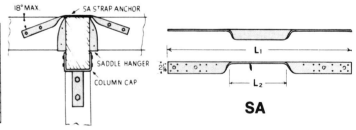

SA

WB WALL BRACING

Type	Material	Size
WB106	16-ga. (galv.)	1¼″ x 9′5⅝″ long
WB126	16-ga. (galv.)	1¼″ x 11′4⅜″ long

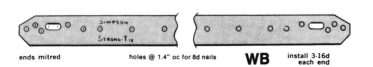

ends mitred   holes @ 1.4″ oc for 8d nails   **WB**   install 3-16d each end

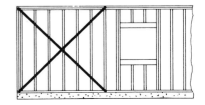

Figure 33-3   (continued)

HUTF JOIST HANGARS

Model No.	Joist Size	DIMENSIONS A	B	H	W	TF	Nail Schedule Header	Joist	Aver. Ult.	I.C.B.O. Loads* Uplift	Norm	Max
HU26TF	2x6	1¼″	2″	5⅜″	1 9/16″	2″	(10)-16d	(4)-10d†	4,800	420	1620	2030
HU28TF	2x8	1¼″	2″	7¼″	1 9/16″	2½″	(10)-16d	(4)-10d†	6,000	420	1620	2030
HU210TF	2x10	1¼″	2″	9¼″	1 9/16″	2½″	(12)-16d	(4)-10d†	7,200	420	1620	2030
HU212TF	2x12	1¼″	2″	11⅛″	1 9/16″	2½″	(14)-16d	(6)-10d†	8,400	630	1860	2320
HU214TF	2x14	1¼″	2½″	13⅛″	1 9/16″	2½″	(16)-16d	(6)-10d†	9,600	630	1860	2330
HU216TF	2x16	1¼″	2½″	15⅛″	1 9/16″	2½″	(18)-16d	(8)-10d†	10,800	840	2070	2580
HU34TF	3x4	1¼″	2″	3½″	2 9/16″	2½″	(8)-16d	(2)-10d	8,270	210	2160	2600
HU36TF	3x6	1¼″	2″	5⅜″	2 9/16″	2½″	(10)-16d	(4)-10d	9,830	420	2390	2990
HU36TF	3x8	1¼″	2″	7¼″	2 9/16″	2½″	(12)-16d	(4)-10d	11,390	420	2390	2990
HU310TF	3x10	1¼″	2″	9¼″	2 9/16″	2½″	(14)-16d	(6)-10d	12,950	630	2630	3320
HU312TF	3x12	1¼″	2½″	11″	2 9/16″	2½″	(16)-16d	(6)-10d	13,760	630	2630	3320
HU314TF	3x14	1¼″	2½″	13″	2 9/16″	2½″	(18)-16d	(8)-10d	14,580	840	3350	4180
HU316TF	3x16	1¼″	2½″	15″	2 9/16″	2½″	(20)-16d	(8)-10d	15,400	840	3350	4180
HU44TF	4x4	1¼″	2″	3½″	3 9/16″	2½″	(8)-16d	(2)-10d	8,270	210	2600	2600
HU46TF	4x6	1¼″	2″	5⅜″	3 9/16″	2½″	(10)-16d	(4)-10d	9,830	420	3160	3210
HU48TF	4x8	1¼″	2″	7¼″	3 9/16″	2½″	(12)-16d	(4)-10d	11,390	420	3160	3600
HU410TF	4x10	1¼″	2″	9¼″	3 9/16″	2½″	(14)-16d	(6)-10d	12,950	630	3400	4130
HU412TF	4x12	1¼″	2½″	11″	3 9/16″	2½″	(16)-16d	(6)-10d	13,760	630	4070	4400
HU414TF	4x14	1¼″	2½″	13″	3 9/16″	2½″	(18)-16d	(8)-10d	14,580	840	4310	4710
HU416TF	4x16	1¼″	2½″	15″	3 9/16″	2½″	(20)-16d	(8)-10d	15,400	840	4310	4710
HU66TF	6x6	1¼″	2″	5⅜″	5½″	2½″	(10)-16d	(4)-16d	9,830	210	3210	3210
HU68TF	6x8	1¼″	2″	7¼″	5½″	2½″	(12)-16d	(4)-16d	11,390	420	3600	3600
HU610TF	6x10	1¼″	2″	9¼″	5½″	2½″	(14)-16d	(6)-16d	12,950	630	4130	4130
HU612TF	6x12	1¼″	2½″	11⅛″	5½″	2½″	(16)-16d	(6)-16d	13,760	630	4400	4400
HU614TF	6x14	1¼″	2½″	13⅛″	5½″	2½″	(18)-16d	(8)-16d	14,580	840	4710	4710
HU616TF	6x16	1¼″	2½″	15⅛″	5½″	2½″	(20)-16d	(8)-16d	15,400	840	4710	4710
HU24-2TF	(2)2x4	1¼″	2″	3½″	3⅛″	2½″	(8)-16d	(2)-10d	8,270	210	2540	2600
HU26-2TF	(2)2x6	1¼″	2″	5⅜″	3⅛″	2½″	(10)-16d	(4)-10d	9,830	420	2780	3210
HU28-2TF	(2)2x8	1¼″	2″	7¼″	3⅛″	2½″	(12)-16d	(4)-10d	11,390	420	2780	3470
HU210-2TF	(2)2x10	1¼″	2″	9¼″	3⅛″	2½″	(14)-16d	(6)-10d	12,950	630	3010	3770
HU212-2TF	(2)2x12	1¼″	2½″	11⅛″	3⅛″	2½″	(16)-16d	(6)-10d	13,760	630	3590	4400
HU214-2TF	(2)2x14	1¼″	2½″	13⅛″	3⅛″	2½″	(18)-16d	(8)-10d	14,580	840	3830	4710
HU216-2TF	(2)2x16	1¼″	2½″	15⅛″	3⅛″	2½″	(20)-16d	(8)-10d	15,400	840	3830	4710
HU210-3TF	(3)2x10	1¼″	2″	9¼″	4 11/16″	2½″	(14)-16d	(6)-16d	12,950	630	4130	4130
HU212-3TF	(3)2x12	1¼″	2½″	11⅛″	4 11/16″	2½″	(16)-16d	(6)-16d	13,760	630	4400	4400
HU214-3TF	(3)2x14	1¼″	2½″	13⅛″	4 11/16″	2½″	(18)-16d	(8-)16d	14,580	840	4710	4710
HU216-3TF	(3)2x16	1¼″	2½″	15⅛″	4 11/16″	2½″	(20)-16d	(8)-16d	15,400	840	4710	4710

HUTF

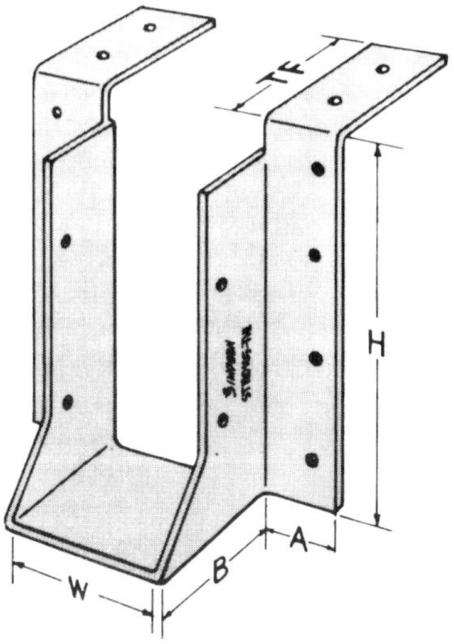

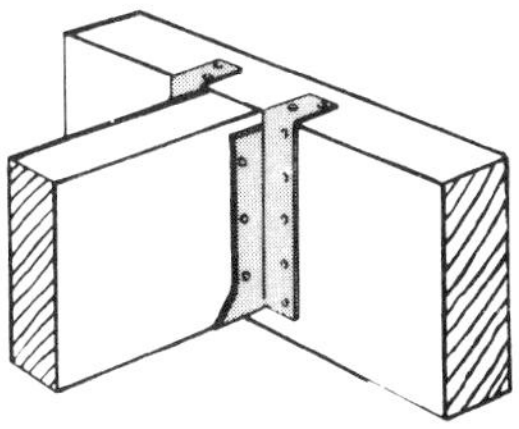

HU JOIST HANGARS

Model No.	Joist Size	DIMENSIONS A	B	H	W	Nail Schedule Header	Joist	Aver. Ult.	I.C.B.O. Loads* Uplift	Normal	Max
HU26	2x4 2x6	1″	2″	3 1/16″	1 9/16″	4-16d	2-10d†	2,600	110	535	670
HU28	2x8	1″	2″	5¼″	1 9/16″	6-16d	4-10d†	3,700	420	805	1010
HU210	2x10	1″	2″	7⅛″	1 9/16″	8-16d	4-10d†	4,900	420	1070	1345
HU212	2x12	1″	2″	9″	1 9/16″	10-16d	6-10d†	6,200	630	1340	1680
HU214	2x14	1″	2½″	10⅛″	1 9/16″	12-16d	6-10d†	8,500	630	1610	2015
HU34	3x4	1¼″	2″	3⅜″	2 9/16″	4-16d	2-10d	2,600	210	535	670
HU36	3x6	1¼″	2″	5⅜″	2 9/16″	8-16d	4-10d	5,020	420	1070	1345
HU38	3x8	1¼″	2″	7⅛″	2 9/16″	10-16d	4-10d	7,430	420	1340	1680
HU310	3x10	1¼″	2″	8⅞″	2 9/16″	14-16d	6-10d	9,850	630	1875	2350
HU312	3x12	1¼″	2½″	10⅝″	2 9/16″	16-16d	6-10d	11,700	630	2145	2690
HU314	3x14	1¼″	2½″	12⅜″	2 9/16″	18-16d	8-10d	13,560	840	2410	3010
HU316	3x16	1¼″	2½″	14⅛″	2 9/16″	20-16d	8-10d	15,420	840	2680	3360
HU44	4x4	1¼″	2″	2⅞″	3 9/16″	4-16d	2-10d	2,600	210	535	670
HU46	4x6	1¼″	2″	4⅞″	3 9/16″	8-16d	4-10d	5,020	420	1070	1345
HU48	4x8	1¼″	2″	6⅝″	3 9/16″	10-16d	4-10d	7,430	420	1340	1680
HU410	4x10	1¼″	2″	8⅜″	3 9/16″	14-16d	6-10d	9,850	630	1875	2350
HU412	4x12	1¼″	2½″	10⅛″	3 9/16″	16-16d	6-10d	11,700	630	2145	2690
HU414	4x14	1¼″	2½″	11⅞″	3 9/16″	18-16d	8-10d	13,560	840	2410	3010
HU416	4x16	1¼″	2½″	13⅝″	3 9/16″	20-16d	8-10d	15,420	840	2680	3360
HU66	6x6	1¼″	2″	5″	5½″	8-16d	4-16d	5,020	420	1070	1345
HU68	6x8	1¼″	2″	6⅝″	5½″	10-16d	4-16d	7,430	420	1340	1680
HU610	6x10	1¼″	2″	8⅜″	5½″	14-16d	6-16d	9,850	630	1875	2350
HU612	6x12	1¼″	2½″	10⅛″	5½″	16-16d	6-16d	11,700	630	2145	2690
HU614	6x14	1¼″	2½″	11⅞″	5½″	18-16d	8-16d	13,560	840	2410	3010
HU616	6x16	1¼″	2½″	13⅝″	5½″	20-16d	8-16d	15,420	840	2680	3360
HU24-2	2x4	1¼″	2″	3 1/16″	3⅛″	4-16d	2-10d	2,600	210	535	670
HU26-2	2x6	1¼″	2″	5 1/16″	3⅛″	8-16d	4-10d	5,020	420	1070	1345
HU28-2	2x8	1¼″	2″	6 13/16″	3⅛″	10-16d	4-10d	7,430	420	1340	1680
HU210-2	2x10	1¼″	2″	8 9/16″	3⅛″	14-16d	6-10d	9,850	630	1875	2350
HU212-2	2x12	1¼″	2½″	10 5/16″	3⅛″	16-16d	6-10d	11,700	630	2145	2690
HU214-2	2x14	1¼″	2½″	12 1/16″	3⅛″	18-16d	8-10d	13,530	840	2410	3010

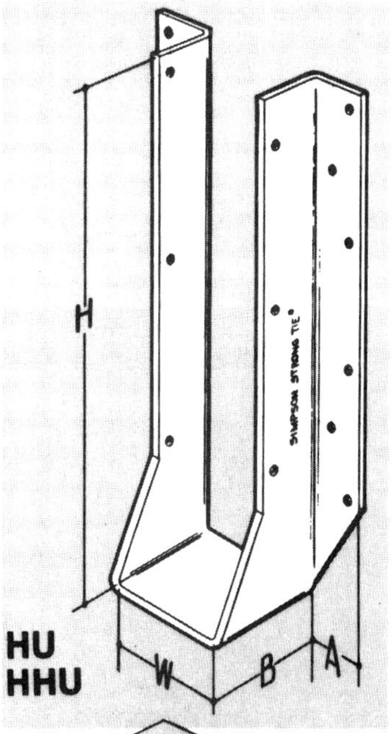

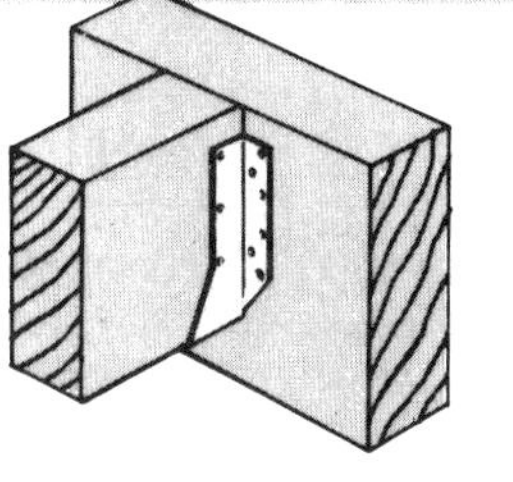

Figure 33-3 (continued)

sheet is shown on the sheet with the foundation plans and on page 247 (Figure 33-2). The letter coded dimensions of the foundation details refer to the U.B.C. (Uniform Building Code) Footing Requirements are shown at the bottom of 10/d4. The foundation recommendations in the top portion of 10/d4 give other information of a general nature. For example, the interior foundations are to be reinforced with one #4 bar at the top and one at the bottom. This corresponds with the two bold dots shown on interior foundation details, such as 3/d1.

Anchors and Ties

Steel hold-downs, tie straps, and framing anchors are used in the West to resist the destructive forces of earthquakes and in the southeast to resist the forces of hurricanes. The details for the Town House, especially those on Drawing d-1, show several types of hardware labeled only by the manufacturer's name and number. Simpson, the manufacturer named on the Town House details, the manufacturer of structural ties and anchors.

The ties, anchors, and other such items to be used in the Town House are shown in Figure 33-3.

The basic methods and materials used for framing on the West coast also differ slightly from those used in other parts of North America. The West has a better supply of tall, straight trees for saw timber; therefore, larger wood beams and posts are common, Figure 33-4. For example, headers over framed wall openings in the Eastern states are usually built of two or more 2 × s. The garage door headers for the Town House are 8 × 16. (See Drawing 3, Garage Floor Framing Plan.)

Another important difference in construction methods is the use of blocking between framing members. Because of the need for extra rigidity in earthquake and high-wind regions, more blocking is required by the building codes in this region. Blocking is placed between rafters and joists at all bearing walls, *purlins* (intermediate supports), and beams. Blocking is shown in section views with a single diagonal line to differentiate it from continuous framing lumber, Figure 33-5.

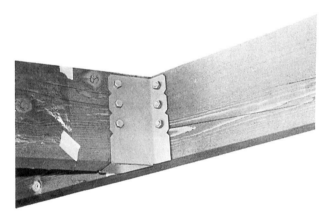

Figure 33-4   Large wooden beams are common on the west coast.

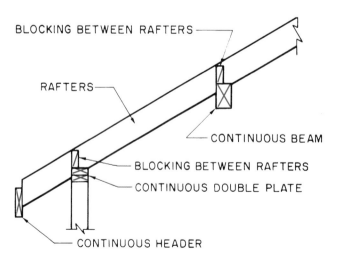

Figure 33-5   Continuous framing is shown by crossing diagonals. Blocking is shown by a single diagonal line.

✓ CHECK YOUR PROGRESS ————————————————————————

Can your perform these tasks?

☐ Given drawings for buildings in different parts of the country, list the differences that result from their different locations.

☐ Explain the purpose of each anchor, hold-down, and tie strap shown on a set of drawings.

☐ List the dimensions of the stem and the footing for monolithic foundations.

☐ Describe the reinforcement in a monolithic foundation.

ASSIGNMENT ——————————————————————————————————

Refer to the Town House drawings (in the packet) to complete the assignment.

1. How deep is the footing under the garage door in Plan B?

2. What reinforcing steel is to be used in the footing at the front of the building in Plan B?

3. How deep is the footing under the front entrance in Plan B?

4. What size is the beam over the opening between the kitchen and dining room in Plan B?

5. How is the kitchen-dining room partition tied to the beam above the opening between these rooms?

6. What is the size and spacing of anchor bolts in the front wall of the house in Plan B?

7. How are the joists above the garage in Plan B tied to the frame wall under the dining room?

8. What is the size and spacing of the anchor bolts in the privacy walls of the courtyards?

9. What size and spacing is the framing of the pot shelf at the front corner the kitchen in Plan A?

10. How are nonbearing partitions tied to roof trusses above in most places?

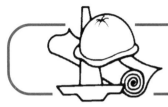

UNIT 34 Party Walls

OBJECTIVES ————————————————

After completing this unit, you will be able to perform the following tasks:

• Identify and explain fire-resistant construction.

• Identify and explain construction for acoustical insulation.

Multifamily buildings are constructed the same as single-family buildings in most respects. The most important differences between these two classes of buildings is the construction of party walls in multifamily dwellings. A *party wall* is a wall that is shared by two separate living units, Figure 34-1. In addition to the usual requirements of a wall, a party wall provides more fire resistance and privacy.

The Town House party walls have stricter fire-resisting requirements in some places than in others. In the Town House drawings, the architect refers to party walls where the fire resistance factor is lower. Where special fire-code requirements must be met, the architect refers to area-separation walls. The terms *party wall* and *area-separation wall* are often used interchangeably. Other architects may use them with reverse meanings.

FIRE-RATED CONSTRUCTION

Fire-rated construction serves two purposes in the event of a fire. It slows the spread of fire, and it maintains structural support longer than non-fire-rated construction. Approved fire caulking or fill must be used at all utility-line fire-rated wall penetrations.

Structural Support

In residential buildings, the structural members in five-rated construction are usually the same as in non-fire-rated construction. These members are protected from fire damage by the nonstructural materials that are used to resist fire spread. Because they are protected from fire damage, these commonly used materials are able to provide structural support longer than unprotected members. Therefore, the important elements of fire-rated residential construction are those that slow the spread of fire.

Slowing Fire Spread

The most obvious way in which a wall, floor, ceiling, or roof can slow the spread of fire is by having a

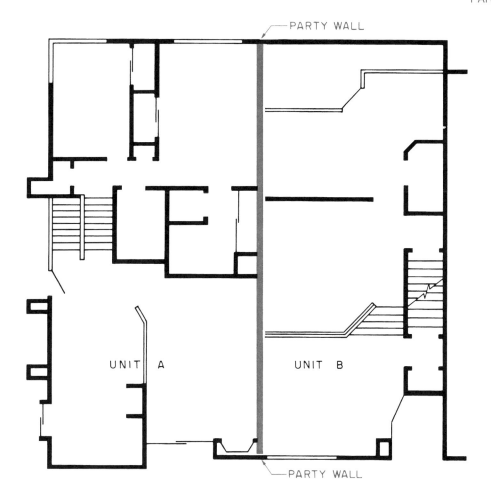

Figure 34-1   A party wall separates two or more units.

fire-resistant surface. Plaster and gypsum wallboard are fire-resistant materials. They do not burn, and they do not easily transmit the heat of fire to the framing members on the other side. Of course, the thicker the wallboard or plaster, the better it resists the flow of heat. For this reason party walls are often required to have double thicknesses of gypsum wallboard on each side, Figure 34-2.

Notice that the wallboard in Figure 34-2 is indicated as type *x*. This is a special fire-rated wallboard. Although all gypsum plaster is noncombustible, standard wallboard breaks down and crumbles in the high heat of a fire. Fire-rated wallboard holds up much longer in a fire.

Fire-rated construction is frequently used to separate garages and mechanical rooms from living spaces. Party walls are also fire rated to prevent a fire from spreading between housing units. To completely separate the housing units, a fire-rated party wall should extend all the way from the foundation through the roof, Figure 34-3.

Most building codes allow an alternative to extending the party wall through the roof. The fire-rated

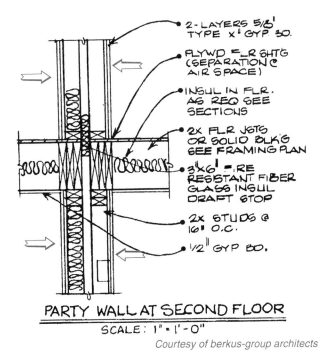

- 2-LAYERS 5/8"
  TYPE X" GYP. BD.
- PLYWD FLR SHTG
  (SEPARATION @
  AIR SPACE)
- INSUL IN FLR.
  AS REQ SEE
  SECTIONS
- 2x FLR JSTS
  OR SOLID BLK'G
  SEE FRAMING PLAN
- 3"x6" FIRE
  RESISTANT FIBER
  GLASS INSUL
  DRAFT STOP
- 2x STUDS @
  16" O.C.
- 1/2" GYP BD.

PARTY WALL AT SECOND FLOOR
SCALE: 1" = 1'-0"

Courtesy of berkus-group architects

Figure 34-2   Party walls often have double layers of gypsum wallboard on one or both surfaces.

Figure 34-3   Fire walls can be extended through the roof for maximum protection.

construction may end at the bottom of the roof as long as the roof is of fire-rated construction, Figure 34-4. This usually means that the roofing material resists fire for as long as the party wall. For example, if the party wall is required to be a one-hour code wall (it resists fire for one hour), the roof must be covered with only material that also resists fire for one hour.

Fire-rated walls must also prevent fire from spreading vertically inside the wall. If left open, the spaces between the studs in a frame wall act like chimneys, allowing flame to spread very quickly from one level to another. To prevent vertical flame spread, stud spaces are not permitted to be more than one story

high. The spaces between the studs are closed off with *firestops* at each level, Figure 34-5. Instead of wood firestops, the wall cavity can be blocked off with fire-resistant fiberglass insulation, Figure 34-2.

Openings are usually avoided in fire-rated walls. Where it is necessary to include a door, it is made of fire-resistant material. The fire rating of the door must comply with the building code. Fire-rated doors are often allowed to have slightly lower rating than the walls in which they are installed. Doors in fire walls are equipped with a self-closing mechanism.

SOUND INSULATION

To provide privacy between the housing units, party walls should not allow the sound from one unit to be heard in the next unit. The measurement of the capability of a building element to reduce the passage of sound is its *sound transmission classification (STC)*, Figure 34-6.

Sound is transmitted by vibrations in any material: solid, liquid, and gas. To slow the passage of sound, a party wall must reduce the flow of vibrations. The materials used in the construction of most walls vibrate relatively well. Also, they transmit these vibrations to the air inside the wall. The air carries the sound to the other side where it is transmitted to the air on the opposite side of the wall. The electrical installer may

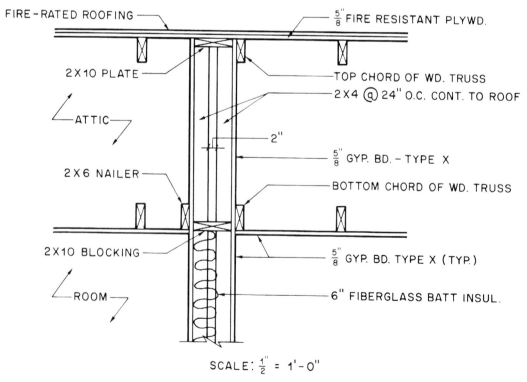

FIRE-RATED ROOFING

$\frac{5}{8}$" FIRE RESISTANT PLYWD.

2 X 10 PLATE

TOP CHORD OF WD. TRUSS

2 X 4 @ 24" O.C. CONT. TO ROOF

ATTIC

2"

$\frac{5}{8}$" GYP. BD. - TYPE X

2 X 6 NAILER

BOTTOM CHORD OF WD. TRUSS

2 X 10 BLOCKING

$\frac{5}{8}$" GYP. BD. TYPE X (TYP.)

ROOM

6" FIBERGLASS BATT INSUL.

SCALE: $\frac{1}{2}$" = 1'-0"

Figure 34-4   Fire-rated party wall in attic space

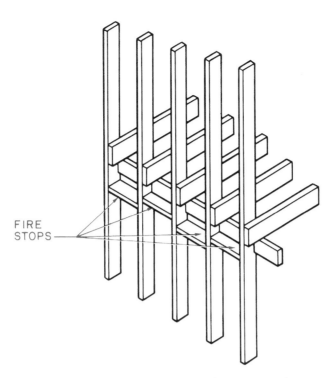

Figure 34-5   Fire stops are installed between studs to prevent vertical drafts inside the wall.

STC RATING	EFFECTIVENESS
25	Normal speech can be understood quite easily.
35	Loud speech can be heard, but not understood.
45	Must strain to hear loud speech.
48	Some loud speech can barely be heard.
50	Loud speech cannot be heard.

Figure 34-6   Sound transmission classes

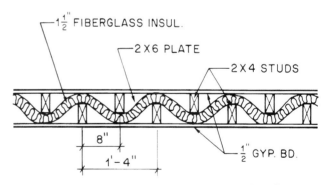

Figure 34-7   Typical sound-insulated wall

be required to utilize fiberglass or some other sound attenuation material around flush device boxes. Raceways may require installations that reduce sound transmission, such as expansion joints, flexible raceway, or raceway offsetting through the party wall.

The sound transmission classification of a wall can be improved greatly by not allowing the studs to contact both surfaces. This may be accomplished in one of two ways. One method is to attach clips made for

sound insulation to the studs; then, fasten the wallboard to these clips. The clips absorb the vibrations. Using the clips and sound-deadening fiberboard results in an STC rating of 52.

An STC of 45 is achieved without clips by using 2 × 4 studs and 2 × 6 plates. The studs are staggered on opposite sides of the wall so that no studs contact both surfaces, Figure 34-7. The STC can be increased to 49 by including fiberglass insulation.

✓ CHECK YOUR PROGRESS ————————————————————

Can your perform these tasks?

☐ Identify party walls in multifamily building.

☐ Identify fire-rated construction on drawings.

☐ List the locations of firestops inside framed walls.

☐ Describe construction to reduce sound transmission.

 ASSIGNMENT ————————————————————

Refer to the Town House drawings (in the packet) to complete the assignment.

1. List the type, thickness, and number of layers of wallboard at each of the following locations in the Town House:

    a. Area separation wall in Plan B dining room

    b. Party wall in Plan B kitchen

    c. Party wall in Plan B garage

    d. Area separation wall in Plan B bedroom #2

    e. Area separation above ceiling in Plan B master bedroom

    f. Party wall in Plan A living room

    g. Garage ceiling under nook in Plan B

2. What is used to stop the vertical spread of fires inside the party walls in the Town House?

3. How is fire prevented from spreading over the top of area separation walls where they meet the Town House roof?

4. What is the total thickness of the party wall at the library in Plan B?

5. What is done to stop the transmission of sound through the air space in the party wall of the library in Plan B?

6. What is the STC rating of the party wall of the library in Plan B?

UNIT 35 Plumbing

OBJECTIVES

After completing this unit, you will be able to perform the following tasks:

- Explain the basic principles of plumbing design.
- Identify the plumbing symbols used on drawings.

All houses include water supply and waste plumbing, and many also have gas plumbing. The water supply plumbing provides fresh hot and cold water to all points of use throughout the house. After the water has been supplied and used, the waste plumbing carries the water and any waste material to the municipal sewer or the septic tank. Waste plumbing is sometimes called *DWV* (drainage, waste, and vent.) The gas plumbing supplies fuel gas to the water heater, furnace, range, and other gas-fired fixtures.

PLUMBING MATERIALS

The materials most often used for plumbing are copper, plastic, cast iron, and black iron. A brief description of each is given in the paragraphs that follow.

Copper is frequently used for plumbing because it resists corrosion. However, it is relatively expensive. Copper pipes and fittings may be threaded or smooth for soldered joints.

Plastics for use in plumbing materials are lightweight, noncorrosive, and easily joined. However, most plastic plumbing materials cannot be used around heat above 200°. Also, plastics are not suitable for some applications where high strength is required, although they are sometimes used for general supply and DWV plumbing.

Cast iron is used extensively for DWV plumbing because of its strength and resistance to corrosion. However, it is seldom used for supply plumbing in residential construction.

Black iron is used extensively for gas piping. Black iron pipes and fittings are threaded, so they can be screwed together. Brass fittings are frequently used to join black iron pipe.

Fittings

A wide assortment of fittings is used for joining pipe, making turns at various angles, controlling the flow of water, and gaining access to the system for service, for example. Most fittings are made of the materials of which pipe is made. Plumbers must be familiar with all types of fittings so they can install their work according to the specifications of the designer.

Couplings, Figure 35-1, are used to join two pipes in a straight line. Couplings are generally used only where a single length of pipe is not long enough.

Union, Figure 35-2, allow piping to be disconnected easily. A union consists of two parts, with one part being attached to each pipe. Then the two parts of the union are screwed together. When it becomes necessary to disconnect the pipe, the two halves of the union are unscrewed.

Elbows, Figure 35-3, are used to make changes in direction of the piping. Elbows turn either 90° or 45°.

Tees and wyes, Figure 35-4, have three openings to allow a second line to join the first from the side. Tees have a 90° side outlet. Wyes have a 45° side outlet.

Cleanouts, Figure 35-5, allow access to sewage plumbing for cleaning. A cleanout consists of a threaded opening and a matching plug. When cleaning is necessary, the plug is removed and a *snake* or *auger* is run through the line. Cleanouts are installed in each straight run of DWV.

Valves, Figure 35-6, are used to stop, start, or regulate the flow of water. The faucets on a sink or lavatory are a type of valve.

Figure 35-1   A coupling is used to permanently join lengths of pipe.

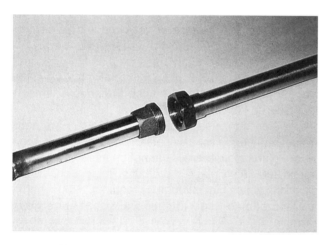

Figure 35-2   A union allows the piping to be disconnected easily.

Figure 35-3   90° elbow and 45° elbow

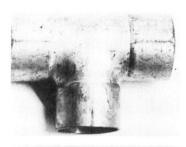

Figure 35-4   Tee and sanitary tee

CLEANOUT

Figure 35-5   A cleanout allows access to the system

Figure 35-6   Each branch of piping should include a shutoff valve.

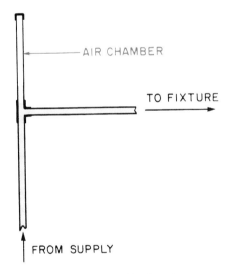

Figure 35-7   Air chamber

DESIGN OF SUPPLY PLUMBING

In most communities, water is distributed through a system of water mains under or near the street. When a new house is constructed, the municipal water department *taps* (makes an opening in) this main. The supply plumbing from the municipal tap to the house is installed by plumbers who work for the plumbing contractor.

The main supply line entering the house must be larger in diameter than the individual branches running from the main to each point of use. There are two ba-

sic reasons for this. First, water develops friction as it flows through pipes, and the greater size reduces this friction in the long supply line. Second, when more than one fixture is used at a time, the main supply must provide adequate flow for both. Generally, the main supply line for a one- or two-family house is $^3/_4$-inch or 1-inch pipe.

At the point where the main supply enters the building, a water meter is installed. The water meter measures the amount of water used. The municipal water department relies on this meter to determine the proper water bill for the building. The main water shutoff valve is located near the water meter.

From the main shutoff, the supply continues to the water heater. Somewhere between the water heater and the meter, a tee is installed to supply cold water to the house. From the main supply, branches are run to each fixture or point of use.

When a valve is suddenly closed at a fixture, the water tends to slam into the closed valve. This causes a sudden pressure buildup in the pipes and may cause the pipes to *hammer* (a sudden shock in the supply piping). To prevent this, an air chamber is installed at a high point in the system. An *air chamber* consists of a short vertical section of pipe that is filled with trapped air, Figure 35-7. When a valve is suddenly closed, the air chamber acts as a shock absorber. Although water cannot be compressed, air can be. When the pressure tends to build up suddenly, the air in this chamber compresses and cushions the resulting shock. Figure 35-8 shows a typical water supply system.

DESIGN OF WASTE PLUMBING

The main purpose of DWV, as stated earlier, is to remove water after it has been used and to carry away solid waste. To accomplish this, a branch line runs from each fixture to the main building sewer. The main building sewer carries the *effluent* (fouled water and solid waste) to the municipal sewer or septic system.

Traps

The sewer contains foul-smelling, germ-ladened gases which must be prevented from entering the house. If waste water simply emptied into the sewer from the pipe, this sewer gas would be free to enter the building. To prevent this from happening, a trap is installed at each fixture. A *trap* is a fitting that naturally fills with water to prevent sewer gas from entering the building, Figure 35-9. Not all traps are easily seen. Some fixtures, such as *water closets* (toilets), have built-in traps, Figure 35-10.

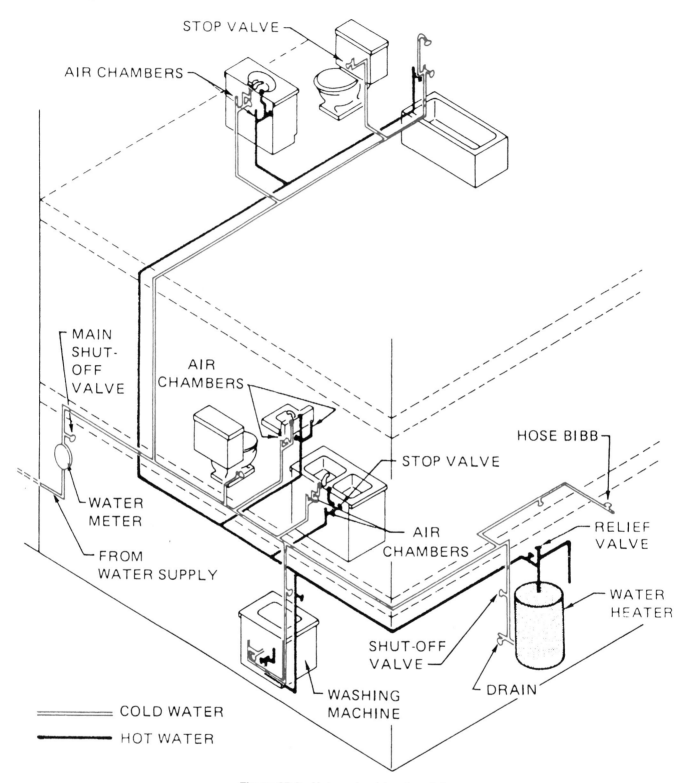

STOP VALVE

AIR CHAMBERS

MAIN
SHUT-
OFF
VALVE

AIR
CHAMBERS

WATER
METER

FROM
WATER SUPPLY

HOSE BIBB

STOP VALVE

AIR
CHAMBERS

RELIEF
VALVE

WATER
HEATER

SHUT-OFF
VALVE

DRAIN

WASHING
MACHINE

COLD WATER

HOT WATER

Figure 35-8   Hot- and cold-water piping

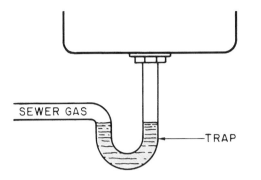

Figure 35-9 A trap fills with water to prevent sewer gas from entering the building.

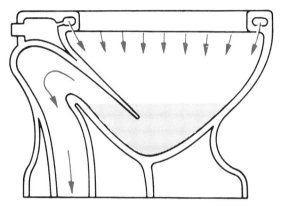

Figure 35-10   A water closet has a built-in trap.

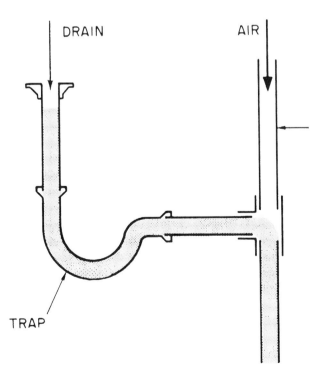

Figure 35-11   Venting a trap allows air to enter the system and prevents siphoning.

Vents

As the water rushes through a trap, it is possible for a siphoning action to be started. (The air pressure entering the fixture drain is higher than that on the other side of the trap. This forces the water out of the trap.)

To prevent DWV traps from siphoning, a vent is installed near the outlet side of the trap. The vent is an opening that allows air pressure to enter the system and break the suction at the trap, Figure 35-11. Because the vent allows sewer gas to pass freely, it must be vented to the outside of the building. Usually all of the fixtures are vented into one main vertical pipe, through the roof, Figure 35-12.

PLUMBING PLANS

For residential construction, the architect does not usually include a plumbing plan with the set of working drawings. The floor plan shows all of the plumbing fixtures by standard symbols. These symbols are easily recognized, because they resemble the actual fixture. The dimensions of the fixtures are provided by the manufacturer on rough-in sheets, Figure 35-13.

If the building and the plumbing are fairly simple, plumbers may prepare estimates and bids, and complete the work from the symbols on the floor plan only. For more complex houses, the plumbing contractor usually draws a plumbing isometric, Figure 35-14, or a special plumbing plan. The drawing set with this textbook includes a plumbing plan and details for plan A of the Town House. This sheet includes more details than would normally be found on a plumbing plan for a single-family housing unit. The extra detail is included here to help you understand the plumbing plan.

Plumbing plans show each kind of piping by a different symbol. Common plumbing symbols are shown in the Appendix. It will help you understand the plumbing plan if you trace each kind of piping from its source to each fixture. For example, trace the gas piping for the Town House. The gas lines can be recognized by the letter G in the piping symbol. The gas supply is shown as a broken line until it is inside the garage. Broken lines are used to indicate that the pipe is underground or concealed by construction. Although it is not noted on this plan, the plumbing contractor should know that the building code requires the gas line to be run in a sleeve where it passes through the foundation and the concrete slab, Figure 35-15. Just inside the garage wall, the broken line changes to a solid line. At this point, a symbol indicates that the solid line (exposed piping) turns down or away. Here, at this point, the gas piping runs above the concrete slab and along the garage wall. A callout on this line indicates that the diameter of the pipe is ³⁄₄-inch. At the back of the garage,

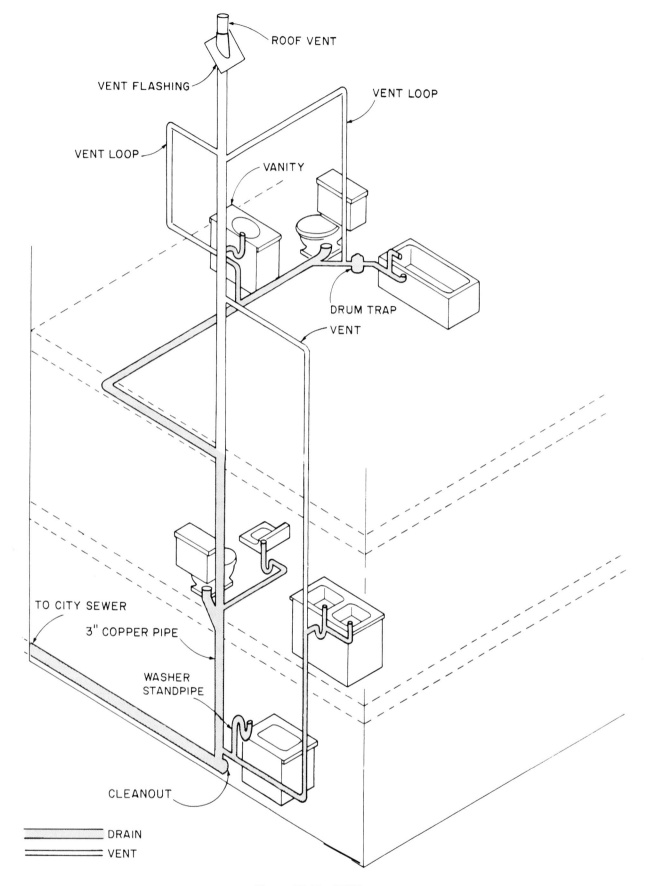

ROOF VENT

VENT FLASHING

VENT LOOP

VENT LOOP

VANITY

DRUM TRAP

VENT

TO CITY SEWER

3" COPPER PIPE

WASHER STANDPIPE

CLEANOUT

DRAIN

VENT

Figure 35-12   DWV system

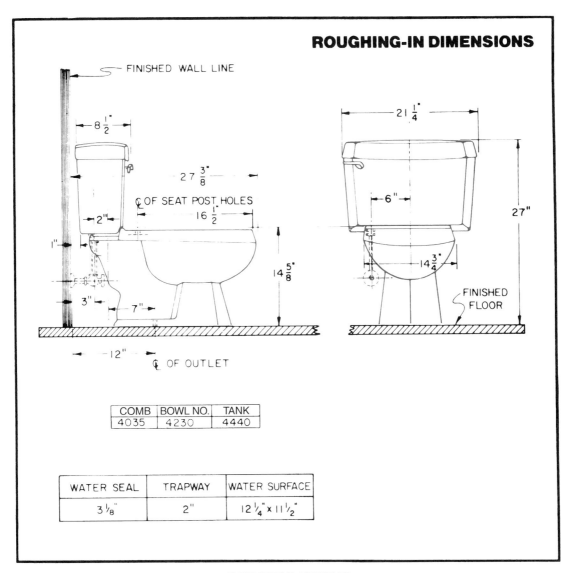

ROUGHING-IN DIMENSIONS

COMB	BOWL NO.	TANK
4035	4230	4440

WATER SEAL	TRAPWAY	WATER SURFACE
$3\frac{1}{8}$"	2"	$12\frac{1}{4}$" x $11\frac{1}{2}$"

SPECIFICATIONS

FIXTURE SPECIFICATION

☐ **U/R 4035 NEW VENUS** — vitreous china floor-mounted siphon jet close-coupled water-saver combination for 12" rough with regular-rim bowl, requires only 3½ gallons per flush — tank furnished with Fluidmaster 400A ballcock and adjustable tilt flush valve — (2) bolt caps and U/R 8085 lift-off seat included.

COLOR SPECIFICATION

☐ Acid Resisting White

☐ Acid Resisting Color _____
(show U/R color desired)

Universal-Rundle

® 303 NORTH STREET • NEW CASTLE PA 16103

Figure 35-13   Typical manufacturer's rough-in sheet

the gas line has a *T*. Both of the outlets of this *T* are ¹/₂-inch in diameter. One side of the *T* supplies the forced air unit (F.A.U.). The other side of the *T* continues around behind the F.A.U. to another *T*, and then to the water heater. The side outlet of the second *T* supplies a log lighter in the fireplace. This branch is drawn on the first floor plumbing plan. Notice that the log-lighter branch is reduced further to ¹/₄-inch diameter.

You should trace each type of plumbing in a similar manner to be sure you understand it. Refer to the details on the drawing for clarification of the complex areas. As you trace each line, look for the following:

- kind of plumbing (hot water, cold water, waste)

- diameter

- fittings

- exposed or concealed

- where line passes through building surfaces

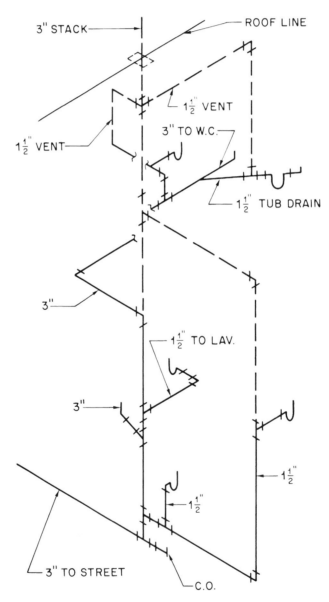

Figure 35-14 Single-line isometric of system shown in Figure 35-12

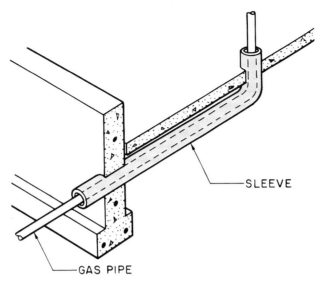

Figure 35-15   Sleeve for running gas piping
under and through concrete

✓ CHECK YOUR PROGRESS ———————————————————————

Can your perform these tasks?

- ☐ List the plumbing fittings shown on a plumbing plan.
- ☐ List the material and size of all piping and fittings for waste plumbing.
- ☐ List the material and size of all piping and fittings for supply plumbing.
- ☐ Use manufacturer's literature to list dimensions for the location of plumbing to fixtures.

ASSIGNMENT ————————————————————————————

Refer to the Town House drawings (in the packet) to complete the assignment.

1. What size pipe supplies the washing machine?

2. What size is the cold water supply to the water heater?

3. What size is the cold water branch to the lavatory in bathroom #2?

4. At what point does the $3/4$-inch cold water branch to the kitchen reduce to $1/2$-inch for the hose bibb?

5. Does the cold water supply turn up or down as it leaves the bathroom area to supply the kitchen area?

6. List each of the fittings that water will pass through after it drains out of the master bathroom lavatory.

7. List each of the fittings that water will pass through after it drains out of the master bathroom lavatory.

8. What size is the waste piping from the water closet in the master bath?

9. What size is the waste piping from the kitchen sink?

10. What size is the waste piping from the washing machine?

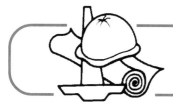

UNIT 36 Heating and Air Conditioning

OBJECTIVES

After completing this unit, you will be able to perform the following tasks:

* Locate and identify the heating and air conditioning equipment shown on the mechanical plan for a house.

* Explain how heat or conditioned air is distributed throughout a building.

FORCED AIR SYSTEMS

One of the most common systems for climate control circulates the air from the living spaces through or around heating or cooling devices. A fan forces the air into large sheet metal or plastic pipes called *ducts.* These ducts connect to openings, called *diffusers,* in the room. The air enters the room and either heats it or cools it as needed.

Air then flows from the room through another opening into the *return duct.* The return duct directs the air from the room over a heating or cooling device, depending on which is needed. If cool air is required, the return air passes over the surface of a cooling coil. If warm air is required, the return air is either passed over the surface of a *combustion chamber* (the part of a furnace where fuel is burned) or a heating coil. Finally, the conditioned air is picked up again by the fan and the air cycle is repeated, Figure 36-1.

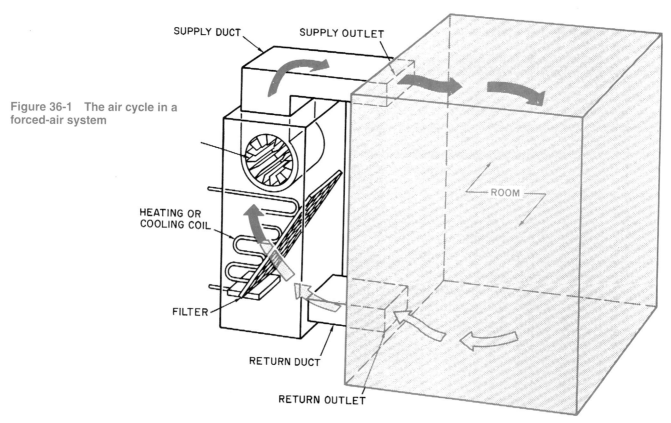

Figure 36-1 The air cycle in a forced-air system

Furnace

If the air cycle just described is used for heating, the heat is generated in a furnace. Furnaces for residential heating produce heat by burning fuel oil or natural gas, or from electric heating coils. If the heat comes from burning fuel oil or natural gas, the *combustion* (burning) takes place inside a combustion chamber. The air to be heated does not enter the combustion chamber, but absorbs heat from the outer surface of the chamber. The gases given off by the combustion are vented through a chimney. In an electric furnace, the air to be heated is passed directly over the heating coils. This type of furnace does not require a chimney.

Refrigeration Cycle

If the air from the room is to be cooled, it is passed over a cooling coil. The most common type of residential cooling system is based on the following two principles:

* As liquid changes to vapor, it absorbs large amounts of heat.

* The boiling point of a liquid can be changed by changing the pressure applied to the liquid. This is the same as saying that the temperature of a liquid can be raised by increasing its pressure and lowered by reducing its pressure.

The principal parts of a refrigeration system are the cooling coil *(evaporator), compressor* (an air pump), the *condenser,* and the *expansion valve,* Figure 36-2.

Keep in mind that common refrigerants can boil (change to a vapor) at very low temperatures—some as low as 21°F below zero. Also remember that a liquid boils at a higher temperature when it is under pressure.

The warm air from the ducts is passed over the evaporator. As the cold refrigerant liquid moves through

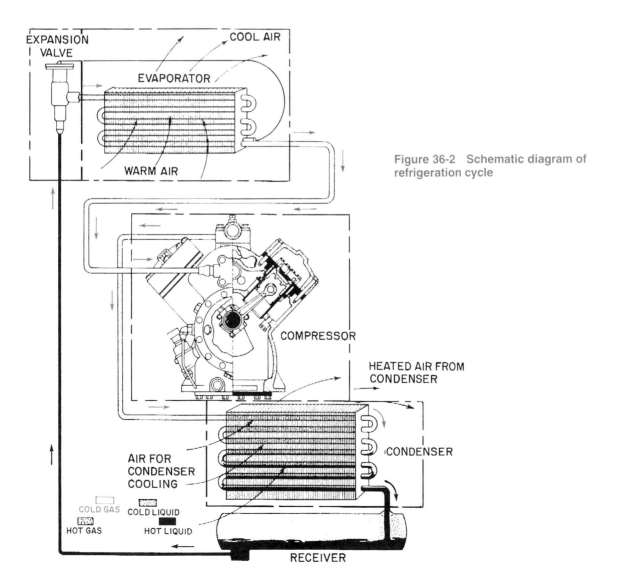

Figure 36-2 Schematic diagram of refrigeration cycle

the evaporator coil, it picks up heat from the warm air. As the liquid picks up heat, it changes to a vapor.

The heated refrigerant vapor is then drawn into the compressor where it is put under high pressure. This causes the temperature of the vapor to rise even more.

Next, the high-temperature, high-pressure vapor passes to the condenser where the heat is removed. In residential systems this is done by blowing air over the coils of the condenser. As the condenser removes heat, the vapor changes to a liquid. It is still under high pressure, however.

From the condenser, the refrigerant flows to the expansion valve. As the liquid refrigerant passes through the valve, the pressure is reduced. This lowers the temperature of the liquid still further, so that it is ready to pick up more heat.

The cold, low-pressure liquid then moves to the evaporator. The pressure in the evaporator is low enough to allow the refrigerant to boil again and absorb more heat from the air passing over the coil of the evaporator.

HOT-WATER SYSTEM

Many buildings are heated by hot-water systems. In a hot-water system, the water is heated in an oil or gas-fired boiler, then circulated through pipes to radiators or convectors in the rooms. The boiler is supplied with water from the fresh waster supply for the house. The water is circulated around the combustion chamber where it absorbs heat.

In some systems, one pipe leaves the boiler and runs through the building and back to the boiler. In this type, called a *one-pipe system,* the heated water leaves the supply, is circulated through the outlet, and is returned to the same pipe, Figure 36-3. Another type, the *two-pipe system,* uses two pipes running throughout the building. One pipe supplies heated water to all of the outlets. The other is a return pipe which carries the water back to the boiler for reheating, Figure 36-4.

Hot-water systems use a pump, called a *circulator,* to move the water through the system. The water is kept at a temperature of 150° –180°F in the boiler. When heat is needed, the thermostat starts the circulator.

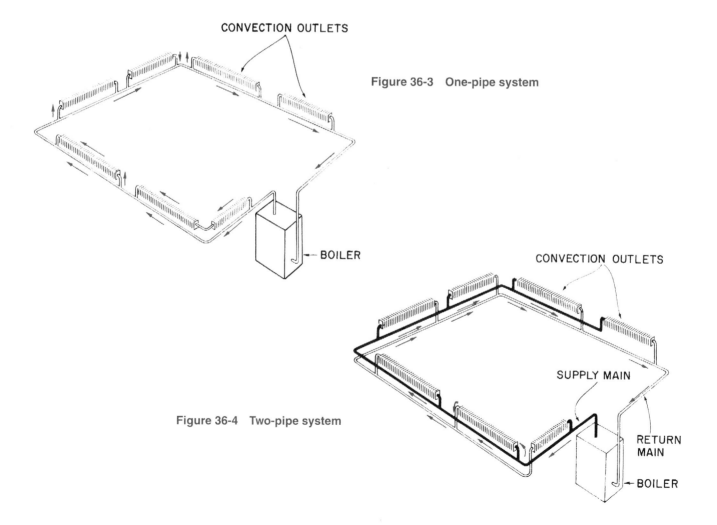

Figure 36-3   One-pipe system

Figure 36-4   Two-pipe system

Solar Collectors

Solar energy can also be used as a source of heat for a hot-water system. All that must be done is to concentrate the heat of the sun on pipes carrying water to the heating system.

A flat-plate solar collector is a box that absorbs the heat of the sun and transfers it to water or antifreeze. The top of the box is made of glass or specially formulated plastic. As the rays of the sun pass through the glass top, they strike the collector plate. The collector plate is made up of a network of tubes that carry the water or antifreeze. Below the collector plate, the box is filled with insulation. At each end of the collector plate a larger tube, called a *manifold,* connects the tubes to the system, Figure 36-5.

In operation, solar flat-plate collectors are placed where the rays of the sun strike their surface in the winter, Figure 36-6. The energy from the sun heats the antifreeze solution in the pipes of the collector. This warmed liquid is pumped to a large tank near the regular boiler. The water entering the boiler is circulated through a coil in the tank. This preheats the water going into the boiler, Figure 36-7. The use of solar collectors can reduce the fuel consumption of a boiler by as much as 80 percent.

ELECTRIC RESISTANCE HEAT

There are a number of heating system designs that rely on electric heating elements located in each

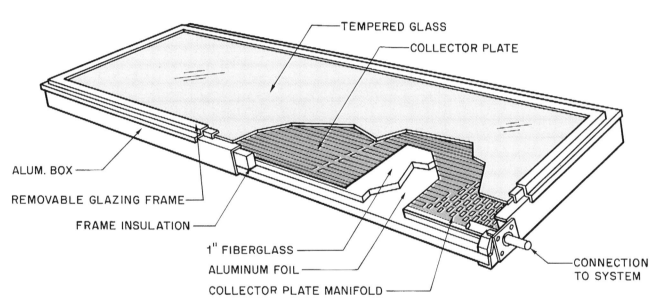

Figure 36-5   Flat-plate solar collector

Figure 36-6   The collectors are mounted where they will get direct sunlight.

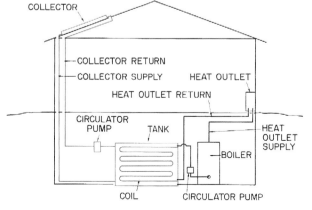

Figure 36-7   Solar-assisted hot-water heating system

room. Some such systems have electric heating elements embedded in the floor or ceiling. In these systems, the surface of the room is heated. Another kind of electric heat has heating outlets similar to those used for hot-water heat.

HEATING AND AIR CONDITIONING EQUIPMENT ON DRAWINGS

As with plumbing, architects do not usually prepare heating and air conditioning drawings for residential construction. The architect specifies the type of heating and air conditioning to be used. The HVAC (heating, ventilating, and air conditioning) contractor prepares required drawings as needed for personal use only.

The mechanical drawing for the Town House includes some limited HVAC information. This is enough information so the other trades will know what they may encounter. This mechanical drawing shows the following:

- Location of the air conditioning compressor
- Location of the central forced-air unit
- How the forced-air unit is vented
- Piping between the compressor and forced-air unit
- Approximate locations of diffusers
- Approximate locations of return-air grilles

✓ CHECK YOUR PROGRESS

Can your perform these tasks?

☐ Describe the type of heating system to be installed.

☐ List the locations and sizes of heating and air conditioning diffusers and return-air grilles.

☐ List the locations of hot-water convection outlets.

☐ Describe the location of major heating and air conditioning equipment.

☐ Trace the flow of water and antifreeze through a solar collector system.

 ASSIGNMENT

Refer to the drawings for the Town House, Plan A.

1. Where is the air conditioning compressor located?

2. Where is the tubing which connects the air conditioning compressor to the forced-air unit?

3. What size tubing connects the compressor to the forced-air unit?

4. How many diffuser outlets supply conditioned air to the rooms?

5. How many return-air grilles are there?

6. Where is the forced-air unit located?

7. What kind of fuel does the forced-air unit use for heating?

8. Where are the sizes of the diffuser outlets given?

UNIT 37   Electrical

OBJECTIVES

After completing this unit, you will be able to perform the following tasks:

- Identify the electrical symbols shown on a plan.

- Explain how the lighting circuits are to be controlled.

CURRENT, VOLTAGE, RESISTANCE, AND WATTS

To understand the wiring in a building you should know how electricity flows. Electricity is energy. To do any work (turn a motor, light a lamp, or produce heat) the electrical energy must have movement. This movement is called *current*. The amount of current is measured in *amperes,* sometimes called *amps.* A single household-type light bulb requires a current of slightly less than 1 ampere. An electric water heater might require 50 amperes.

The amount of force of pressure causing the current to flow affects the amount of current. The force behind an electric current is called *voltage.* If 115 volts causes a current flow of 5 amperes, 230 volts will cause a current flow of 10 amperes.

The ease with which the current is able to flow through the device also affects the amount of current. The ease or difficulty with which the current flows through the device is called the *resistance* of that device. As the resistance goes up, the current flow goes down. As the resistance goes down, the current flow goes up.

The amount of work the electricity can do in any device depends on both the amount of current (amps) and the force of the current (volts). Electrical work is measured in *watts.* The number of watts of power in a device can be found by multiplying the number of amperes by the number of volts. Stated another way the current flowing in a device can be found by dividing the number of watts by the voltage. For example, how much current flows through a 1,500-watt heater at 115 volts? 1,500 divided by 115 equals about 13 amperes. Figure 37-1 shows the current, wattage, and voltage of some typical electrical equipment.

Figure 37-1   Current, voltage, and power ratings of some typical electrical devices

DEVICE	AMPERES	VOLTS	WATTS
Ceiling light fixture	1.3	115	150
Vacuum Cleaner	6.1	115	700
Radio	0.4	115	4
Clock	0.4	115	4
Dishwasher	8.7	115	1,000
Toaster	13	115	1,500
Cook Top	32	230	7,450
Oven	29	230	6,600
Clothes Dryer	25	230	5,750
Washing Machine	10	115	1,150
Garbage Disposal	7.4	115	850

CIRCUITS

In order for current to flow, it must have a continuous path from the power source, through the electrical device, and back to its source. This complete path is called a *circuit,* Figure 37-2.

Many circuits include one or more switches. A switch allows the continuous path to be broken, Figure 37-3. By using two 3-way switches, the circuit can be controlled from two places, Figure 37-4. When the circuit is broken by a switch, a broken wire, or for any other reason, it is said to be *open.*

Any material that carries electric current is called a *conductor.* In Figure 37-2 each of the wires is a conductor. When two or more wire conductors are bundled together, they make a cable, Figure 37-5.

In larger buildings the wiring is frequently installed by pulling individual wires through steel or plastic pipes,

called *conduit.* In houses it is more common to use cables containing the needed wires plus one ground conductor. The ground conductor does not normally carry current. The *ground,* as it is usually abbreviated, connects all of the electrical devices in the house to the ground. If, because of some malfunction, the voltage reaches a part of the device that someone might touch, the ground protects them from a serious shock. The current that might otherwise flow through the person follows the ground conductor to the earth. The earth actually carries this current back to the generating station.

Additional protection against serious shock can be provided by using a *ground-fault circuit interrupter* (GFCI or GFI.) A GFI is a device that measures the flow of current in the hot (supply) conductor and the neutral (return) conductor. If a faulty device allows some of the current to flow through a person rather than the

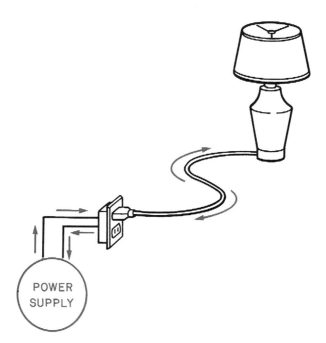

Figure 37-2   A complete circuit includes a path from the supply to the device and back again.

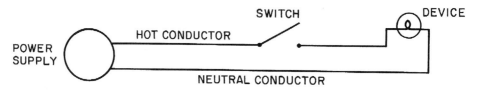

Figure 37-3   A switch is used to break (or open) the circuit.

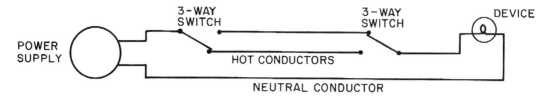

Figure 37-4   Three-way switches allow a device to be controlled from two locations. Notice that if either switch is activated, the device will be energized.

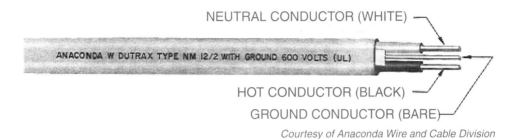

Courtesy of Anaconda Wire and Cable Division

Figure 37-5   This cable has two circuit conductors and one ground conductor.

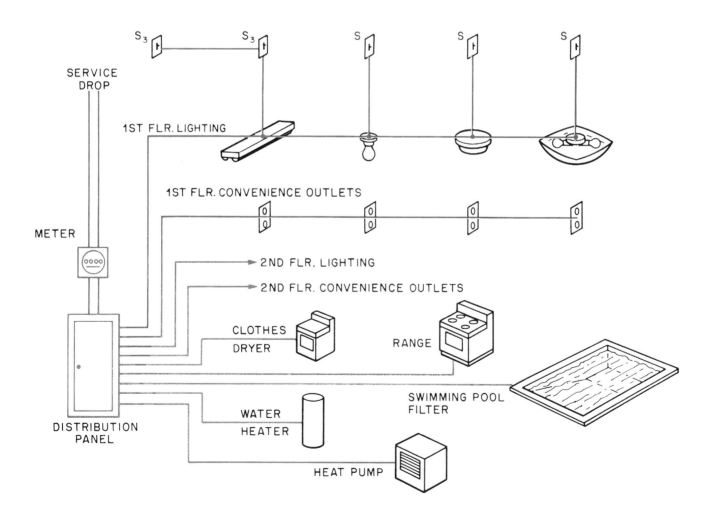

Figure 37-6   The electrical service is split up into branch circuits at the distribution panel.

neutral conductor, the GFI stops all current flow immediately. GFIs are so effective that the *National Electric Code®* requires their use on circuits for outlets installed outdoors, in kitchens, bathrooms, garages, and near any other water hazards.

The electrical service entrance was discussed earlier in Unit 11. The service feeder cable ends at a distribution panel. From the distribution panel, the electrical system is split up into several branch circuits, Figure 37-6. Each branch circuit includes a circuit breaker or fuse. The circuit breaker or fuse opens the circuit if the current flow exceeds the rated capacity of the circuit. Branch circuits for special equipment such as water heaters and air conditioners serve that piece of equipment only. Branch circuits for small appliances and miscellaneous use may serve several outlets. Branch circuits for lighting are restricted to lighting only, but a single circuit may serve several lights. Lighting circuits also include switches to turn the lights on and off.

The National Fire Protection Association publishes the *National Electrical Code®* which specifies the design of safe electrical systems. Electrical engineers and electricians must know this code which is accepted as the standard for all installations. The following are among the items it covers:

- Kinds and sizes of conductors

- Locations of outlets and devices

- Overcurrent protection (fuses and circuit breakers)

- Number of conductors allowed in a box

- Safe construction of devices

- Grounding

- Switches

The specifications for the structure indicate such things as the type and quality of the equipment to be used, the kind of wiring, and any other information that is not given on the drawings. However, electricians must know the *National Electrical Code®* and any state or local codes that apply because specifications sometimes refer to these codes.

ELECTRICAL SYMBOLS ON PLANS

The drawings for residential construction usually include electrical information on the floor plans. Only the symbols for outlets, light fixtures, switches, and switch wiring are included. The exact location of the device may not be dimensioned. The position of the device is determined by the electrician after observing the surrounding construction. It should also be noted that all wiring is left to the judgement of the electrician and the regulations of the electrical codes. Switch wiring for light fixtures is included only to show which switches control each light fixture. Switch wiring is shown by a broken line connecting the device and its switch, Figure 37-7.

In rooms without a permanent light fixture, one or more convenience outlets may be split wired and controlled by a switch. In split wiring one-half of the outlet is always hot; the other half can be opened by a switch, Figure 37-8.

Figure 37-7 Switch legs on a plan

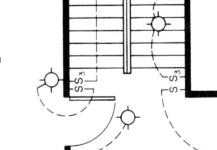

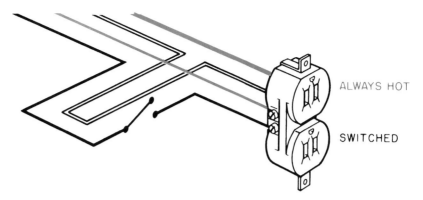

Figure 37-8   Split-wired outlet

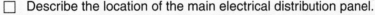

✓ CHECK YOUR PROGRESS

Can your perform these tasks?

- [ ] List all the convenience outlets, special outlets, light fixtures, and switches shown on plans.
- [ ] Describe the location of the main electrical distribution panel.

ASSIGNMENT

Refer to the drawings for the Town House, Floor Plan A, including the garage.

1. How many light fixtures are shown on the floor plan? (Include the garage.)

2. How many switches are shown on the floor plan? (Include the garage.)

3. How many duplex outlets are shown?

4. Briefly describe the location of each split-wired outlet and the switch or switches that control each.

5. List five pieces of equipment shown on the floor plan that probably require separate branch circuits.

6. How many outlets are to have ground-fault circuit interrupters included in their circuits?

7. What is the location of the switch, or switches, that control(s) the light over the stairs to the bedroom level?

8. What is the location of the switch, or switches, that control(s) the light fixture over the stairs to the garage?

9. Where is the smoke detector located?

10. Where is each of the two telephone outlets?

PART III Test

A. For each of the items in Column I, indicate the function listed in Column II that it performs.

I		II	
1.	Diffuser	a.	Absorbs the heat from room air
2.	Combustion chamber	b.	Reduces the pressure of refrigerant
3.	Evaporator	c.	Connects the water passages in a flat-plate collector
4.	Condenser	d.	Joins two pipes, so they can be easily disconnected
5.	Compressor	e.	Outlet where heated or cooled air enters a room
6.	Expansion valve	f.	Contains the fire in a furnace
7.	Circulator	g.	Prevents hammering in supply plumbing
8.	Collector manifold	h.	Removes heat from refrigerant
9.	Air chamber	i.	Pumps water or antifreeze through a heating system
10.	Union	j.	Increases pressure and temperature of refrigerant

B. For each of the symbols in Column I indicate the object in Column II which it represents.

I	II
1.	a. Electrical outlet for special equipment
2.	b. Pipe turning down or away
3. C.O.	c. Switch wiring
4.	d. Convenience outlet
5.	e. Valve
6.	f. Pipe turning up or toward
7.	g. Hot water piping
8.	h. Wall switch
9. S	i. Wye fitting with a cleanout
10.	j. Tee fitting

C. Refer to the illustrated Hidden Valley drawing index and partial site plan to answer the following questions.

1. In which building is unit number 56 located?
2. On which drawing sheet is the foundation plan for unit number 56?
3. On which drawing sheet are the foundation details for Plan-B units?
4. List all of the drawing sheets that are *not* needed to construct Building IV.
5. Of the buildings shown, which would be built first?

table of contents

Courtesy of berkus-group architects

D. Refer to the Town House drawings to answer these questions.

1. In Plan B, which rooms have their floors at the same elevation as the kitchen?
2. In Plan B, what separates the living room and dining room?
3. In Plan B, as you climb the stairs from the first floor, what room do you enter?
4. In Plan B, as you climb the stairs from the garage, what room do you enter?
5. In Plan B, what is the distance from the bottom of the garage stairs to the finished floor in the library?
6. What is the width of Plan B in the living room area measuring to the inside face of the framing?
7. What size are the anchor bolts in the foundation of Plan B?
8. In the front part of the foundation plan for Plan B, there is a 24" square by 12" deep pad for a pier which supports a wood beam. What does this beam support?
9. In Plan B, what supports the floor joists under bedroom #2 at the end nearest the front of the building?
10. In Plan B, what prevents the party wall in the nook area from wracking?
11. In Plan B, where the bedroom floor joists rest on the dining room-kitchen partition, how is the blocking between the joists fastened to the top plate of the wall?
12. What is the wall surface material at the area separation wall in the dining room of Plan B?
13. What is the total thickness, including surface material, of the party wall in the garage stairway in Plan B?
14. What size lumber (thickness × width × length) would be ordered for the joists in the Plan A kitchen floor?
15. In plan A what is the finished size (thickness × width × length) of the floor joists under bedroom #2?
16. What is the length of the cantilevered part of the floor joists under bedroom #2 of Plan A?
17. What material is used for the wood sill on the foundation of Plan A at the exterior corner of the kitchen?
18. In Plan A, what kind and size fasteners are used to anchor the railing outside the dining room entrance?
19. In Plan A, how far is the air-conditioner compressor pad from the forced air unit?
20. In Plan B, who many three-way switches are required?
21. In Plan A, how many ceiling light fixtures are there, including luminous soffits?
22. In Plan A, how many plumbing vents penetrate the roof?
23. What diameter pipe supplies the hose bibb at the front of the kitchen?
24. In Plan A, how many cleanouts are included in the waste piping?
25. In Plan A, where is the hot-water shutoff closest to the shower in bathroom #2?

Part IV

COMMERCIAL
CONSTRUCTION

Part IV presents a thorough examination of the information found on prints for heavy commercial construction. The materials and methods used for large buildings are different from those used in light-frame construction. Also, the drawing set for a commercial project usually includes many more sheets than are found in the drawing sets for smaller buildings. To understand these drawings and make practical use of them, you will need to understand the organization of the drawings and how the heavier materials are described on the drawings.

Part IV also covers air conditioning and heating, plumbing, and electrical drawings in much more detail than earlier sections of the textbook. For those who work in the mechanical or electrical trades, the importance of understanding the drawings for these systems is obvious. However, all of the trades work in the same spaces and the work of one trade effects the other trades. Estimators, superintendents, inspectors, and many other construction professions also require an understanding of all of the construction trades.

The School Addition is an excellent building for inclusion in this book. It uses structural steel framing, reinforced concrete foundations, and a varied assortment of other materials typically associated with heavy construction; yet, unlike most large commercial buildings, the School Addition is small enough, so that most of the drawing can be packaged with this book. Some drawings have been eliminated, either because they pertain to other work done on existing school building or because they do not provide new opportunities for the study of commercial construction drawings. All of the School Addition drawings are numbered as they were in the complete set, and the text book makes frequent reference to the drawing numbers. The Material Keynotes, which are explained in this part of the textbook, are printed on the tabbed pages at the end.

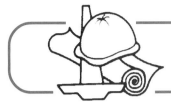

UNIT 38 Commercial Construction

OBJECTIVES

After completing this unit, you will be able to perform the following tasks:

- Describe the major differences between light–frame construction and heavy commercial construction.

- Describe common structural steel shapes and their designations.

- Explain the differences between foundations for heavy structures and those for light–frame buildings.

ARCHITECTURAL STYLE

Large buildings, such as the School Addition described in the drawings with this book, are usually designed in a different style from one- and two-family homes or even larger light-frame buildings, such as the Town House. Homes are designed to satisfy the tastes and expectations of the families who will live in them, Figure 38–1. For centuries homes have been rectangular structures, with sloping roofs, horizontal siding or masonry veneer, and traditional architectural trim. Recently, contemporary architecture has begun to break away from these traditional design elements and homes have had sharper lines, a variety of siding materials, and less classic trim. Still, the design of most homes

Figure 38-1   Usually, one- and two-family homes are architecturally different from larger commercial buildings.

Figure 38-2 Generally, medium to large commercial buildings are very functional in design.

is something recognized and expected as the design for a house. Some light commercial buildings, usually those that employ residential construction methods like light-frame construction and composition roof shingles, are designed with these classic home-design elements.

Most larger commercial buildings, however, are too large to employ the same construction methods and have different architectural styles than houses, Figure 38–2. For example, commercial buildings usually have a larger roof area, so it is not practical to slope the roof. This means that not only does the flat roof change the appearance of the building, but it necessitates the use of different roof framing and roof covering materials. Because of their size and because they are usually constructed as a business investment, commercial buildings are designed to be as maintenance free as possible. The wood or vinyl molding that decorates the exterior of many houses would require too much maintenance for a commercial office building. This is not to say that a commercial building cannot be architecturally pleasing. In fact, because more money is generally available to design and construct large commercial buildings, they often tend to be extremely attractive and viewed as works of art. However, the aesthetic qualities in a commercial building are generally built into the

overall design of the building, rather than applied after the shell is erected.

Larger buildings require greater strength. The weight of the materials required to build the School Addition is many times the weight of the materials to build even a large house. Also, the external loads (wind, snow, and so forth) on a larger building are greater than those on a smaller building. All of these greater loads require stronger structural members, more precisely engineered construction methods, and fastenings that transmit the loads from one structural member to the next. This usually dictates the use of structural steel and reinforced concrete for the structural elements of the building—the frame and foundation.

STRUCTURAL STEEL

Steel is made by alloying small amounts of carbon and other elements with iron. The amount of carbon in the steel determines it's toughness. Other elements, such as chromium, copper, nickel, or titanium are used to produce specific properties. Stainless steel, for example, is made by alloying chromium with the steel. Structural steel is mild steel (a small amount of carbon is used) which is rolled into specific shapes used for construction. The most common structural shapes are shown in Figure 38–3. W shapes are used for most beams and columns in steel-frame buildings. The designator for a structural steel member includes a symbol or capital letter to indicate the basic shape, one or more numbers to indicate the size, and a weight-per-foot designation. For example, drawing S200 for the School Addition, Figure 38–4, shows a W6 × 25 column. This is a **wide**–flange shape, with a **6**-inch web, weighing **25** pounds per foot. Standard dimensions for structural steel shapes are published by the American Institute for Steel Construction (AISC) *Manual of Steel Construction*.

Structural steel members are most often joined by welding. Bolts are sometimes used to attach dissimilar materials, such as wood and steel framing members. Where bolts are to be used, they are usually shown either by symbols or a drawing of a bolt on the details, Figure 38–5. Detail J–14 of the drawings for the School Addition (Figure 38–5) introduces a technique that is often used on drawings for commercial construction. Instead of describing all of the information on the detail

Shape	Letter Designation	Size Designation
 Wide Flange	W	depth of web × weight per foot
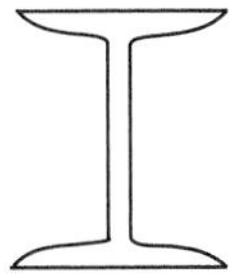 Standard Beam	S	depth of web × weight per foot
 Channel or Miscellaneous Channel	C or MC	depth × weight per foot
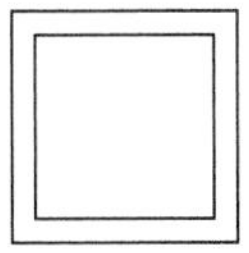 Structural Tube	TS	width × depth
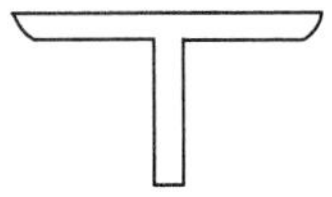 Tee	T	depth × weight per foot
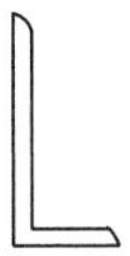 Angle	L	vertical leg × horizontal leg × thickness

38-3   **Common structural shapes and their designations.**

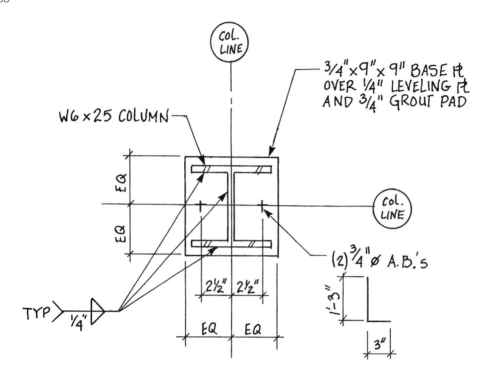

TYPICAL
BASE PLATE DETAIL
(28 LOCATIONS)

38-4   Structural steel designation on a drawing.

drawing itself, many of the drawing callouts refer to a material legend. The Material Legend, which is shown in its entirety on pages 339 through 346, identifies all of the material items in the building by a legend number. J–14 shows four items identified only as 5.10D. The Material Legend identifies these as anchor bolts.

Welding Symbol

Welds are depicted in drawings by the use of standard welding symbols, Figure 38–6. The American Welding Society (AWS) has developed a standard system for welding symbols. Each part of the welding symbol has a specific meaning. The arrow indicates what steel parts are to be welded. The term *welding symbol* should not be confused with *weld symbol*. The welding symbol is the complete symbol, including ar-

row, reference line, and any information added to it. The weld symbol indicates the type of weld to be made. The type of weld used most often in structural steel for construction is called a *fillet weld* and is indicated by a triangle. Other types of welds include *plug welds, groove welds,* and *butt welds*, Figure 38–7. The reference line is always drawn parallel to the bottom edge of the drawing. The lower side of the reference line is called the *arrow side*. Any information, such as the type of weld, that is shown below the reference line applies to the side of the assembly to which the arrow points. The upper side of the reference line is called the *other side*. Information above the reference line applies to the side of the assembly away from the arrow, Figure 38–8. The tail of the arrow is used to specify welding processes, electrode materials, and any other specifications that cannot be shown elsewhere on the welding symbol.

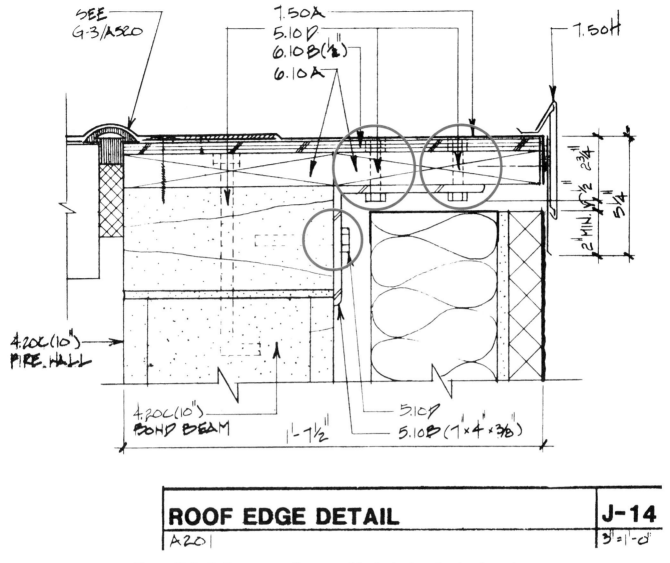

SEE
G-3/A520

7.50A
5.10D
6.10B(½")
6.10A

7.50H

2¾"

2" MIN.

½"

5¼"

4.20C(10")
FIRE WALL

4.20C(10")
BOND BEAM

1'-7½"

5.10D
5.10B(7"×4"×3/8")

ROOF EDGE DETAIL

A201

J-14

3"=1'-0"

Figure 38-5   Bolts are sometimes used for fastening structural members.

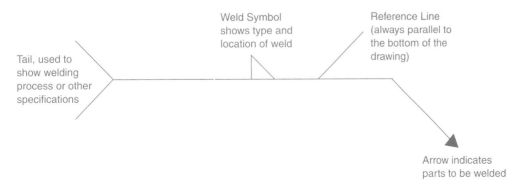

Tail, used to
show welding
process or other
specifications

Weld Symbol
shows type and
location of weld

Reference Line
(always parallel to
the bottom of the
drawing)

Arrow indicates
parts to be welded

Figure 38-6   The basic welding symbol.

FILLET	PLUG OR SLOT	SPOT OR PROJEC- TION	STUD	SEAM	BACK OR BACKING	SUR- FACING	SCARF (FOR BRAZED JOINT)
◿	▭	○		⊖	◡	◡◡	∥
◺	▭	○ / ○	⊗	⊖ / ⊖	◠		//

FLANGE		GROOVE						
EDGE	COR- NER	SQUARE	V	BEVEL	U	J	FLARE- V	FLARE- BEVEL
⊥⊥	⊥⊥	∥	∨	⩔	⋃	⋌	⟉	⟊
⊤⊤	⊤⊤	∥	∧	⩘	⋂	⋉	⟍	⟎

NOTE· _ _ _ _ REPRESENTS REFERENCE LINE WHICH IS NOT SHOWN AS A SOLID LINE FOR THIS PURPOSE.

38-7   Weld symbols used to show various types of welds. (Bennett & Siy: *Blueprint Reading for Welder, 5th Edition*, copyright 1993 by Delmar Publishers Inc.)

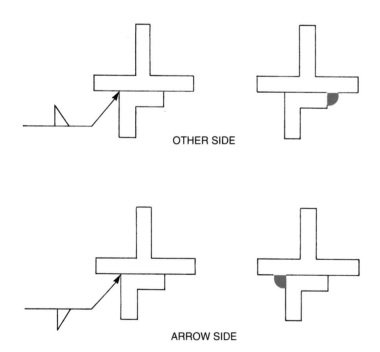

OTHER SIDE

ARROW SIDE

Figure 38-8   Arrow-side, other-side significance.

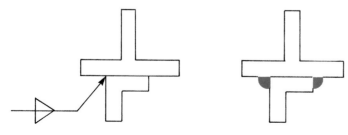

BOTH SIDES

Figure 38-9 Field weld symbol.

Shop Fabrication

Many structural sub assemblies are more easily or more accurately fabricated in a shop environment. For example, if a special beam is to be installed with a welded sub assembly on its ends, the sub assembly would probably be fabricated in the steel shop so the beam can be delivered to the site ready for erection. A large structural steel frame might require as many *fabrication drawings* (sometimes called *shop drawings*) as structural drawings. Usually, the field contractors do not deal with shop drawings and they are not covered in this book. However, it is important to recognize notes on the structural drawings that indicate a difference between field erection and shop fabrication. One example of such information is the field weld symbol, Figure 38–9. The field weld symbol is a small flag at the break in the arrow of the welding drawing. This symbol indicates the weld is to be done in the field, not in the fabrication shop.

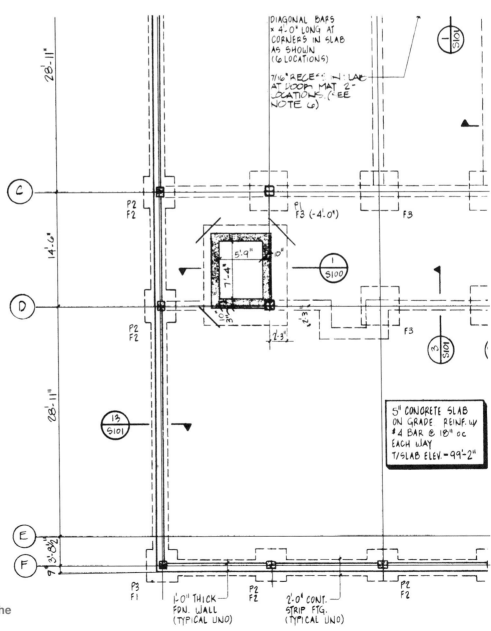

Figure 38-10 The layout of the foundation is shown on the foundation plan.

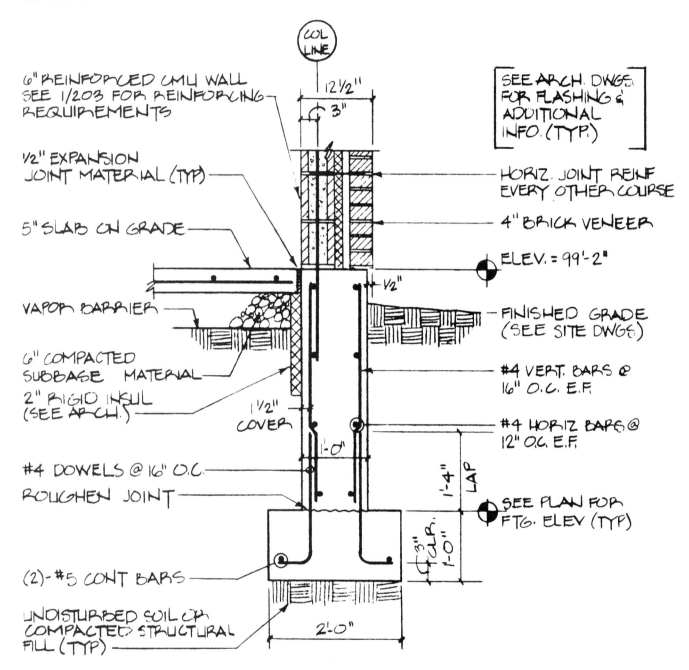

6" REINFORCED CMU WALL
SEE 1/203 FOR REINFORCING
REQUIREMENTS

1/2" EXPANSION
JOINT MATERIAL (TYP)

5" SLAB ON GRADE

VAPOR BARRIER

6" COMPACTED
SUBBASE MATERIAL

2" RIGID INSUL
(SEE ARCH.)

#4 DOWELS @ 16" O.C.

ROUGHEN JOINT

(2)-#5 CONT BARS

UNDISTURBED SOIL OR
COMPACTED STRUCTURAL
FILL (TYP)

COL
LINE

12 1/2"

3"

SEE ARCH. DWGS.
FOR FLASHING &
ADDITIONAL
INFO. (TYP.)

HORIZ. JOINT REINF
EVERY OTHER COURSE

4" BRICK VENEER

ELEV. = 99'-2"

FINISHED GRADE
(SEE SITE DWGS)

#4 VERT. BARS @
16" O.C. E.F.

#4 HORIZ BARS @
12" O.C. E.F.

SEE PLAN FOR
FTG. ELEV (TYP)

1/2"

1 1/2"
COVER

1'-0"

1'-4" LAP

3" CLR. 1'-0"

2'-0"

TYPICAL WALL SECTION

13
S101

SCALE: 3/4"=1'-0"

Figure 38-11   Foundation detail.

FOUNDATIONS FOR COMMERCIAL CONSTRUCTION

Houses and small frame buildings usually have very simple foundation systems. The superstructure does not impose a heavy load on the foundation, so a simple *haunched slab* or *inverted-T foundation* is all that is required to support the building in most soil conditions. As buildings become larger and more complex, they place a greater load on the foundation system. Also, larger buildings tend to be designed with much of the load being transmitted through a relatively small number of columns, instead of being spread over a sill member by closely spaced studs. To support these loads, the foundations of heavy buildings are carefully engineered by structural engineers and are more heavily reinforced than the foundations for smaller buildings.

Generally, continuous wall foundations are similar to those for smaller buildings. They consist of a *spread footing* which supports a continuous concrete wall. Because of the greater size of the structure, the footing and wall may be more massive for a heavy building and will probably require more and larger-diameter reinforcing steel. The overall layout of the foundation is shown on a foundation plan, Figure 38–10. The dimensions and the reinforcement of the footings and walls is planned by engineers to carry the necessary loads, so it is likely that there will be several detail drawings to further describe the foundation systems at different points in the building, Figure 38–11. Some of the reinforcement may be bars that protrude only a short distance out of the footing and into the wall. Reinforcing bars used in this way are called *dowels* and are used to hold the wall from shifting on the footing.

Column footings are located by the column centerlines. All of these footings must be centered under the columns they support unless specifically noted otherwise. The outline of the footing is shown on the foundation plan using dashed lines. This indicates that the footing is actually hidden from view by the concrete slab. Column footing dimensions are usually shown in a schedule, Figure 38–12. There are usually several column footings of the same size. The mark-schedule system allows the drafter to show all of the similar footings with a single entry on the schedule.

Piles are long poles or steel members that are driven into the earth to support columns or other grade beams. Pile foundations support their loads by means of several piles in a cluster, topped with a pile cap made of reinforced concrete. Pile foundations are not as common as spread footings and continuous walls. *Grade beams* are reinforced concrete beams placed at grade (ground) level to provide a bearing surface for the superstructure.

FOOTING SCHEDULE

MARK	SIZE	REINF.	REMARKS
F1	3'-0"x 3'-0"x 12"	(4)#5 BARS EA. WAY	
F2	4'-0"x 4'-0"x 12"	(5)#5 BARS EA. WAY	
F3	4'-6"x 4'-6"x 12"	(5)#6 BARS EA. WAY	
F4	3'-0"x 4'-6"x 24"	(5)#5 SHORT BARS (7)#5 LONG BARS	OVERPOUR EXISTING BUILDING FOOTING (SIM. TO 14/S101)

Figure 38-12 Footing schedule.

✓ CHECK YOUR PROGRESS ─────────────────────────────

Can you perform these tasks?

☐ Explain several differences between typical light-frame construction and larger commercial construction.

☐ Describe common structural shapes from their identification on drawings.

☐ Explain how the foundations for large buildings are different from the foundations for single-family homes.

 ASSIGNMENT ────────────────────────────────────

1. List five features of the School Addition that would be very different on a light-frame building and explain the differences.

2. Describe each of the structural steel designations listed below:

   W14 × 82
   MC12 × 45
   L6 × 3 1/2 × 5/16

3. What is the structural steel shape and depth of the steel beam shown supporting the roof bar joists in detail 3/S202?

4. Draw a sketch showing where the welds would be made for each of the following symbols:

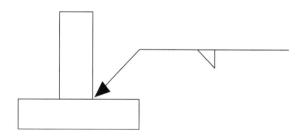

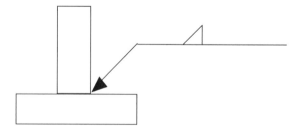

5. Describe the information given in the welding symbol on detail 3/S202.

6. Based on the information shown on sheet S100, would you say that the School Addition has a pier foundation, grade beams, or a continuous wall foundation with spread footings?

7. How many pieces of reinforcing steel are shown in the footing on Wall Section 13/S101? Describe each, including its size and purpose.

8. How many 4'-6" × 4'-6" × 12" column footings are shown on the foundation plan for the School Addition?

9. There are several pairs of dashed (hidden) lines running vertically through the interior of the foundation plan. A note on one of these pairs of lines indicates that the slab is to be thickened. Why is it thickened at these points? How much thicker is it at these points than the surrounding area? What additional reinforcement is provided in the thickened portion of the slab?

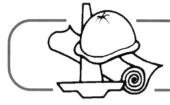

UNIT 39 Coordination of Drawings

OBJECTIVES

OBJECTIVES

After completing this unit, you will be able to perform the following tasks:

- Interpret the indexes and keys normally found on drawings for commercial construction.

- Use the number system for referencing details, sections, and other drawings.

- Explain the use of a grid system to locate columns and piers.

- Use the plans, elevations, and details to describe the layout of the rooms and spaces in a large commercial building.

INDEXES

A set of working drawings for a large commercial building might fill more than one hundred sheets. The School Addition, although it is a small by comparison with many large buildings, is part of a renovation and expansion program that spans three buildings and is described on sixty-six drawing sheets. A Key Plan in the lower right corners of sheets A101 and A102 shows the addition in relationship to the rest of the Junior/Senior High School building. Finding one's way around these sixty-six sheets would be a difficult task without some kind of an index. As with most commercial construction drawing sets, the cover page of the drawing set includes the index, Figure 39–1.

It is not reasonable to include all sixty-six drawing sheets, plus the cover page, with this textbook because the textbook would have increased by more than two hundred pages. Therefore, the drawing packet with this text includes all of the drawings that are necessary to completely understand the additions to the Junior/Senior High School, but could not be reprinted on the pages of the text. The Index, found in Figure 39–1, is complete with all original drawing numbers. The drawings contained in the drawing packet are indicated by underlining their drawing numbers. Parts of other sheets are included as illustrations in this textbook. Drawings for civil (site work) that are not needed for an understanding of the structure and drawings for work on other school property are not included.

The *Drawing Index* includes the number of the drawing and the drawing title. *Drawing numbers* are made up of one or two letters and numerals. The letters indicate the basic type of construction shown on the drawing. Most architects use the letter A to indicate architectural drawings, S to indicate structural drawings, C for civil, and so on. The numbers indicate the sequence of the drawings in that section. Some architects indicate levels of drawings by using a series within a range to indicate a level. For example, 100 through 199 might be reserved for foundations; 200 through 299 might be floor plans and elevations showing the overall superstructure; and 300 through 399 might be roof framing. This system is usually quite easy to spot by studying the index briefly.

INDEX OF DRAWINGS

	GENERAL CONSTRUCTION CONTRACT
A000	MASTER KEYNOTES
G001	SITE PLAN (JR/SR HIGH SCHOOL)
G002	KEY PLANS (JR/SR HIGH SCHOOL)
G003	KEY PLANS (BERNE ELEMENTARY & WESTERLO ELEMENTARY
C100	SITE DEMOLITION PLAN (JR/SR HIGH SCHOOL & BERNE ELEMENTARY
C101	LAYOUT PLAN (JR/SR HIGH SCHOOL)
C102	GRADING PLAN (JR/SR HIGH SCHOOL)
C103	PLANS & DETAILS (JR/SR HIGH SCHOOL)
C201	PARTIAL SITE PLAN, SOUTH (JR/SR HIGH SCHOOL)
C202	PARTIAL SITE PLAN, NORTH (JR/SR HIGH SCHOOL)
C203	PLAN AND DETAILS - SEWAGE DISPOSAL (JR/SR HIGH SCHOOL & BERNE ELEMENTARY)
C204	PLAN AND DETAILS; NEW WELL (JR/SR HIGH SCHOOL & BERNE ELEMENTARY)
C205	DRAINAGE PROFILES AND DETAILS (JR/SR HIGH SCHOOL & BERNE ELEMENTARY)
C206	DETAILS SEWAGE DISPOSAL (JR/SR HIGH SCHOOL & BERNE ELEMENTARY)
C207	PLANS AND DETAILS (JR/SR HIGH SCHOOL & BERNE ELEMENTARY)
C208	SECTIONS AND DETAILS (JR/SR HIGH SCHOOL & BERNE ELEMENTARY)
A001	ROOM FINISH SCHEDULE, DOOR SCHEDULE AND DETAILS (JR/SR HIGH SCHOOL)
A101	FIRST FLOOR PLAN (JR/SR HIGH SCHOOL)
A102	SECOND FLOOR PLAN (JR/SR HIGH SCHOOL)
A103	CAFETERIA & BOILER ROOM PLAN, DEMOLITION PLAN & DETAILS (JR/SR HIGH SCHOOL)
A104	FLOOR PLAN & DETAILS (BUS GARAGE)
A201	EXTERIOR ELEVATIONS & BUILDING SECTIONS (JR/SR HIGH SCHOOL)
A301	STAIR PLANS & SECTIONS (JR/SR HIGH SCHOOL)
A302	STAIR DETAILS (JR/SR HIGH SCHOOL)
A303	RAMP & ELEVATOR (JR/SR HIGH SCHOOL)
A501	WALL SECTIONS (JR/SR HIGH SCHOOL)
A502	WALL SECTIONS (JR/SR HIGH SCHOOL)
A503	PLAN DETAILS (JR/SR HIGH SCHOOL)
A504	WALL DETAILS (JR/SR HIGH SCHOOL)

Figure 39-1 Index from cover page of school drawings.

A510	WINDOW TYPES AND DETAILS (JR/SR HIGH SCHOOL)
A511	EXTERIOR ELEVATIONS (WESTERLO ELEMENTARY)
A512	WINDOW TYPES + DETAILS (WESTERLO ELEMENTARY)
A520	ROOF PLAN + DETAILS (JR/SR HIGH SCHOOL)
A610	INTERIOR ELEVATIONS + TOILET ROOM FLOOR PLAN (JR/SR HIGH SCHOOL)
S100	FOUNDATION PLAN + NOTES (JR/SR HIGH SCHOOL)
S101	FOUNDATION DETAILS (JR/SR HIGH SCHOOL)
S200	SECOND FLOOR FRAMING PLAN + DETAILS (JR/SR HIGH SCHOOL)
S201	ROOF FRAMING PLAN + DETAILS (JR/SR HIGH SCHOOL)
S202	STEEL FRAMING DETAILS (JR/SR HIGH SCHOOL)
S203	MISCELLANEOUS STEEL DETAILS + MASONRY REINFORCING DETAILS (JR/SR HIGH SCHOOL)
	ROOF RECONSTRUCTION CONTRACT
A521	ROOF PLAN + DETAILS (JR/SR HIGH SCHOOL)
A522	ROOF PLAN + DETAILS (BERNE ELEMENTARY)
A523	ROOF PLAN + DETAILS (BERNE ELEMENTARY)
	ASBESTOS ABATEMENT CONTRACT
ASB-1	ASBESTOS ABATEMENT - PLANS AND DETAILS (JR/SR HIGH SCHOOL)
ASB-2	PARTIAL PLAN - BOILER ROOM (JR/SR HIGH SCHOOL)
ASB-201	ASBESTOS ABATEMENT - PLANS AND DETAIL (WESTERLO ELEMENTARY)
	PLUMBING CONTRACT
P-1	PARTIAL PLAN - ADDITION, FIRST FLOOR AND SECOND FLOOR (JR/SR HIGH SCHOOL)
P-2	PARTIAL PLAN - CAFETERIA EXPANSION; DETAILS (JR/SR HIGH SCHOOL)
	HEATING, VENTILATING + AIR CONDITIONING CONTRACT
H-1	FIRST + SECOND FLOOR PLAN ADDITION; SYMBOL LEGEND (JR/SR HIGH SCHOOL)
H-2	PARTIAL PLAN BOILER ROOM; REMOVALS AND NEW WORK (JR/SR HIGH SCHOOL)
H-3	PARTIAL PLAN - CAFETERIA EXPANSION; DETAILS (JR/SR HIGH SCHOOL)
H-4	SCHEDULES AND DETAILS (JR/SR HIGH SCHOOL)

Figure 39-1   (continued)

H-5	DETAILS (JR/SR HIGH SCHOOL)
H101	PARTIAL PLAN - BOILER ROOM; DETAILS: SCHEDULES (WESTERLO ELEMENTARY)
H201	PARTIAL PLAN, BOILER ROOM, DETAILS, SCHEDULES (WESTERLO ELEMENTARY)
H202	DETAILS AND SCHEDULES (WESTERLO ELEMENTARY)
	ELECTRICAL CONTRACT
E-1	FIRST + SECOND FLOOR PLAN, ADDITION LIGHTING (JR/SR HIGH SCHOOL)
E-2	FIRST + SECOND FLOOR PLAN, ADDITION UTILITIES (JR/SR HIGH SCHOOL)
E-3	PARTIAL PLANS CAFETERIA + COMPUTER LABS (JR/SR HIGH SCHOOL)
E-4	BASEMENT PLAN, PART PLANS BOILER ROOM, DETAILS (JR/SR HIGH SCHOOL)
E-5	FIRST FLOOR PLAN, DETAILS (JR/SR HIGH SCHOOL)
E-6	SECOND FLOOR PLAN, DETAILS (JR/SR HIGH SCHOOL)
E-7	SCHEDULES (JR/SR HIGH SCHOOL)
E101	GROUND FLOOR PLAN; DETAILS; SCHEDULES (WESTERLO ELEMENTARY)
E102	FIRST FLOOR PLAN; DETAILS; SCHEDULES (WESTERLO ELEMENTARY)
E103	SECOND FLOOR PLAN; DETAILS; SCHEDULES (WESTERLO ELEMENTARY)

Figure 39-1 (continued)

MATERIAL KEYING

In light-frame construction of small buildings it is customary for callouts on the drawings to label the material used. On larger projects, however, it is customary for a material keying system to be used so that the drawings themselves are not cluttered by excessive notation. The following, which is reprinted from the cover page of the School Addition drawings, explains the system:

MATERIAL KEYING SYSTEM:

A KEYNOTING SYSTEM IS USED ON THE DRAWINGS FOR MATERIALS REFERENCES AND NOTES. REFER TO THE KEYNOTE LEGEND ON THE DRAWINGS FOR THE INFORMATION WHICH RELATES TO EACH KEYNOTE SYMBOL ON THE RESPECTIVE DRAWING.

EACH KEYNOTE SYMBOL CONSISTS OF A NUMBER FOLLOWED BY A LETTER SUFFIX. THE NUMBER RELATES TO THE SPECIFICATION SECTION WHICH GENERALLY COVERS THE ITEM THAT IS REFERENCED, AND THE LETTER SUFFIX COMBINED WITH THE NUMBER CREATES A KEY-

NOTE SYMBOL WHICH IDENTIFIES THE SPECIFIC REFERENCE NOTATION USED ON THE DRAWING. THE LETTER SUFFIX DOES NOT RELATE TO ANY CORRESPONDING REFERENCE LETTER IN THE SPECIFICATION.

THE ORGANIZATION OF THE KEYNOTING SYSTEM ON THE DRAWINGS, WITH THE KEYNOTE REFERENCE NUMBERS RELATED TO THE SPECIFICATIONS SECTIONS NUMBERING SYSTEM, SHALL NOT CONTROL THE CONTRACTOR IN DIVIDING THE WORK AMONG SUBCONTRACTORS OR IN ESTABLISHING THE EXTENT OF WORK TO BE PERFORMED BY ANY TRADE.

Figure 39–2 illustrates how this keynote system works. Notice that many of the callouts on this detail are a decimal number followed by a letter. That is the keynote symbol. Referring to the Index of Drawings, Figure 39–1, we find that Sheet A000 is Master Keynotes. The complete legend of Master Keynotes, as it appeared on Sheet A000, is reprinted on pages 339 through 346. The keynotes that pertain to a particular drawing sheet are also copied on the right end of the drawing sheet. In Figure 39–2 the keynotes on the right side of the figure pertain to the drawings that were

MATERIAL KEYING LEGEND
(SEE SPECIFICATIONS')

3.30 CONCRETE
SEE STRUCTURAL DRAWINGS*
3.30A CAST-IN-PLACE CONCRETE

4.20 UNIT MASONRY
4.20C CONCRETE MASONRY UNITS

5.10 STRUCTURAL STEEL
SEE STRUCTURAL DRAWINGS*
5.10A STEEL BEAM

5.20 STEEL JOISTS
SEE STRUCTURAL DRAWINGS*

5.50 MISCELLANEOUS METALS

5.50A MISCELLANEOUS STEEL—
SHAPE AND SIZE AS NOTED

5.50B STEEL ANGLE

5.50C 1-¼" NOMINAL STEEL PIPE
(1.66" O.D.)
POST 4'-0" O.C. MAXIMUM

5.50D 10 GAUGE METAL PAN RISERS
& TREADS

5.50E MC 12 × 10.6 STRINGER

5.50F 3-½" × 2-½" × ¼"—LLV—PAN
SUPPORT AT 3'-0" O.C.
MAXIMUM

5.50G 1-½" × 1-½" × ¼" STEEL
ANGLE CARRIERS

5.50U 1-¼" NOMINAL STEEL PIPE

5.50AM (1.66" O.D.)
HANDRAIL/GUARDRAIL

5.50AP 1" NOMINAL STEEL PIPE
HANDRAIL
(1.315" O.D.)

5.50AY STEEL WALL BRACKET,
JULIUS BLUM #662, OR EQUAL

5.50AY STEEL PLATE, SIZE AS NOTE

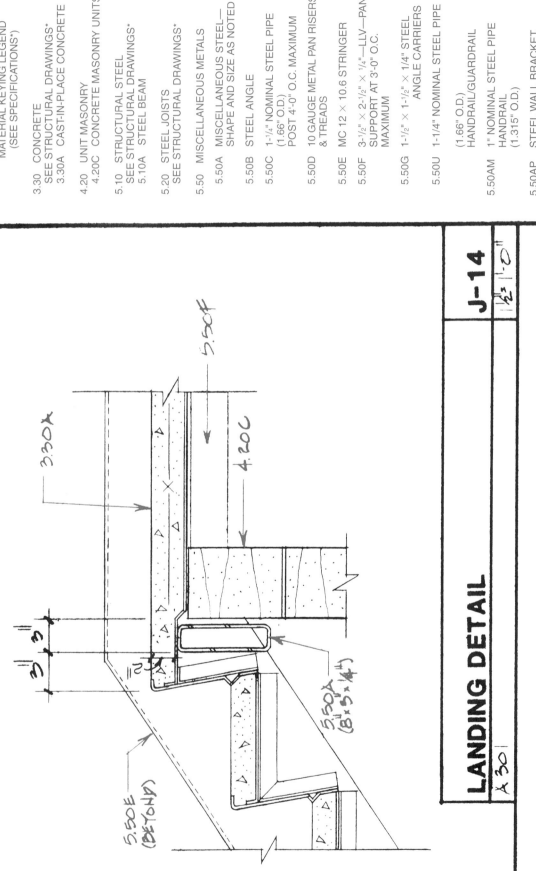

LANDING DETAIL

A 301   J-14   ½" = 1'-0"

Figure 39-2   Material keynote system.

together on Sheet A302, where J–14 was located. The top right callout on J–14 has a keynote symbol, 3.30A. Referring to the Material Keying Legend, 3.30A is cast-in-place concrete. This keynote replaces the customary callout on the drawing that would indicate that the landing and tread surfaces are cast-in-place concrete. More specific information, such as the grade of concrete and type of finish would be indicated in Section 3 of the specifications. Moving counterclockwise around the drawing, the next keynote symbol is 5.50E and is referenced to the stair stringer which is *beyond* the main part of the drawing. According to the Material Keying Legend, 5.50E is covered in Section 5.50 of the specifications, Miscellaneous Metals, and is MC12 × 10.6 stringer. (Unit 38 explained the meaning of this structural steel designation.)

Many of the material keynotes are referenced on several of the drawing sheets. In that case, they are reproduced on each of those sheets, as well as on the Master Legend on Sheet A000. All of the whole sheets with this book include the material keying legends as they were intended by the architects. It was not practical to print material keying legends on every page of the textbook that contains drawings, so it will be necessary to refer to the Master Legend on pages 339 through 346.

REFERENCE SYMBOLS USED ON DRAWINGS

A large building, like the School Addition requires a large number of sections and details to describe all aspects of the structure. The symbols for section views and the symbols for detail drawings are similar in appearance and format, but have one significant difference. The symbol, appearing on a plan or other large-scale drawing to indicate existing detail is commonly indicated as a circle containing the detail number above a horizontal line. Below the horizontal line is the sheet number on which the detail appears. Some element of the symbol, such as a rectangular flag, indicates that it is a detail drawing symbol. Section view symbols use the same scheme for specifying section number and sheet number, but use a pointer instead of a rectangular flag to indicate the direction in which the section is viewed. Figure 39–3 is reprinted from the cover page for the School Addition. In this case, Section B–2, which would be found on Sheet A100, is drawn with a view of the building from the top of the page toward the bottom. (The Index shows that the A series of drawings includes A000 and A101 through A610, but no A100. Apparently, A100 is just a hypothetical sheet for the sake of explaining how symbols are used.) An example from the drawings that are used with this text-

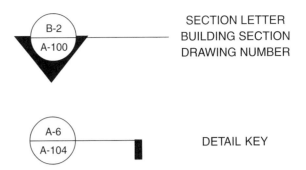

Figure 39-3   Detail and section-view symbols.

book can be seen on the First Floor Plan, Sheet A101. At the extreme right and left ends of the floor plan there is a section view symbol, with its pointers directed toward the top of the drawing (north, according to the symbol in the lower right corner of the drawing). This symbol refers to Section View J–14 on Sheet A201. Section J–14 is in the top right of Sheet A201 and is a full cross-section view of the building as seen by a person standing on the south side of the building and looking north. Looking back at the First Floor Plan shows the detail drawing symbol just above the section view symbol at the left end of the building. The direction of the rectangular flag indicates that it is also drawn as though viewed while facing north. That detail, which is at a larger scale ($^3/_4$" = 1'–0") than the building sections ($^1/_8$" = 1'–0"), is number A–4 on Sheet A502. Detail A–4 include further detail symbols, such as one about half way up the wall referring to E–2 on A504, an expansion joint detail, at 3" = 1'–0".

Interior elevations, which are often used to show the locations of fixtures, built-in furniture, and special architectural details, are usually referenced with their own type of symbols on floor plans. The symbol for an interior elevation also shows the direction from which the elevation is viewed and it includes the number of the elevation and the sheet on which that elevation drawing appears. The symbol in Figure 39–4 references elevation drawing L–10, which would be found on Sheet A103. The direction the triangular shape of the symbol points is the direction in which the elevation is viewed.

INTERIOR ELEVATION
L-10/A103

Figure 39-4   Elevation reference symbol.

STRUCTURAL GRID COORDINATE

Buildings with columns of reinforced concrete or structural steel as a main element of the building frame are laid out on a structural grid. The plans are drawn with a system of horizontal and vertical reference lines, so that all columns and details can be referenced to this grid. Vertical lines are numbered and horizontal lines are lettered or the system is reversed, with horizontal lines numbered and vertical lines lettered, Figure 39–5. Columns are located by grid lines through their centers. The locations of major architectural components, such as walls, are located by dimensions that are referenced to the grid lines. The dimensions from the grid lines to the faces of exterior walls are usually given at the corners of the building. Interior partitions and other architectural elements may be located by dimensioning to the faces of these exterior walls, which are referenced to the structural grid.

MENTAL WALK THROUGH

As with each of the buildings discussed earlier in this textbook, one of the first steps you should take to familiarize yourself with the School Addition is a *mental walk through*. Start at the front entrance to the first floor, in the lower right corner of the First Floor Plan. Enter door number 101 and step into a stairwell marked stair #1. At the far end of the stairwell is door #102, which opens into a corridor. If you turn right in the corridor, you can walk up a ramp into the existing building, ten inches higher than the first floor of the addition. Across the main corridor of the addition are the doors into two toilet rooms. A symbol in front of those doors indicates that Elevation A–8 on A610 will show us what the toilet room doors look like. As we continue our walk through, we will come to many more elevation symbols. Look at each of the elevations on your walk through to help you form a mental picture of the addition. You will

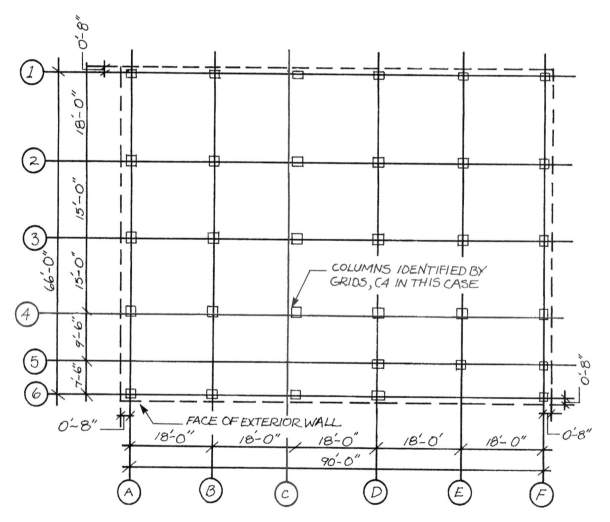

Figure 39-5   Structural grid coordinate system.

also see several detail and section symbols. It will not be necessary to check each of these to form a mental picture of the addition, but you should look up a couple of each, just to make sure you are comfortable locating them.

Moving to the west in the main corridor, we come to a door (103) on our left. This is the entrance to a storage room. A dimension inside the storage room indicates that it is 13' – 8½" wide. The exterior wall (south wall) of the storage room is on grid line F (see the left end of the floor plan). The wall with the door is on grid line D, which is 32' – 7½" from grid line F. We could determine the actual face-to-face dimension between the north and south walls by looking at the details and sections, but that is not necessary for this mental picture of the building. Notice that, where the east and west walls of the storage room intersect with the wall of the corridor, there are two heavy dashed crosses, indicating columns. Most walls are on column lines. These are on vertical grid lines 6 and 7.

Walk through the janitor closet (JAN. CL.), each of the classrooms, and all corridors and stairwells on your mental walk through. Only a few highlights are described in this textbook, but after you have followed the walk through described here, you should take a second walk through, noticing and studying everything until you can visualize it.

Along both sides of the corridor are several references to Material Keynote 10.50A. Find that keynote on the legend at the right end of the drawing sheet or on the Master Keynote Legend.

Through door 105 we enter classroom 103. On the floor plan, room 103 is only separated from room 104 by a single dashed line. That warrants investigation. A detail symbol directs us to A–6 on sheet A504. That is a large-scale section view of a track detail. Looking again at the floor plan, we see a material keynote reference near the lower end of the dashed line: 10.62A. According to the legend, 10.62A is a folding partition, so rooms 103 and 104 are separated by a folding partition.

Across the corridor from classroom 103 is stair #2, which goes to the second floor. At the extreme left end of the corridor is a large rectangular area identified by keynote 14.20A, an elevator. Take the elevator to the second floor.

When we step off the elevator we are on the second floor. Our walk through is continued on Sheet A102, Second Floor Plan. The second floor of the addition is almost identical to the first floor. It would be foolhardy to take it for granted that the two levels are identical, but a quick survey shows that they have the same rooms in approximately the same layout. The rooms on the second floor are numbered in the 200 series to indicate that they are on the second floor.

✓ CHECK YOUR PROGRESS

Can you perform these tasks?

- ☐ Find and understand the information given in the index to drawings and the materials specifications on drawings.

- ☐ Cross reference between sections, details, and other drawing sheets on the basis of the numbering system on the drawings.

- ☐ Locate columns according to their grid coordinates.

- ☐ Find your way around plans and elevations well enough to form a mental picture of the building.

ASSIGNMENT

Refer to the drawings of the School Addition (included in the packet) to complete this assignment.

1. List the sheet numbers for all of the structural drawings for the addition.

2. On what sheet would you look to find wall details?

3. On what sheet would you find a framing plan for the roof?

4. What is indicated by Keynote 10.80C?

5. What keynote number is used to indicate plywood?

6. On the First Floor Plan, at about the mid point of the north wall, a circle is drawn over a triangular flag. The notation inside the circle is J–6/A201. Explain what this symbol indicates and where further information can be found.

7. According to the building elevations and several wall details, there is a precast concrete sill or band which is about waist high on the exterior walls. What is the dimension from the top of the foundation to the bottom of this sill or band?

8. What is the height of the concrete block wall on the north wall of the first floor corridor?

9. What is the material on the lower surface of the canopy over the entrance at door 101?

10. What are the dimensions of the liquid marker board and tackboards on the west wall of classroom 107?

11. What are the dimensions of the mirrors in the toilet rooms?

12. Describe the location of column D3, relative to the floor plans.

13. What is the center-to-center spacing between columns F7 and F9?

14. What are the overall outside dimensions of the precast concrete band, such as is used near the base of typical exterior walls?

15. There is a door from a classroom (112) in the existing building into the storage room in the northeast corner of the addition. The door is made of what material?

16. Describe the differences between the rooms adjacent to the elevator on the first floor and on the second floor.

17. Where do you find the reflected ceiling plan?

18. Is the electrical service entrance section inside the existing building or outside the existing building?

19. How many light switches with motion sensors are there? Where are they located?

20. Which lighting fixtures are scheduled to have electronic ballasts?

UNIT 40 Structural Drawings

OBJECTIVES

After completing this unit, you will be able to perform the following tasks:

- Describe the footings for columns and walls, including dimensions and reinforcement.

- Interpret the information found on a foundation plan, including dimensions of foundation walls, reinforcement of foundations, and the locations of the various elements.

- Describe each column, beam, and lintel shown on the structural drawings.

- Interpret structural details and sections.

FOUNDATIONS FOR COMMERCIAL BUILDINGS

The foundations for large, commercial buildings perform the same functions as those for light-frame buildings. The foundation supports the loads (weight) imposed on the superstructure and spreads those loads over a large enough area so that the earth can support it uniformly. In most commercial construction the foundation system is comprised of spread footings and stem walls and pads which act as footings under columns. These are the same elements that are found in most one- and two-family homes. The biggest difference in the foundation for a commercial building and that for a small house is the thickness of the concrete and the amount of reinforcing steel. There are apt to be more detail drawings for the foundation of a larger building, because of the need to describe the different sizes and shapes of footings and the reinforcement at many points in the foundation. The same knowledge is required to read the foundation plan for a large building as for a smaller one.

The drawings consist of a plan that shows the layout of the foundation and the major dimensions and detail drawings that describe reinforcement, expansion joints, and variations in design at special locations. The placement of column footings is indicated by referencing the structural grid coordinates. The structural grid indicates where the centerline of the column will be, but it is very important that the footing be placed accurately on this center. If the column does not rest on the center of the footing, the footing may tip due to the uneven pressure on the soil below, Figure 40–1.

Figure 40-1 The column load must be centered on the footing. (Willis: *Blueprint Reading for Commercial Construction,* by Delmar Publishers Inc.)

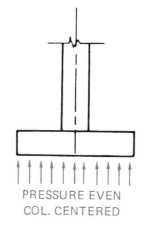

PRESSURE EVEN
COL. CENTERED

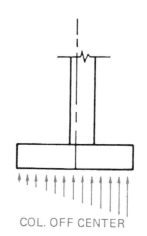

COL. OFF CENTER

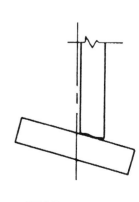

FOOTING TIPPED

STRUCTURAL STEEL FRAMING

A structural steel building frame consists of *columns* (vertical members) and *beams* (horizontal members.) The largest beams, called *girders*, attach to the columns, Figure 40–2. *Joists* are intermediate beams and are supported at their ends by the girders. *Lintels* are the beams that support the weight above an opening, such as a door, window, or nonstructural panel. If the first floor is a concrete slab, it is described on the foundation plans and details. The second and higher floors are described on framing plans, as is the roof frame. Many buildings have several floors that are framed alike. In this case a note might indicate that a framing plan is typical.

Framing plans for structural steel are drawn on the structural grid coordinate system, Figure 40–3. Beams are shown by a single-line symbol with an accompanying note to indicate the structural shape to be used. The lines indicating beams stop short of the symbol for the girder or column when the beam is framed into the supporting member and does not continue over it. The abbreviation *do*, which stands for ditto, indicates that the specification for the first member in a series is to be repeated for all members. A number in parentheses at the end of the designation is used to indicate the elevation of the top of the member. This may be the elevation or the distance above or below the floor line, Figure 40–4.

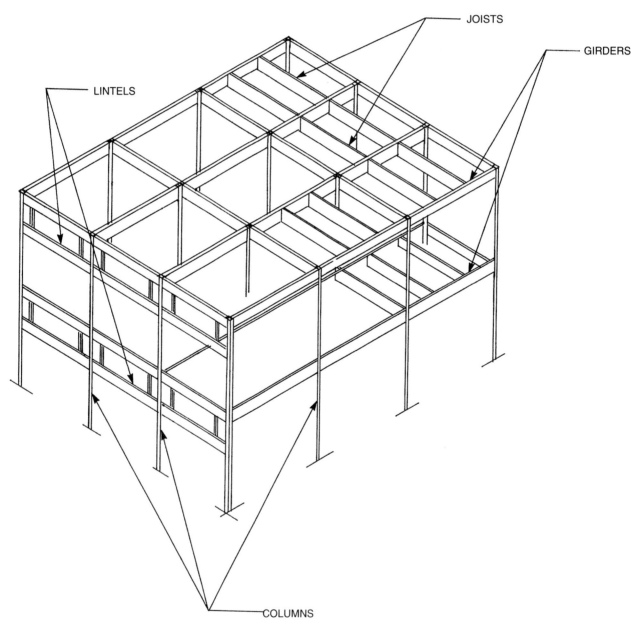

Figure 40-2 Major parts of a building frame.

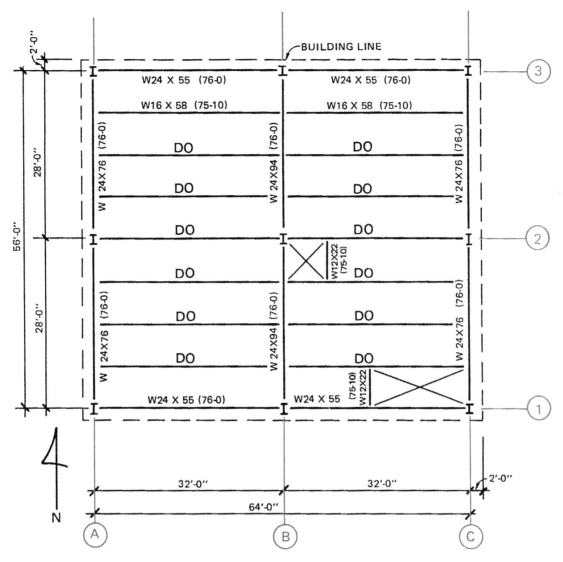

Figure 40-3   Structural coordinates are shown on framing plans and floor plans.

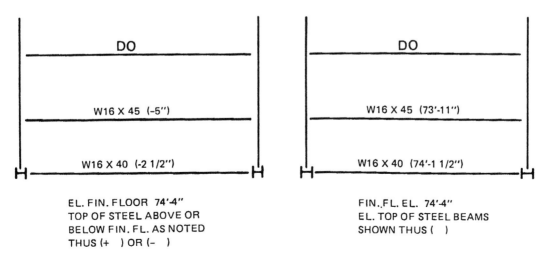

EL. FIN. FLOOR 74'-4"
TOP OF STEEL ABOVE OR
BELOW FIN. FL. AS NOTED
THUS (+   ) OR (–   )

FIN. FL. EL. 74'-4"
EL. TOP OF STEEL BEAMS
SHOWN THUS (   )

Figure 40-4   Notations on the drawings indicate the relative elevations of beams.

Joists are frequently *open-web steel joists*, sometimes called *bar joists*, Figure 40–5. Bar joists are manufactured in H, J, and K series, depending on the grade of steel used and their strength requirements. K series are stronger than H series and H series are stronger than J series. A bar joist designation includes a number to indicate depth, a letter to indicate strength series, and a number to indicate chord size. For example, the designation 16H6 indicates a 16 inch deep H series joist with number 6 chords.

The actual lengths of members are not shown on the general contract drawings. This information is shown

Figure 40-5   Open web steel joist. (Willis: *Blueprint Reading for Commercial Construction,* by Delmar Publishers Inc.)

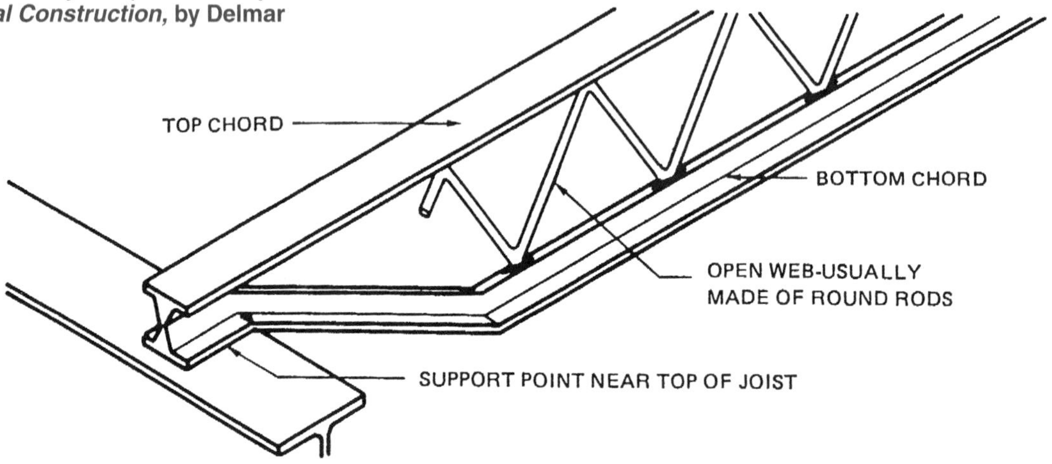

TOP CHORD

BOTTOM CHORD

OPEN WEB-USUALLY MADE OF ROUND RODS

SUPPORT POINT NEAR TOP OF JOIST

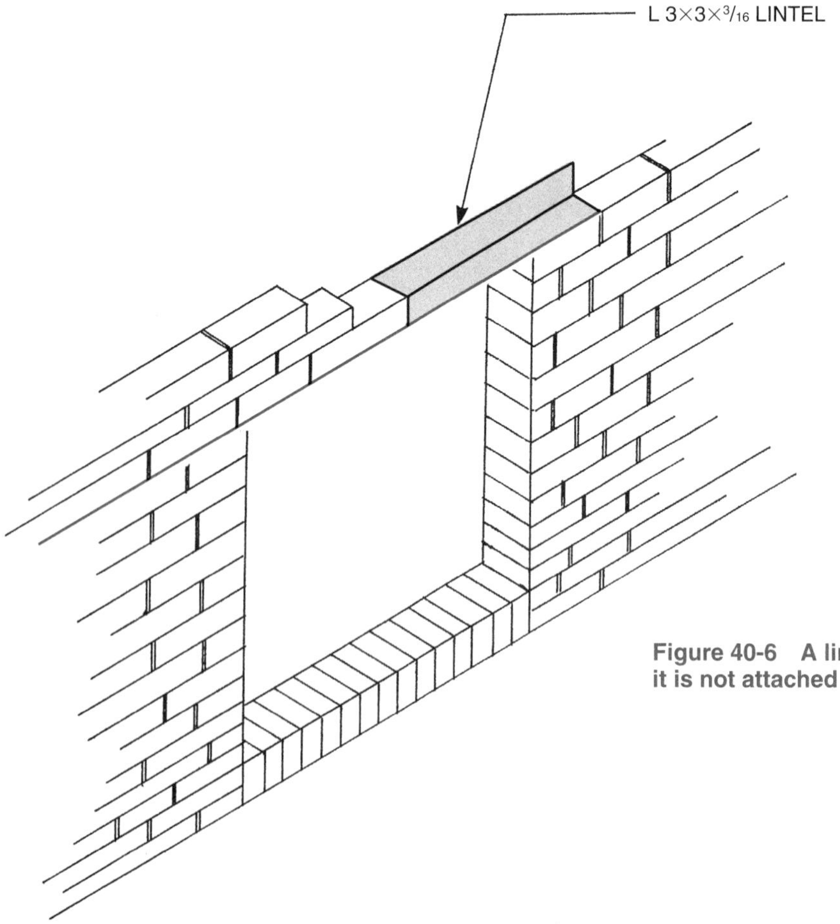

L 3×3×³/₁₆ LINTEL

Figure 40-6   A lintel may be considered loose steel if it is not attached to the frame.

on shop drawings that are drawn by the steel fabricator sometime after the construction drawings are completed. It is an easy matter, however, to find the span of the member by looking at the framing plan. Connections are shown on details and sections.

Lintels are sometimes categorized as *loose steel.* In masonry construction, a steel lintel is placed above each opening to support the weight of the masonry above the opening, Figure 40–6. These lintels are not attached to the steel building frame, so they are called loose steel. If there are several lintels alike, they are normally shown on the plans by a symbol, then described more fully in a lintel schedule, Figure 40–7.

MASONRY REINFORCEMENT

Masonry materials have great compressive strength. That is, they resist crushing quite well, but they do not have good tensile strength. The mortar in masonry joints is especially poor at resisting the forces that tend to pull it apart, such as a force against the side of a wall or a tendency for the wall to topple. Masonry joint reinforcement is done by embedding specially made welded-wire reinforcement in the joints, Figure 40–8. Greater strength can be achieved by building the masonry wall with reinforcement bars in the cores of the masonry units, then filling those cores with concrete. Concrete used for this purpose is called *grout.* Reinforcing steel is also embedded in masonry walls to tie structural elements together. For example, it is quite common for rebars to protrude out of the foundation and into the exterior masonry walls. The strength of a masonry wall can be increased considerably by the use of bond beams, Figure 40–9. A *bond beam* is made by placing a course of U-shaped masonry units at the top of the wall. Reinforcing steel is placed in the channel formed by the U shape, then the channel is filled with grout. The result is a reinforced beam at the top of the wall.

LINTEL SCHEDULE				
MARK	MATERIAL	TYPE	MAS. OPNG	REMARKS
L1	WT8x13	⊥	SEE ARCH.	ATTACH ONE END TO COL. SEE 1/S200
L2	WT8x13	⊥	"	END BEARING BOTH ENDS
L3	WT8x13	⊥	"	CONTINUOUS W/CENTER SUPPORT ON T.S. 4x4
L4	WT4x9	⊥	"	AT ALL BELOW WINDOW UNIT VENTILATOR OPENINGS
L5	(2)∠6x4x5/16 W/ 1/4"x9½ CONT. BOT.℔	⊥⊥	"	LINTEL IN FIRE WALL WELD BOT.℔ TO ANGLES.
L6	(2)∠6x4x5/16	⊥L	"	IN EXIST. BLDG. WALL
L7	(3)∠4x3½x5/16	⊥⊥L	"	AT CAFETERIA HVAC UNIT WALL OPENINGS COORDINATE WITH G-7/H3 AND A103(2-LOCATIONS)

Figure 40-7   Lintel schedule.

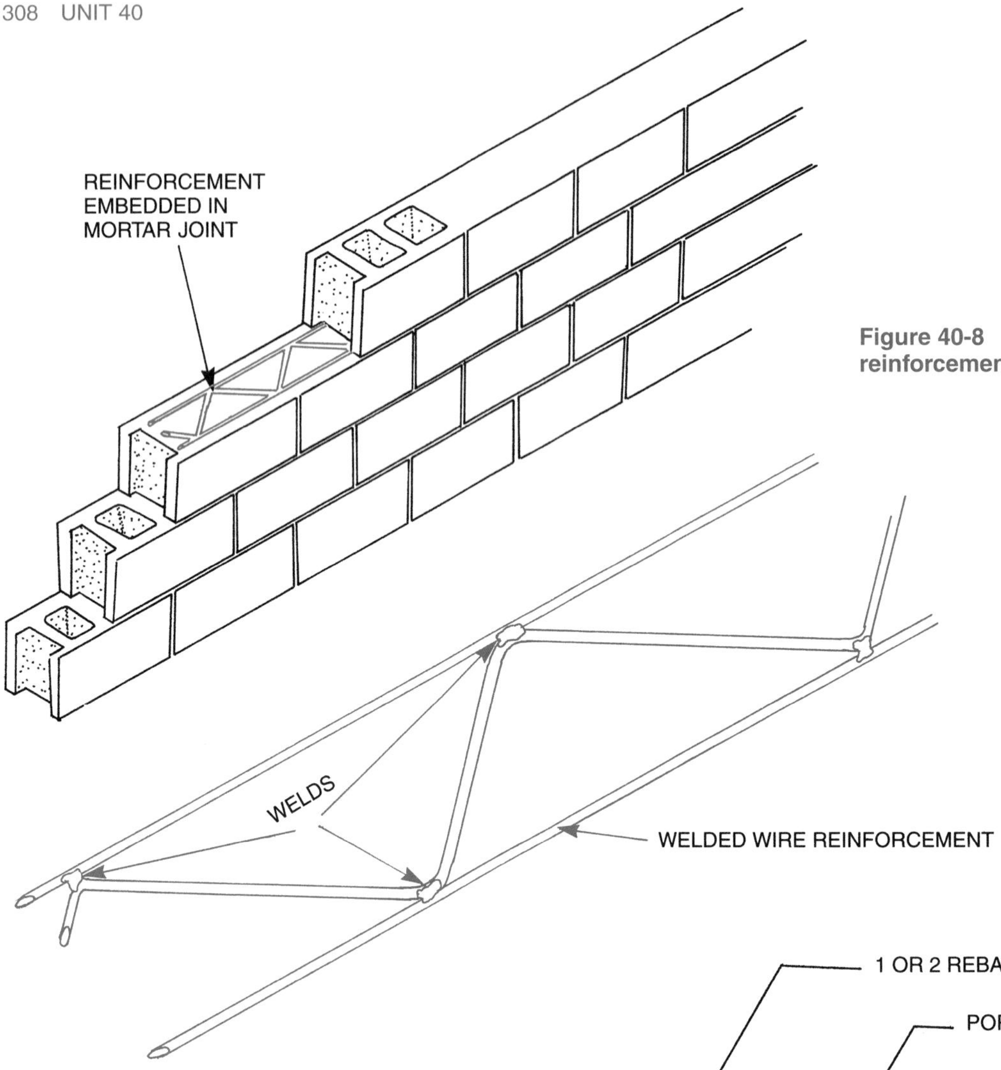

REINFORCEMENT
EMBEDDED IN
MORTAR JOINT

Figure 40-8   Masonry joint reinforcement.

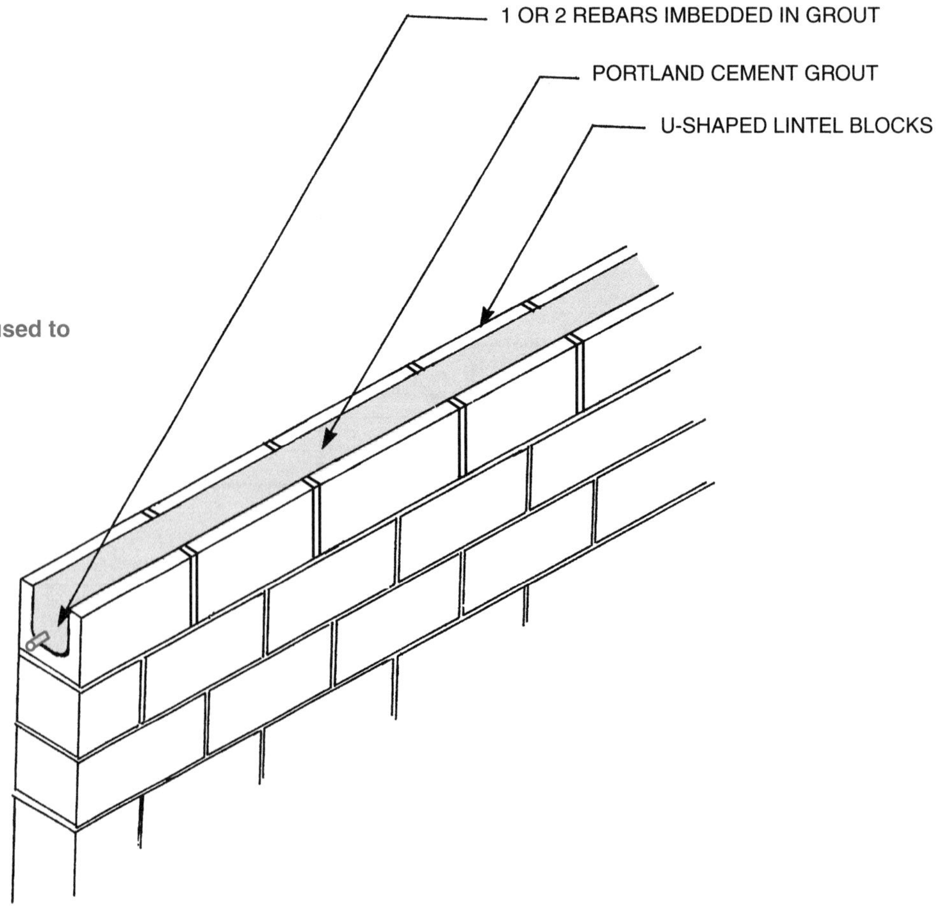

WELDS

WELDED WIRE REINFORCEMENT

1 OR 2 REBARS IMBEDDED IN GROUT

PORTLAND CEMENT GROUT

U-SHAPED LINTEL BLOCKS

Figure 40-9   A bond beam is used to strengthen the top of the wall.

✓ CHECK YOUR PROGRESS ────────────────────────────────

Can you perform these tasks?

☐ Using the information on drawings, describe the footings for a building.

☐ Using the information on drawings, give complete information about the foundation for a building.

☐ Describe the columns, beams, and lintels shown on drawings.

ASSIGNMENT ───

Refer to the Drawings of the School Addition (included in the packet) to complete this assignment.

1.   What are the dimensions of the footing for the northwestern most column?

2.   What is the width and depth of the footing at the west end of the addition?

3.   What is the elevation of the top of the floor in the elevator pit?

4.   How many lineal feet of #5 reinforcement bars are needed for the footing under the east wall of the addition?

5.   What size and kind of material is used to prevent the foundation wall from moving on the footings? How much of this material is needed for the west end of the addition?

6.   How many pieces of what size reinforcing steel are to be used in the footing for the column between the entrances to classrooms 103 and 104?

7.   How closely is the vertical reinforcement spaced in a typical section of the foundation wall?

8.   What is the overall length of each piece of rebar used to secure the interior masonry partitions to the concrete slab?

9.   What is the spacing of the dowels used to secure the exterior masonry walls to the foundation?

10. Describe how the corridor walls are secured to the columns.

11. How many pieces of W18 x 50 steel are used in the construction of the addition?

12. What size and shape structural steel supports the north ends of the floor joists under classroom 203?

13. What size and shape structural steel supports the ends of the girder in the second floor at the front (side with the door) of the elevator shaft?

14. What is the nominal depth of the joist in the second floor corridor?

15. What is the elevation of the top of the second floor girder between D5 and D6, relative to the second finish floor elevation of 110'-3"?

16. What is the shape and size of the member that supports the masonry above the windows in classroom 108?

UNIT 41   Mechanical Drawings

After completing this unit, you will be able to perform the following tasks:

- Identify and briefly describe the major pieces of HVAC equipment to be used in a building.

- List the sizes of pipes and fittings shown on mechanical plans and explain the major functions of each.

- Describe the sizes and shapes of air-handling ducts shown on mechanical drawings.

HEATING VENTILATING AND AIR CONDITIONING PLANS

The heating, ventilating, and air conditioning system for a building larger than a single-family house is designed by an engineer who specializes in this work. The drawings for this work are usually drawn on a basic outline of the floor plan for the building. This allows the mechanical drawings to show where HVAC and plumbing equipment is in relation to rooms, walls, floors, and so on. The coordination of structural, mechanical, and electrical work is often an issue of some concern on construction projects, so it is important to understand where and how pipes will pass between beams or through walls. For the sake of this book, we are limiting the mechanical drawings to those for HVAC and plumbing and we will look at the HVAC drawings first.

Most commercial buildings are heated by burning fuel (natural gas or fuel oil) and distributing the resulting heat in the form of warm air, hot water, or steam. Warm-air heating systems require ducts throughout the building to carry the warm air to the occupied spaces and to return the cool air to the source. A warm-air system has the advantage of allowing for inexpensive air conditioning. The same ducts can be used to distribute either warm air in winter or cool air in summer. The ducts, however, require much more space in ceilings and walls and are harder to route through and between structural elements than are pipes. When hot water or steam are used to transport the heat to the occupied space, two pipes are required: one to supply the hot water or steam and one to return the cooler medium to its source. The basic type of system is readily apparent to anyone looking at the mechanical plans. The plans for an air handling system include ductwork. The plans for a hydronic system include more pipes and fittings.

UNIT VENTILATORS

At first look, the heating plans for the School Addition might appear foreign and difficult to understand, but like most of the drawings you have studied before this point, these drawings are easy to understand when taken one step at a time. Notice that each of the rooms in the addition has a rectangle on the outside wall with a callout of UV–1, UV–2, or something similar. These indicate that this building is heated by unit ventilators. Figure 41–1 is the Schedule of Unit Ventilators. (In the

actual drawing package it was #H–7 on Sheet H–4.) A *unit ventilator* is a unit that combines a means of mixing room air and outside air to ventilate the room while also warming it. The warmth is provided by the hot water that is piped in from the boiler. Figure 41–2 explains

MARK	MAKE	MODEL	FAN				V/PH/HZ	HEATING				
								HOT WATER				
			TOT CFM	MIN O.A.	RPM	HP		EAT	EWT	GPM	MBH	△P
UV–1	AAF	AV–4000	1000	270	650	1/8	120/1/60	50	180	2.0	36	0.99
UV–2	AAF	AV–5000	1250	270	650	1/8	120/1/60	50	180	2.0	42	1.27
UV–3	AAF	AV–3000	750	250	650	1/8	120/1/60	40	180	2.0	28	0.79
UV–4	AAF	AH–3000	750	270	650	1/8	120/1/60	40	180	2.0	36	0.79
UV–5	AAF	AH–3000	750	270		1/2	120/1/60	40	180	2.0	36	0.79

UNIT VENTILATOR SCHEDULE **H-7**

TOT CFM total cubic feet per minute through the unit
MIN O.A. minimum outside air flow into the unit
RPM revolutions per minute of the fan
HP fan motor horsepower
V/PH/HZ fan motor volts, phase rating (1 or 3), and Hertz (frequency or cycles)
EAT entering air temperature

EWT entering water temperature
GPH gallons per minute of water through the unit
MBH thousands of BTUs per hour (a measure of heat output)
△P delta pressure (drop in water pressure through the system

Figure 41-1   Unit ventilator schedule.

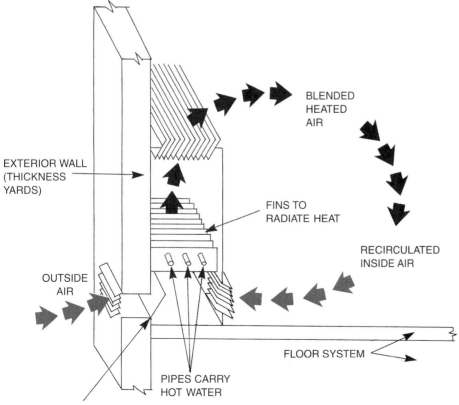

Figure 41-2   Operation of a unit ventilator.

the principles of a unit ventilator. The heating controls include a function that controls the balance of recirculated inside air and fresh outside air. During the day, when the building is in use, the damper is opened to allow more fresh air to be circulated through the room. At night, the damper is closed, so that less cool air enters the room, resulting in an energy savings. The amount of air that is drawn in from the outside when the dampers are closed as much as possible is indicated on the drawings and on the schedule as C.F.M. (cubic feet per minute). Other information on the schedule describes such things as the total amount of air that passes through the unit, the water temperature intended by the designer, the speed of the fan, and so on.

This school does not include air conditioning. If it did, the unit ventilators would be supplied with cold water when cooling is called for. The same system of balancing inside and outside air would be used for energy conservation and fresh air, but instead of passing the air over heated pipes it would be passed over cold pipes. The Unit Ventilator Schedule would include information for air conditioning.

The unit ventilators are only a few feet long, yet the plans show a continuous string of equipment of some sort along the exterior walls of the classrooms. Although they are not labelled anywhere on the plans, the Construction Notes refer to bookcases. The manufacturers of unit ventilators also manufacture bookcases and filler pieces that are designed to fit with the ventilators and fill the rest of the wall space. These bookcases have a false back, with space for pipes to and from the ventilators.

In some spaces, such as stairways and storage closets, the heaters are not labeled as UV (unit ventilators.) These are cabinet unit heaters (CUH) that mount on the surface of the wall and do not have ventilation capability and fan coil units (FCU) that are intended to provide more heat than ventilation. Figure 41–3 shows the schedules for the units marked as CUH, UH, and FCU.

HEATING PIPING

Once the basic type of heating system has been determined, the next step is to understand the piping that carries water to and from the heating units. It is easy to spot the pipes on the heating plan. A solid line indicates supply pipes and a dashed line indicates return pipes. Pipe sizes are indicated by callouts throughout the plans. Fittings, such as valves, elbows, and tees are represented by symbols. There are standard piping symbols, but each drafter uses a few special representations of his own, so a symbol legend is included with the drawings.

Pipes are easier than ducts to coordinate with structural elements because the direction of the pipe can be changed easily and frequently. Pipes might run horizontally in a ceiling for some distance, then loop around a column and drop down to a unit ventilator. It is not practical to show the layout of all pipes with separate plan and riser views. The common method for showing pipes where both horizontal and vertical layout is involved is to use isometric schematics. Plumb-

MARK	MAKE	MODEL	STYLE & ARRANGEMENT	STEAM		WATER				CFM	RPM	EAT	MBH	ELECTRICAL	
				PSIG	IB/HR	EWT	Δ T	GPM	Δ P"					V/ø/HZ	HP
CUH–1	STERLING	RWI–1130–04	RECESSED WALL, INVERTER			180	34	2.0	0.44	420	1050	60	33.8	120/1/60	1/10

CABINET UNIT HEATER SCHEDULE

MARK	MAKE	MODEL	TYPE	WATER			MBH	CFM	EAT	LAT	MOTOR		
				ENT.	GFM	Δ P					RPM	H.P.	V/ø/HZ
UH–1	MODINE	HS–18L	HORIZONTAL	180	1.1	0.4	9.4	364	60	84	1550	16MHP	120/1/60

UNIT HEATER SCHEDULE

MARK	MAKE	MODEL	CFM & TOT./F.A.	EXT. S.P.	ELECTRICAL		HEATING				
					WATTS	VOLTS/PH/HZ	EAT	EWT	GPM	MBH	Δ P
FCU–1	AAF	SFG–FFA–3000	350	0	110	120/1/60	50	180	1.0	18	0.1

FAN COIL UNIT SCHEDULE

Figure 41-3   Schedules of cabinet unit heaters, unit heaters, and fan cabinet unit heaters.

ing isometrics were discussed in Unit 39. Sheet H–1 of the School Addition drawings combine plan views and isometric schematics. The long runs of pipes are shown in plan view to indicate where the pipes are in relation to the walls. Where the pipes need to drop down from the ceiling to the level of the ventilators, they are shown in isometric views, Figure 41–4.

For the sake of explanation, the following discussion traces the hot water supply pipes, starting on the left, where the supply comes from. The first fitting is a tee, where a line branches toward the bottom of the drawing, then an elbow toward the right or outside wall. A circle at the end of the line is marked with Construction Note 2, which reads, "Risers to unit on second floor. Offset for clarity." This indicates that the pipe rises up to the second floor unit, but it is not positioned on the drawing where the actual pipe would be in the building. If the riser symbol (small circle) was drawn in the wall, where the pipe should be, the symbol would be lost in all the other lines of the drawing. Continuing on the main line, just beyond the tee, is an elbow where the pipe turns down toward the level of the first-floor unit. This is the isometric portion of the schematic. At the level of the unit ventilators, the pipes are arranged

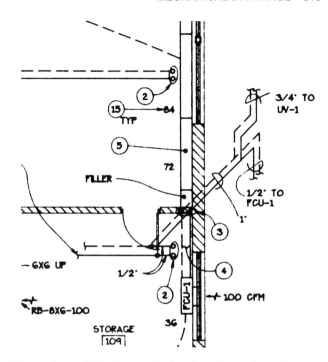

Figure 41-4 This isometric drawing shows how the heating pipes drop down from the ceiling to the level of the ventilators.

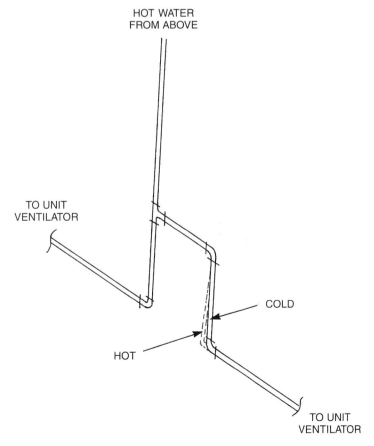

Figure 41-5 Noises are eliminated by allowing pipes to expand and contract.

in a special way to eliminate noises that might result from expansion and contraction as the water temperature changes, Figure 41–5. The ends of the pipes are shown with break lines, which indicate that the pipes actually continue, to their obvious destinations: fittings on the unit ventilators.

AIR HANDLING EQUIPMENT

The unit ventilators introduce fresh air into the school addition. A separate ventilation system removes stale air from each room of the school. All of the rooms in the addition except the smallest closets and the stairways are provided with a system of ducts and a fan to ensure air circulation. A *louvered grille* in the room receives the air and channels it into a duct which carries it to the roof, where a fan exhausts it to the outside.

Figure 41–6 explains the designations for outlet and inlet grilles. The ducts that carry the air to the roof vent are shown by their outlines and do not indicate their sizes. A square or rectangle with a diagonal line through it indicates where the duct rises to the next floor or the roof. Several notes indicate that ducts rise up to EF-# on roof. EF is an abbreviation for exhaust fan. Some of the exhaust fans occupy several square feet of space on the roof, so they cannot be placed too close together.

Trace the flow of air as it is exhausted from the janitor's closet on the first floor. The path starts with a louver marked RB (return) – 8 X 6 (cross sectional size in inches) – 75 (cubic feet per minute nominal air flow). Between the janitor's closet and the storage room, the air enters a 6 X 6 duct which goes up through a chase (space between two walls) to the second floor. At the second floor it is joined by the exhaust air from the second floor janitor's closet. On the Second Floor Plan there is also a symbol indicating a fire damper "W/ Access Door At Floor." This is a damper (metal gate) that closes off the duct if it gets hot in a fire, Figure 41–7. This is necessary, because state codes require fire-rated construction and an open duct is a great place for fire to spread. The access door at the floor allows for maintenance and repairs to the damper. Notes at both levels indicate that the duct between floors is 6 X 6. Above the second floor louver, the duct is 8 X 8 and at the second floor ceiling it is offset an unspecified distance into the storage room, where it turns up to EF–4 on the roof. The reason for the offset is because it would

otherwise be to close to the other duct such that there would not be enough room for both exhaust fans on the roof.

PLUMBING

The plumbing for the School Addition consists of DWV (drainage, waste, and vent) plumbing to remove effluent, hot and cold water supply, and fixtures. The DWV system is a little more complex for a building this size than for a house because the large flat roof collects quite a bit of water in a rainstorm and there must be a way to get rid of it. Therefore, the Second Floor Plumbing Plan shows roof drains, marked R.D. and storm drain piping, indicated by a heavy line. Callouts along the storm drain indicate the size of the pipe, its pitch, and give information about how the storm drain is to be coordinated with structural components of the building. On the First Floor Plumbing Plan the storm drain is represented by a heavy dashed line, indicating that it is below grade.

Notice that the *storm drain* and the *sanitary drain* are separate systems. The storm drain only needs to carry rain water away from the building. The sanitary drain must connect to a sewer or septic system. Although the storm drain and the sanitary drain lines cross near the janitor's closet on the first floor, they are not connected.

In the region where this building was constructed, drainage pipes outside the building lines are installed by the site contractor, not the plumbing contractor. Therefore, the plumbing drawings only show the storm drain and sanitary drain to a point 5'-0" outside the building lines. A note where each line passes through the building wall indicates the *invert elevation* (elevation of the bottom of the inside of the pipe).

The hot and cold water piping is similar to that found in homes, except that the hot water is a two-pipe system. A return hot water pipe allows the hot water to circulate continuously, ensuring that hot water is readily available at the point of use without having to empty all of the cool water from the supply pipe first. Hot water supply and hot water return can be distinguished from each other by the number of short dashes in the symbol. These are explained on the Symbol Legend. The hot water return, which is usually smaller than the supply, maintains circulation to a point close to where most of the hot water will be used. It is often not practical to continue a hot-water return pipe to every fixture in an area with several fixtures like the school addition toilet rooms. Figure 41–8 is riser diagrams for the fixture connections in the School Addition. Notice that there is no hot-water return shown on these diagrams, because the return is connected to the supply before the supply reaches the first fixture.

Figure 41-6   Explanation of grille designations.

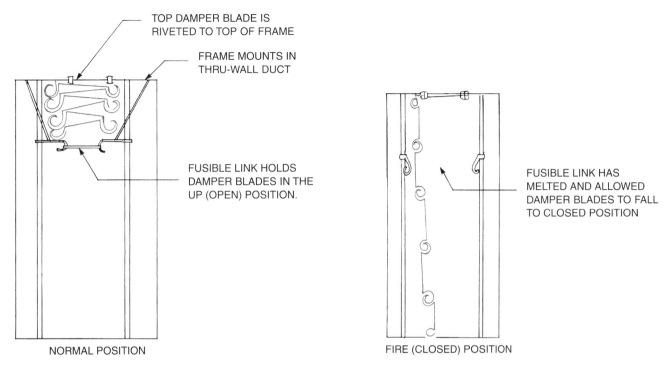

TOP DAMPER BLADE IS
RIVETED TO TOP OF FRAME

FRAME MOUNTS IN
THRU-WALL DUCT

FUSIBLE LINK HOLDS
DAMPER BLADES IN THE
UP (OPEN) POSITION.

NORMAL POSITION

FUSIBLE LINK HAS
MELTED AND ALLOWED
DAMPER BLADES TO FALL
TO CLOSED POSITION

FIRE (CLOSED) POSITION

Figure 41-7   Fire damper for ductwork.

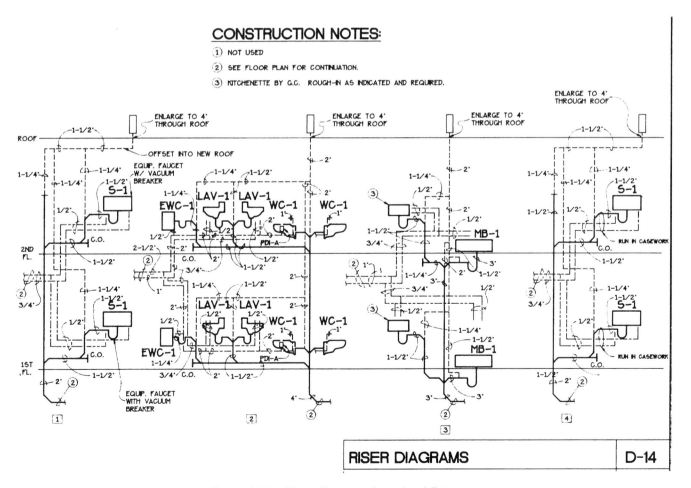

CONSTRUCTION NOTES:

1 NOT USED

2 SEE FLOOR PLAN FOR CONTINUATION.

3 KITCHENETTE BY G.C. ROUGH-IN AS INDICATED AND REQUIRED.

RISER DIAGRAMS   D-14

Figure 41-8   Riser diagrams for school fixtures.

✓ CHECK YOUR PROGRESS

Can you perform these tasks?

☐ Locate information on schedules of equipment and relate that information to the plans.

☐ Understand the symbols and abbreviations used for air conditioning, heating, and plumbing drawings.

☐ Explain the sizes and shapes of ducts to be used for air handling and ventilation.

☐ Trace the path of piping and identify the sizes of pipes and fittings used.

ASSIGNMENT

Refer to the drawings of the School Addition (included in the packet) to complete this assignment.

1. How many unit ventilators are shown on the first floor?

2. How is the machinery room on the first floor behind the elevator heated?

3. List in order the sizes of pipe that the water that flows through from the point where it enters the addition to get to the heating unit in classroom 210 and then back to the point where it leaves the addition.

4. What are the dimensions of the grille where stale air is vented out of classroom 209?

5. List in order the sizes of the ductwork that exhaust air passes through from the time it leaves classroom 209 until it is outside the building.

6. List in order the sizes of the ductwork that exhaust air passes through from the time it leaves the girls toilet room on the first floor until it exits the building.

7. How many gallons per minute of water are expected to flow through the heating unit in stair 2?

8. Why are two roof drains shown on the First Floor Plumbing Plan of this two-story building?

9. What is designated as EWC–1 on the plumbing plans?

10. What is the vertical distance between the storm drain and the sanitary drain at the point where they cross?

11. What is the diameter of the pipe that is used for the storm drain where it goes from the second floor to the first floor?

12. What size pipe is used for the domestic hot water return line?

13. Where does the domestic hot water return line tee out of the domestic hot water supply line?

14. Where would you shut off the water cold water supply to the kitchenette unit in room 101 without shutting off the cold water to the toilet rooms on the same floor?

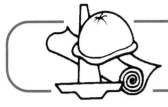

UNIT 42   Electrical Drawings

OBJECTIVES

After completing this unit, you will be able to perform the following tasks:

- Explain the information found on a lighting plan.
- List the equipment served by an individual branch circuit using electrical plans, riser drawings, and schedules.
- Explain the information on a schematic diagram.

DESIGN

Architects typically hire consultants to design the electrical services in buildings. The electrical consultant prepares the electrical drawings. These drawings are usually identified by an E prefix on the sheet number.

The electrical installer should carefully investigate all of the contract drawings, especially the architectural drawings. They should specifically verify exact locations of electrical equipment including switches, receptacles, panels, feeders, etc. It is the electrical installer's responsibility to understand structural drawings where electrical equipment such as large conduit, switchgear, transformers, etc., is to be mounted from structural members. This type of installation is most common in commercial and industrial projects. The electrical installer must be able to read and understand the drawings for other construction installers (plumbing, fire sprinkler, HVAC, etc.) so that potential space and clearance conflicts can be resolved before conduits and electrical are installed.

ELECTRICAL DRAWINGS

Electrical drawing deals predominantly with circuits. Knowledge and understanding of the basic circuits used in electric building construction is necessary for those working with and interpreting electrical drawings. The four basic methods of showing electrical circuits are:

1. Plan views
2. One-line diagrams
3. Riser diagrams
4. Schematic diagrams

PLAN VIEWS

The electrical floor plan of a building shows all the exterior walls, interior partitions, windows, doors, stairs, cabinets, etc. along with the location of the electrical items and their circuitry. Unit 37 covered most of the construction related electrical basics.

Power Circuits

The *power circuit* electrical floor plan shows electrical outlets and devices and includes duplex outlets, specialty outlets, telephone, fire alarm, etc. The conventional method of showing power circuits is to use long dash lines when specified to be installed in the slab or underground, and solid when concealed in ceilings and walls. Short dash lines indicate exposed wiring. Solid lines are often used in commercial installations when the raceway is to be exposed (surface mounted). Slash marks through the circuit lines are used to indicate the number of conductors. Full slash marks are the circuit conductors, a longer full slash is the neutral and half slash marks are the ground wires. When only two wires are required, no slash marks are used. The typical circuit line for residential construction indicates a No. 14 AWG wire size. The typical wire size for commercial construction is No. 12 AWG. The arrows indicate "home runs" to the designated panel. The panel and circuit number designations are adjacent to the arrowheads. Figure 42-1 lists the typical circuiting symbols. The electrical installer will normally group home runs in a raceway (conduit) with combinations similar to the following:

- Three-phase systems with three circuit wires and a common neutral in a raceway.

- Single-phase systems with two circuit wires and a common neutral in a raceway.

Note—To reduce the harmonic problems caused by solid state devices (nonlinear loads), many electrical designers are requiring separate neutrals with each phase conductor.

Some equipment, such as television, fire alarm, clock, PA, and sound might not be connected with wiring lines on the plan. This indicates that wiring for that equipment is not part of the general circuit wiring and is probably not to be fed through the same panels as the rest of the electrical equipment. For example, on the School Addition the PA system and the cloks are fed from the sound system console and do not involve the general circuit wiring.

Lighting Circuits

There is usually a *lighting circuit* electrical floor plan for each level or major space within a building. This floor plan shows light fixtures, emergency lighting, security lighting, special lighting control (i.e., photocell, motion detector, etc.). Typically included is a reflected ceiling plan, Figure 42-2. That means that it is a plan view, but shows what is on the ceiling as though it was reflected onto the floor plan, Figure 42-2. The reflected ceiling plan shows each light fixture with a circle, square, or rectangle that approximates the shape of the fixture. The lighting circuitry on the School Addition is shown with three types of wiring circuit lines, solid for unswitched, dotted for switched, and line-dash-line for the motion sensor circuits. The conventional method of showing lighting circuits is the same as previously explained in the paragraph on *Power Circuits.*

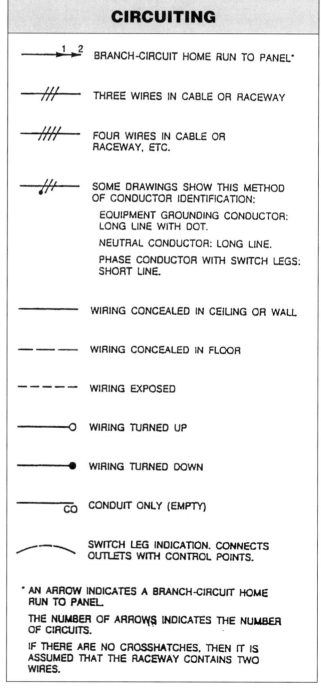

Figure 42-1 Circuiting symbols

CEILING OF TWO ROOMS

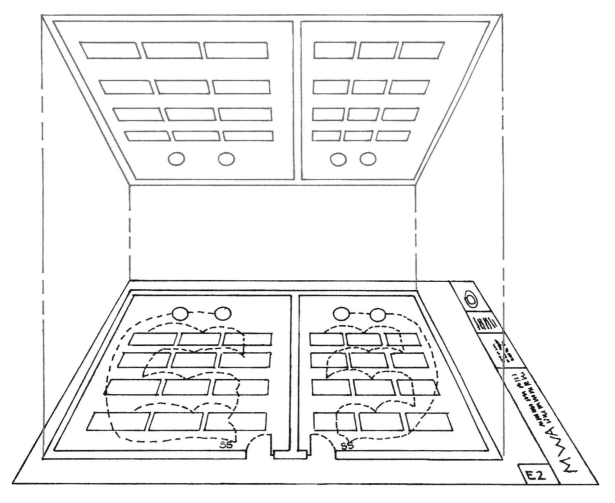

REFLECTED LIGHTING PLAN

Figure 42-2   Reflected ceiling plan

Motion sensors are an interesting feature of the lighting design for the School Addition. The motion sensors are shown as small circles containing the letter M. A motion sensor detects any movement in the area of the sensor, so that in a room controlled by a motion sensor, if there is no movement in the room for a period of time, the sensor opens its contacts, like turning a switch off. A motion detector can be used together with a switch so that the lights are controlled normally by the switch, but if someone forgets to turn the lights off, the motion detector will do it for them.

A low voltage lighting control system requires a *relay* that is activated by the light switch. The relay opens or closes the higher-voltage fixture, Figure 42-3. The relays on the School Addition are indicated by the letter R in a small square. Notice that every lighting switch is connected to a relay.

Symbols

Electric *symbols* are used to simplify the drafting and later the interpreting of the drawings. Electrical symbols *are not standardized* throughout the industry. Most drawings will have a symbol legend or list. You must be knowledgeable of the symbols specifically used on each project, since designers modify basic symbols to suit their own needs. Many symbols are similar (circle, square, etc.). The addition of a line, dot, shading, letters, numbers, etc. gives the specific meaning to the symbol. Learning the basic form of the various symbols is the best starting point in developing the ability to interpret the drawings and their related symbol meanings. Figure 42-4 lists the most common and recommended electrical symbols.

The School Addition drawing E-1, Symbol Legend A-4 contains an electrical symbol list for the project. From

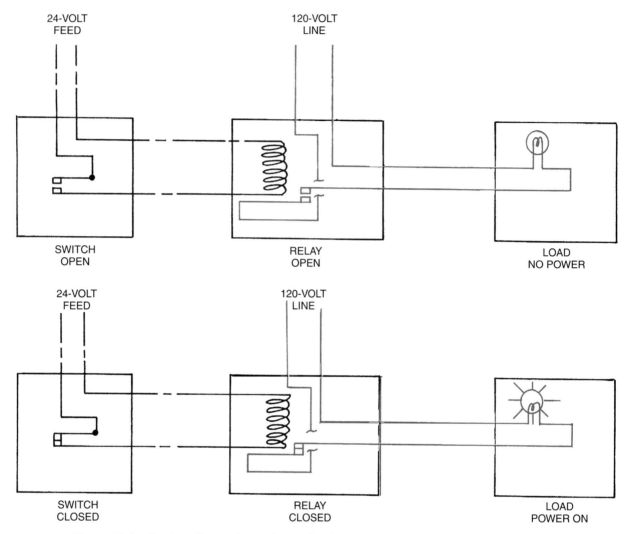

24-VOLT
FEED

120-VOLT
LINE

SWITCH
OPEN

RELAY
OPEN

LOAD
NO POWER

24-VOLT
FEED

120-VOLT
LINE

SWITCH
CLOSED

RELAY
CLOSED

LOAD
POWER ON

Figure 42-3 A relay allows a low-voltage circuit to start and stop a higher-voltage circuit

the symbol list, it can be seen that there are duplex receptacles, switches, telephone outlets, special purpose outlets, and fire alarm devices mounted at various heights. Then there is the General Note in the Symbol Legend that specifies that all mounting heights are to be verified and "modified as directed." This is an example of why the installer must become familiar with the drawings and specifications far in advance of the installation scheduled time. The installer must request clarification or direction and give the designers reasonable time to clarify the questionable specified instructions.

Review

As previously covered in Part I Drawings, Units 1 through 8, the four basic types of drawings are: plan, elevations, sections, and details. The electrical floor plan of a building shows all the exterior walls, interior partitions, windows, doors, stairs, cabinets, etc. along with the locations of the electrical outlets, light fixtures, branch circuits, feeders, panelboards, equipment, and other details necessary to make a complete installation.

The power wiring and the lighting layout are generally shown on separate floor plans. Electrical symbols are used to show the locations of electrical outlets, devices, lighting, equipment, special systems, fire alarms, etc. on the building floor plans.

SINGLE LINE DIAGRAMS

The electrical service *single line diagram* on the School Addition is shown on drawing E-2, Single Line Diagram E-17. The electrical power is brought into the school by way of the service entrance section (S.E.S.) The note "—coordinate with the local utility requirements prior to rough-in" indicates that the electrical contractor is to furnish and install the three 4-inch conduits and that the utility company wants to inspect the

Electrical Reference Symbols

ELECTRICAL ABBREVIATIONS
(Apply only when adjacent to an electrical symbol.)

Central Switch Panel	CSP
Dimmer Control Panel	DCP
Dust Tight	DT
Emergency Switch Panel	ESP
Empty	MT
Explosion Proof	EP
Grounded	G
Night Light	NL
Pull Chain	PC
Rain Tight	RT
Recessed	R
Transfer	XFER
Transformer	XFRMR
Vapor Tight	VT
Water Tight	WT
Weather Proof	WP

ELECTRICAL SYMBOLS

Switch Outlets

Single-Pole Switch	S
Double-Pole Switch	S_2
Three-Way Switch	S_3
Four-Way Switch	S_4
Key-Operated Switch	S_K

Switch and Fusestat Holder	S_{FH}
Switch and Pilot Lamp	S_P
Fan Switch	S_F
Switch for Low-Voltage Switching System	S_L
Master Switch for Low-Voltage Switching System	S_{LM}
Switch and Single Receptacle	S
Switch and Duplex Receptacle	S
Door Switch	S_D
Time Switch	S_T
Momentary Contact Switch	S_{MC}
Ceiling Pull Switch	Ⓢ
"Hand-Off-Auto" Control Switch	HOA
Multi-Speed Control Switch	M
Push Button	•

Receptacle Outlets

Where weather proof, explosion proof, or other specific types of devices are to be required, use the upper-case subscript letters. For example, weather proof single or duplex receptacles would have the uppercase WP subscript letters noted alongside of the symbol. All outlets should be grounded.

Single Receptacle Outlet	
Duplex Receptacle Outlet	
Triplex Receptacle Outlet	
Quadruplex Receptacle Outlet	

Figure 42-4   Recommended electrical symbols

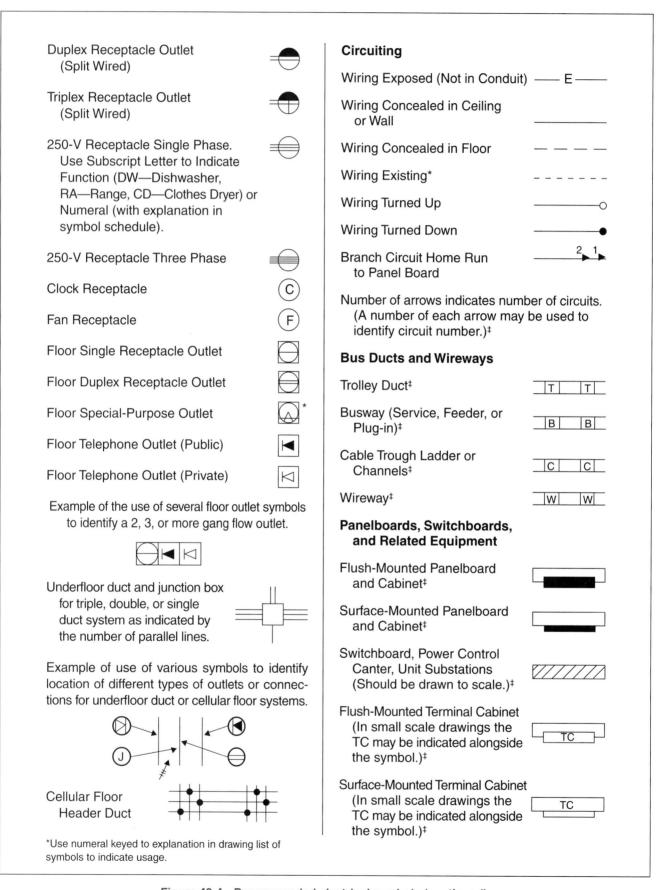

Duplex Receptacle Outlet
(Split Wired)

Triplex Receptacle Outlet
(Split Wired)

250-V Receptacle Single Phase.
Use Subscript Letter to Indicate
Function (DW—Dishwasher,
RA—Range, CD—Clothes Dryer) or
Numeral (with explanation in
symbol schedule).

250-V Receptacle Three Phase

Clock Receptacle

Fan Receptacle

Floor Single Receptacle Outlet

Floor Duplex Receptacle Outlet

Floor Special-Purpose Outlet

Floor Telephone Outlet (Public)

Floor Telephone Outlet (Private)

Example of the use of several floor outlet symbols
to identify a 2, 3, or more gang flow outlet.

Underfloor duct and junction box
for triple, double, or single
duct system as indicated by
the number of parallel lines.

Example of use of various symbols to identify
location of different types of outlets or connections for underfloor duct or cellular floor systems.

Cellular Floor
Header Duct

*Use numeral keyed to explanation in drawing list of
symbols to indicate usage.

Circuiting

Wiring Exposed (Not in Conduit)

Wiring Concealed in Ceiling
or Wall

Wiring Concealed in Floor

Wiring Existing*

Wiring Turned Up

Wiring Turned Down

Branch Circuit Home Run
to Panel Board

Number of arrows indicates number of circuits.
(A number of each arrow may be used to
identify circuit number.)‡

Bus Ducts and Wireways

Trolley Duct‡

Busway (Service, Feeder, or
Plug-in)‡

Cable Trough Ladder or
Channels‡

Wireway‡

**Panelboards, Switchboards,
and Related Equipment**

Flush-Mounted Panelboard
and Cabinet‡

Surface-Mounted Panelboard
and Cabinet‡

Switchboard, Power Control
Canter, Unit Substations
(Should be drawn to scale.)‡

Flush-Mounted Terminal Cabinet
(In small scale drawings the
TC may be indicated alongside
the symbol.)‡

Surface-Mounted Terminal Cabinet
(In small scale drawings the
TC may be indicated alongside
the symbol.)‡

Figure 42-4   Recommended electrical symbols (continued)

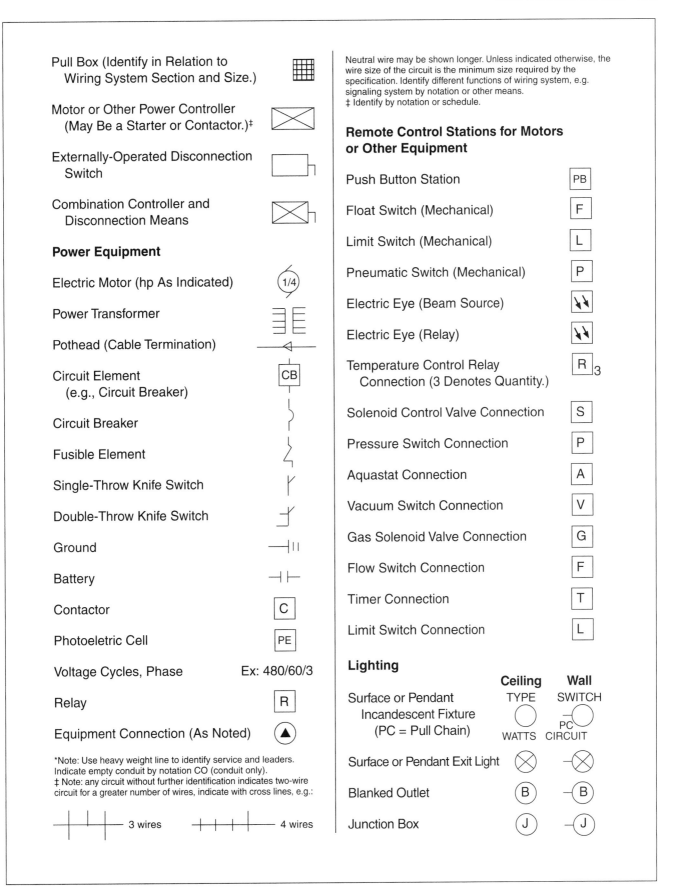

Pull Box (Identify in Relation to Wiring System Section and Size.)

Motor or Other Power Controller (May Be a Starter or Contactor.)‡

Externally-Operated Disconnection Switch

Combination Controller and Disconnection Means

Power Equipment

Electric Motor (hp As Indicated)

Power Transformer

Pothead (Cable Termination)

Circuit Element (e.g., Circuit Breaker)

Circuit Breaker

Fusible Element

Single-Throw Knife Switch

Double-Throw Knife Switch

Ground

Battery

Contactor

Photoeletric Cell

Voltage Cycles, Phase   Ex: 480/60/3

Relay

Equipment Connection (As Noted)

*Note: Use heavy weight line to identify service and leaders. Indicate empty conduit by notation CO (conduit only).
‡ Note: any circuit without further identification indicates two-wire circuit for a greater number of wires, indicate with cross lines, e.g.:

3 wires   4 wires

Neutral wire may be shown longer. Unless indicated otherwise, the wire size of the circuit is the minimum size required by the specification. Identify different functions of wiring system, e.g. signaling system by notation or other means.
‡ Identify by notation or schedule.

Remote Control Stations for Motors or Other Equipment

Push Button Station

Float Switch (Mechanical)

Limit Switch (Mechanical)

Pneumatic Switch (Mechanical)

Electric Eye (Beam Source)

Electric Eye (Relay)

Temperature Control Relay Connection (3 Denotes Quantity.)

Solenoid Control Valve Connection

Pressure Switch Connection

Aquastat Connection

Vacuum Switch Connection

Gas Solenoid Valve Connection

Flow Switch Connection

Timer Connection

Limit Switch Connection

Lighting

Surface or Pendant Incandescent Fixture (PC = Pull Chain)

Surface or Pendant Exit Light

Blanked Outlet

Junction Box

Figure 42-4   Recommended electrical symbols (continued)

Recessed Incandescent Fixtures

Surface or Pendant Individual
    Fluorescent Fixture

Surface or Pendant Continuous-   A
    Row Fluorescent Fixture   Fixture No.
    (Letter Indicating Controlling Switch)   Wattage

Bare-Lamp Fluorescent Strip*

*In the case of continuous-row bare-lamp flourescent strip
above an area-wide diffusing means, show each fixture run
using the standard symbol; indicate area of diffusing means
and type by light shading and/or by light shading and/or
drawing notation.

Electric Distribution or Lighting System, Aerial

Pole‡

Steel or Parking Lot Light
    and Bracket‡

Transformer‡

Primary Circuit‡

Secondary Circuit‡

Down Guy

Head Guy

Sidewalk Guy

Service Weather Head‡

Electric Distribution or Lighting System, Underground

Manhole‡   M

Handhole‡   H

Transformer Manhole   TM
    or Vault‡

Transformer Pad‡   TP

Underground Direct, Burial Cable
    (Indicate type, size, and number
    of conductors by notation
    or schedule.)

Underground Duct Line
    (Indicate type, size, and
    number of ducts by cross-
    section identification of each
    run by notation or schedule.
    Indicate type, size, and number
    of conductors by notation or
    schedule.)

Street Light Standard Feed From
    Underground Circuit‡

‡ Identify by notation or schedule.

Signaling System Outlets

Institutional, Commercial, and Industrial Occupancies

I. Nurse Call System Devices
    (Any Type)

    Basic Symbol

    (Examples of individual item identi-
    fication. Not a part of standard.)

Nurses' Annunciator   1
    (Adding a number after it
    indicates number of lamps,
    e.g., +①24.)

Call Station, Single Cord,   2
    Pilot Light

Call Station, Double Cord,   3
    Microphone Speaker

Corridor Dome Light, 1 Lamp   4

Transformer   5

Any Other Item on Same System   6
    (Use Numbers as Required.)

II. Paging System Devices
    (Any Type)

    Basic Symbol

Figure 42-4   Recommended electrical symbols (continued)

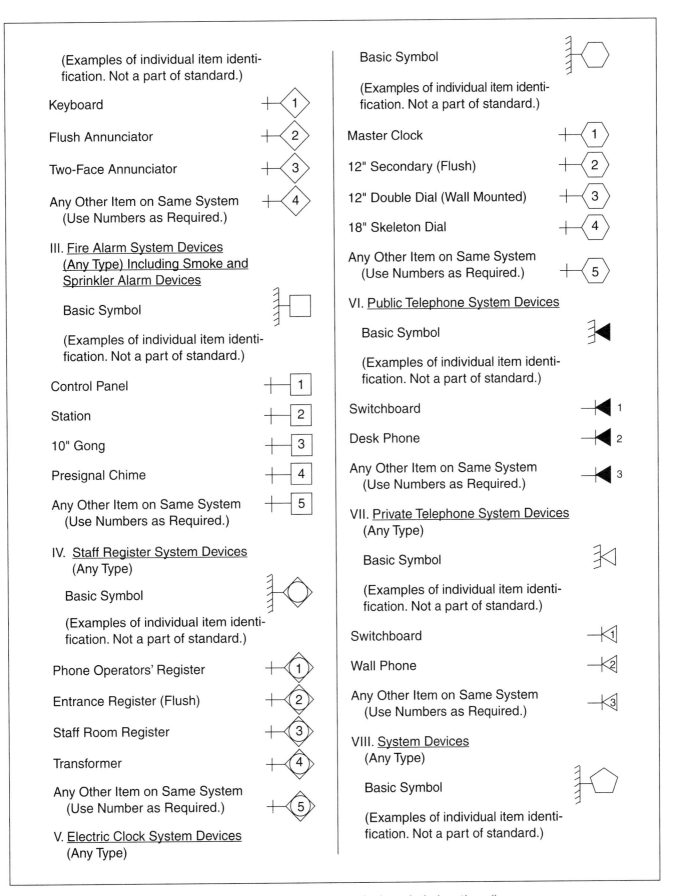

(Examples of individual item identification. Not a part of standard.)

Keyboard 1

Flush Annunciator 2

Two-Face Annunciator 3

Any Other Item on Same System 4
(Use Numbers as Required.)

III. Fire Alarm System Devices
(Any Type) Including Smoke and
Sprinkler Alarm Devices

Basic Symbol

(Examples of individual item identification. Not a part of standard.)

Control Panel 1

Station 2

10" Gong 3

Presignal Chime 4

Any Other Item on Same System 5
(Use Numbers as Required.)

IV. Staff Register System Devices
(Any Type)

Basic Symbol

(Examples of individual item identification. Not a part of standard.)

Phone Operators' Register 1

Entrance Register (Flush) 2

Staff Room Register 3

Transformer 4

Any Other Item on Same System 5
(Use Number as Required.)

V. Electric Clock System Devices
(Any Type)

Basic Symbol

(Examples of individual item identification. Not a part of standard.)

Master Clock 1

12" Secondary (Flush) 2

12" Double Dial (Wall Mounted) 3

18" Skeleton Dial 4

Any Other Item on Same System 5
(Use Numbers as Required.)

VI. Public Telephone System Devices

Basic Symbol

(Examples of individual item identification. Not a part of standard.)

Switchboard 1

Desk Phone 2

Any Other Item on Same System 3
(Use Numbers as Required.)

VII. Private Telephone System Devices
(Any Type)

Basic Symbol

(Examples of individual item identification. Not a part of standard.)

Switchboard 1

Wall Phone 2

Any Other Item on Same System 3
(Use Numbers as Required.)

VIII. System Devices
(Any Type)

Basic Symbol

(Examples of individual item identification. Not a part of standard.)

Figure 42-4   Recommended electrical symbols (continued)

conduit installation prior to back filling the trench where the conduits are installed; however, this should be verified at a pre-construction meeting with the utility company. There are times when an electrical service, or electrical distribution raceway would require being concrete encased for protection. This type of concrete encased underground raceway system is called a *duct bank,* Figure 42-5.

It does not specifically indicate who furnishes and installed the utility feeders in the three 4-inch conduits. This could be an expensive item and should be clarified with the utility company prior to estimating and bidding the project.

The service entrance section to comply with the *National Electrical Code®* (*NEC®*) may have up to, but must not exceed, six main disconnecting units (switches or circuit breakers). Four main fused switches are shown in the "EXISTING" portion of the single line drawing. Note that this single line drawing shows only a portion of the complete existing service entrance section. The service entrance section indicates that it is located outside by the "NEMA 3R ENCLOSURE" definition. NEMA, National Electric Manufacturers Association, generates specifications and is recognized as a design standard for electrical boxes, devices, and equipment. "NEMA 3R" is a rain tight designation. The "NEW" portion of the single line drawing shows the electrical distribution to be installed in this School Addition. The new elevator, electrical panels L10 and L11 are fed from existing panel MDP. These new electrical loads are:

1. 60-amp. circuit breaker with four #4 AWG, one #10 ground and one #10 isolated ground in a 1¼-inch conduit feeding the elevator.

2. 100-amp. circuit breaker with four #1 AWG, one #6 ground in a 1½-inch conduit feeding panel L11.

3. 100-amp. circuit breaker with four #1 AWG, one #6 ground in a 1½-inch conduit feeding panel L10.

A complete building electrical floor plan is not a part of the School Addition drawings and the "EXISTING" electrical distribution plan on drawing E-2 shows only a portion of the total electrical distribution section. This required notes "6," "7," and "8" to be added to drawing E-2, Single Line Diagram E-17, giving the contractor the lengths of the feeders for the elevator, panel L10, and panel L11 which are to be included in this School Addition. In a typical design, a complete building electrical plan is provided indicating the locations of all existing and new electrical equipment. The feeder length would be determined from that drawing.

RISER DIAGRAMS

A *riser diagram* is so named because it usually shows the path of wiring or raceway (conduit) from one level of a building to another and it rises from one floor to the next. A riser diagram does not give information about where equipment is to be located in a room or area. Riser diagrams are used because they are particularly easy to understand. Therefore, they do not require much explanation.

A *power riser diagram,* Figure 42-6, shows a typical building's electrical service and related components. This figure is not the same electrical service as the

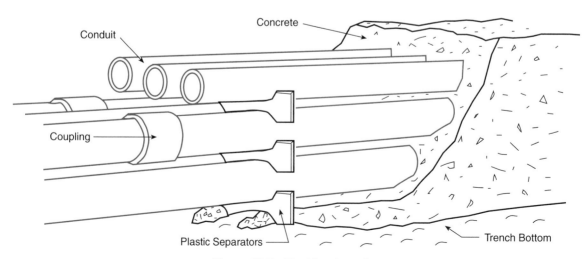

Figure 42-5   Duct bank system

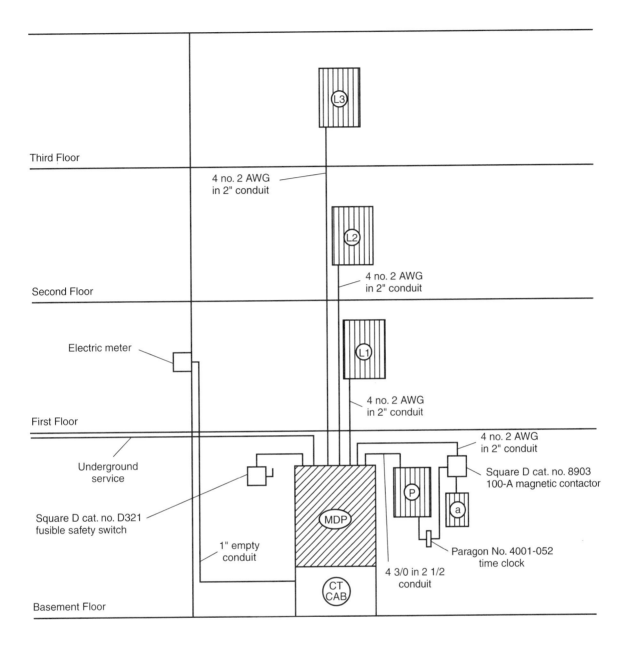

Figure 42-6   Typical riser diagram showing an overview of the building's electrical service and related components

School Addition Single Line Diagram, E-17 on drawing E-2, but comparing the two diagrams shows how a power riser diagram greatly simplifies the interpretation of an installation drawing.

A *special riser diagram* is used for many systems that include:

1. Fire alarm

2. Security

3. Telephone

4. Clock

5. Signal

- Bell
- Call (nurse, emergency, etc.)
- Water sprinkler

The *fire alarm riser diagram,* Figure 42-7, shows the new School Addition fire alarm system with ¾-inch conduits to Ramp Area 113. The School Addition drawing E-2 shows three (3) ¾-inch conduits with the note "Three (3) ¾-inch existing conduits from the Fire Alarm Control Panel (150 feet)." These three conduits are to be used for the School Addition fire alarm connection

Figure 42-7 Fire alarm riser diagram for the School Addition

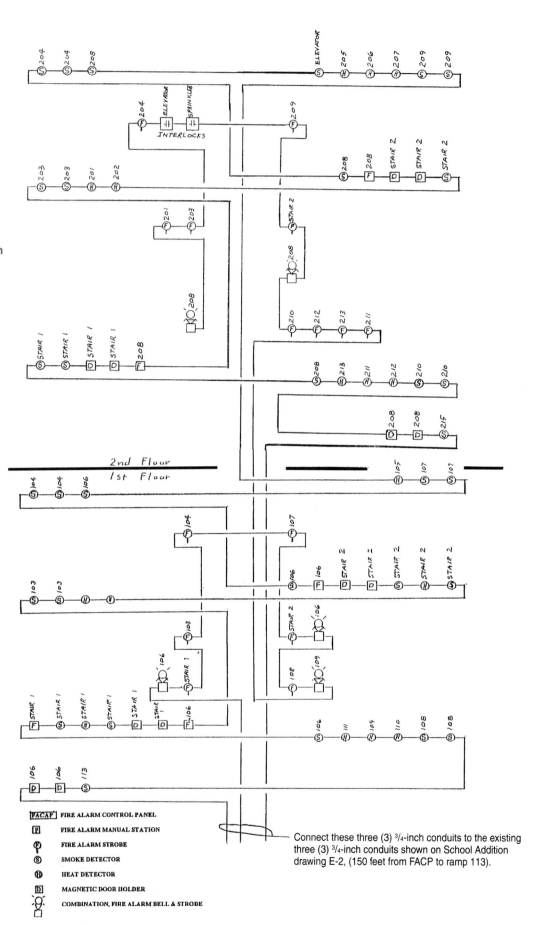

FACP | FIRE ALARM CONTROL PANEL

F | FIRE ALARM MANUAL STATION

P | FIRE ALARM STROBE

S | SMOKE DETECTOR

H | HEAT DETECTOR

D | MAGNETIC DOOR HOLDER

| COMBINATION, FIRE ALARM BELL & STROBE

Connect these three (3) 3/4-inch conduits to the existing three (3) 3/4-inch conduits shown on School Addition drawing E-2, (150 feet from FACP to ramp 113).

to the existing fire alarm system. The original fire alarm control panel was sized to accommodate this School Addition; however, you should verify that the existing *special systems* (fire alarm, security, clock, etc.) will accommodate the additional requirements when adding to or modifying the existing system(s).

The *telephone riser diagram,* Figure 42-8, shows an existing 1¼-inch conduit, 150 feet in length, from Ramp Area 113 to the existing main telephone terminal cabinet. This conduit is to be extended to the telephone terminal board in Room 102, First Floor Plan E-7, drawing E-2. The telephone riser diagram shows telephone conduit to be installed in the area above the suspended ceiling from the telephone terminal board to the outlet locations as shown on drawing E-2 and Outlet Detail A-11.

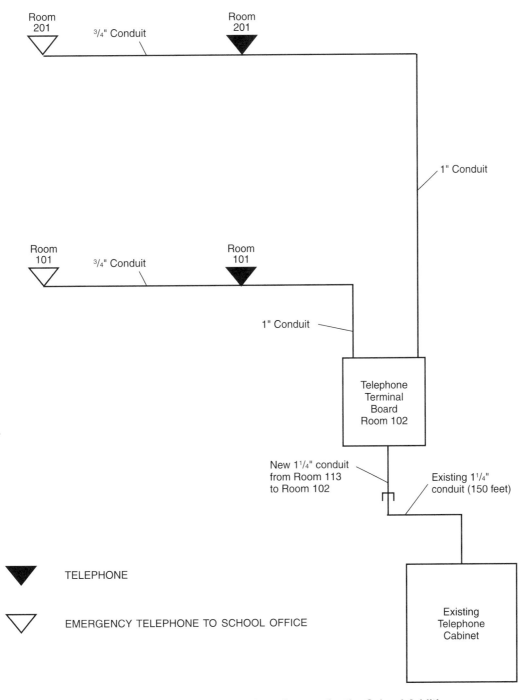

Figure 42-8 Telephone riser diagram for the School Addition

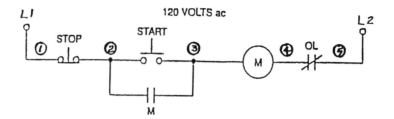

Schematic

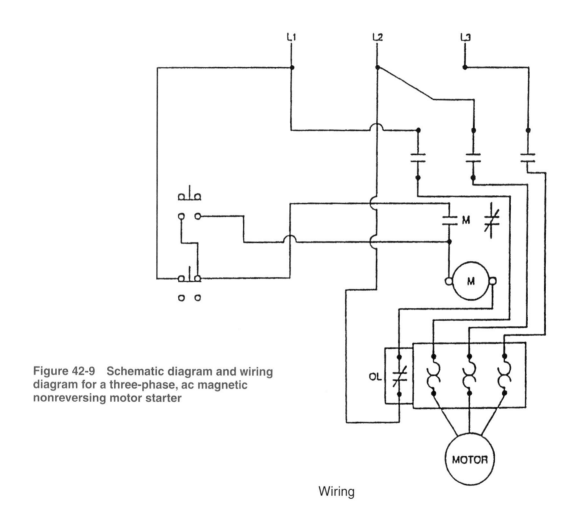

Figure 42-9   Schematic diagram and wiring diagram for a three-phase, ac magnetic nonreversing motor starter

Wiring

SCHEMATIC DIAGRAMS

A schematic wiring diagram is a drawing that uses symbols and lines to show how the parts of an electrical assembly or unit are connected. A schematic does not necessarily show where parts are actually located, but it does explain how to make electrical connections. Several of the schematics with the School Addition show connections to wires labeled G, N, and H. These stand for *ground, neutral,* and *hot.* Schematics are commonly drawn for electrical equipment that involves internal wiring—everything from washing machines to computers. A basic motor control schematic, Figure 42-9, shows a three-phase power source (L1, L2, and L3) through the starter contacts (M) and the overloads (OL) powering the motor. The starter control is taken from phase L1 through a stop push-button, a start push-button paralleled with a latching contact "M" to the starter coil "M," and finally through the normally closed overload contacts back to phase L2. The connection label numbers (①, ②, ③, etc.), shown on the schematic aid the electrician in troubleshooting. Several different labeling methods are used, but they all follow the same principles, and if you understand one method, you will be able to understand the other methods.

SCHEDULES

An electrical schedule is used to systematically list equipment, loads, devices, and information. Schedules organize the information in an easily understood form and can be a valuable method for communicating the design requirements to the contractor and his installers.

The *fixture schedule* lists complete information about each fixture type shown on the lighting plan. The following is a list of the type of information that is typically shown on a fixture schedule:

- Mark—The label used to indicate a fixture type. The mark is written on or next to each fixture on the plan.

- Make—The identification of the manufacturer being used to establish the specific design requirements needed to aesthetically and functionally light the various rooms or areas.

- Volts—It is becoming increasingly common for light fixtures to be powered by up to 277/480 volts, while the *switches* might use only 24 volts (low voltage lighting control), to protect the person using the switch.

- Watts—It is necessary to know the wattage of the lamps (bulbs) in each fixture, so that the space will have the amount of illumination intended by the electrical designer.

- Lamp Type and Quantity—Lamp manufacturers have similar systems of designating lamp characteristics, Figures 42-11 and 42-12.

- Notes or Remarks—This column is for information that does not clearly belong under the other headings.

The School Addition Fixture Schedule A-9, Floor Plans E-7 and E-14 on drawing E-1, shows type "FA" light fixture to be the light source for the classrooms. The School Addition fixture schedule, Figure 42-10, shows the fixture to be manufactured by Williams, the catalog number to be 1222-RWKA125 with three F32T8/SP30 lamps rated at 120 volts with an 183-watt

load. Also, the remarks column indicates some special wiring and specific wiring and specific ballast type requirements. The first and second floor "Reflected Ceiling Plans," E-15 and E-16 on drawing E-7, indicate light fixture type "FA" to be a 2' x 4' light fixture mounted end-to-end.

Fifteen other light fixture types are shown on the fixture schedule and on the drawings and are identified by their corresponding mark or label. All are rated at 120 volts except one emergency battery powered "Remote Lamp," rated at 12 volts. This information is typically given in the specifications but is more readily presented by a schedule.

A *panel schedule* identifies the panels by their mark or label. They are shown by this same designation on the electrical floor plan. The panel schedule for the School Addition lists each of the branch circuits served by a panel, the calculated load for that branch circuit, the voltage of that circuit, and the number of poles and trip rating for each circuit breaker. A panel schedule also might include:

- Type (surface or flush)

- Panel main buss amperes, volts, and phases

- Main circuit breaker/main lugs only

- Breaker frame sizes

- Items fed and/or remarks

On the lighting and power plans, we saw that each device is connected to a branch circuit identified by a panel number (label) and circuit number. Those numbers correspond with the numbers on the panel schedule.

When a commercial project has a kitchen, you should have a *kitchen equipment schedule.* If the drawings do not have one, it is very helpful to make one for the installer. The kitchen equipment schedule should include:

- Equipment number or designation

- Description of each equipment item

- The load in horsepower or kilowatts

- Volts

- Wire size

- Conduit size

- Protection in amperes

- How each equipment item will be furnished (furnished by others or contractor furnished)

- Installation requirements

- Remarks column for any detailed specific information required

| MARK | MAKE | MODEL | VOLTS | WATTS | LAMP | | REMARKS |
					TYPE	NO./FIXT.	
A	STONCO	VWXL11GC	120	100	100WA19	1	CAST GUARD.
FA	WILLIAMS	1222-RWKA125	120	183	F32T8/SP30	3	SPLIT WIRE IN TANDEM PAIRS - WATTAGE IS FOR 2 FIXTURES W/ TOTAL OF 3 2-LAMP BALLASTS. ELECTRONIC BALLAST.
FA1	WILLIAMS	1222-RWKA125	120	92	F32T8/SP30	3	ELECTRONIC BALLAST.
FB	WILLIAMS	2922-KA	120	61	F32T8/SP30	2	ELECTRONIC BALLAST.
FC	KIRLIN	96617-45-46-61	120	36	PLC13W/27K	2	HPF BALLAST.
FC1	KIRLIN	96617-46-61-SM	120	36	PLC13W/27K	2	HPF BALLAST, SURFACE MOUNT; COLOR AS SELECTED
FD	WILLIAMS	8222	120	61	F32T8/SP30	2	ELECTRONIC BALLAST.
FE	WILLIAMS	1262-RWKA125	120	84	F17T8/SP30	4	ELECTRONIC BALLAST.
FF	TERON	EE26-P-H	120	36	PL13/27	2	HPF BALLAST.
FG	WILLIAMS	2122-IM	120	61	F32T8/SP30	2	ELECTRONIC BALLAST.
FH	WILLIAMS	EPG-R272RWKA-125	120	61	F32T8/SP30	2	TO MATCH EXISTING
M1	KIRLIN	SS-51277-24-43 (35W)-45-46-FR	120	40	35W HPS	1	HPF BALLAST.
M2	STONCO	PAR250LX	120	300	250W HPS	1	HPF BALLAST EQUIP WITH MOUNTING BRACKET ARM.
EXITS	LITHONIA	WLES SERIES	120	65	LED	1	SEE SHEET E-2 FOR LOC. ARROWS & MOUNTING AS IND EQUIP W/ BATTERY BACK-UP
EMERG.LGT. BATT.PACK	EXIDE	B200	120/12	150	H1212		
REMOTE LAMP	EXIDE	H1212	12	24	H1212	2	

Figure 42-10   Fixture schedule.

A *receptacle schedule* is valuable when a number of special or specific receptacle types are found on the electrical drawings. If one is not provided, you should make one to expedite the installation time required and reduce the chance of installation error. A receptacle schedule should include:

- The symbol designation used by the designer

- Amperage rating

- Number of wires and poles

- Voltage rating

- NEMA type

- The configuration of the blades or slots

- A manufacturer catalog number reference

- Special information (duplex, single, 3 phase, etc.)

Note: Receptacle information may be found on an *equipment schedule.*

The local plans review and the utility company typically require a *connected load schedule.* This type of schedule includes:

- Type of load

- Building or area designation

- Size of load (kilowatts or horsepower)

- Total electrical load by type or area

- Notes explaining any special methods used in the load calculations

SPECIFICATIONS

The drawings and the specifications are the items that establish the intended design and the construction

How to Read Ordering Guide

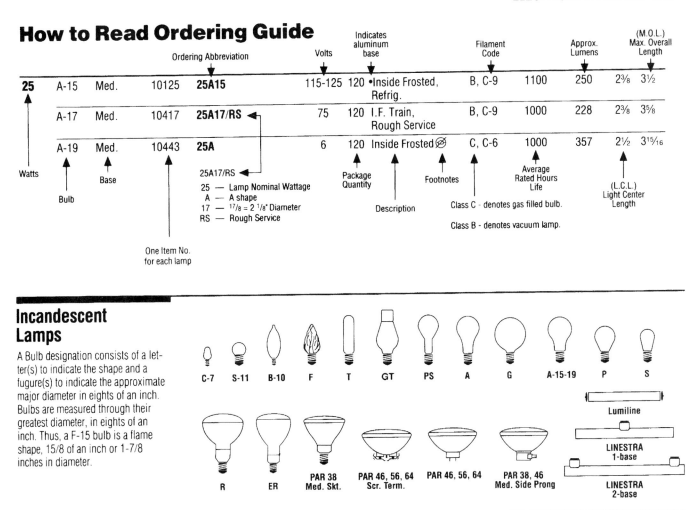

Incandescent Lamps

A Bulb designation consists of a letter(s) to indicate the shape and a figure(s) to indicate the approximate major diameter in eights of an inch. Bulbs are measured through their greatest diameter, in eights of an inch. Thus, a F-15 bulb is a flame shape, 15/8 of an inch or 1-7/8 inches in diameter.

Figure 42-11 Incandescent lamp designations (Osram Sylvania Inc.)

How to Read Ordering Guide

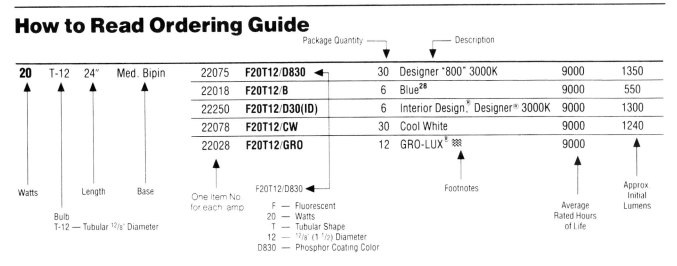

Figure 42-12 Flourescent lamp designations (Osram Sylvania Inc.)

FLUORESCENT LAMPS

The bulb shape and size of a fluorescent lamp are expressed by means of a code consisting of the letter "T" (which designates that the bulb is "tubular" in shape) followed by a number which expresses the diameter of the bulb in eighths of an inch. They vary in diameter from T-5 (⅝ inch) to T-17 (2⅛ inches). In nominal overall length, fluorescent lamps range from 6 to 96 inches, which is always measured from back of lampholder to back of lampholder. For example, the actual overall length of the 40-watt rapid start T-12, 48 inch lamp is 47¾ inches. Circline lamps, which are circular, are available in four sizes: 6½ inches, 8 inches, 12 inches and 16 inches outside diameter. There are also U shaped fluorescent types (Curvalume™) with T-8 and T-12 bulbs. U shaped types are measured for the distance between the ends. The overall length is measured from the face of the bases to the outside of the glass bend.

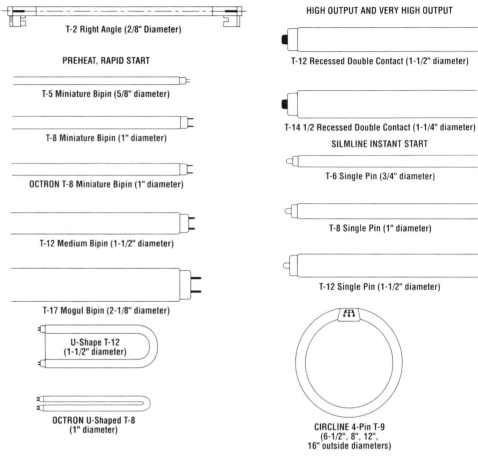

Figure 42-12   (Continued)

requirements required by the owner, architect, and engineers. The contractor and his installers must review the contract documents for conflicts and/or discrepancies between the contract, specifications, and the drawings. The drawings should be reviewed for conflicts between sections (architectural, mechanical, plumbing, structural, civil, HVAC, electrical, etc.). The drawings in your specific section should be reviewed for conflicts sheet by sheet.

The electrical specifications give the quality of materials intended to be used and the installation and testing requirements. Sometimes a specific manufacturer or catalog number is specified with no substitution or "equal" allowed. This usually inflates the cost of the project. There is no standard specification. The contractor and his installers should read through all project specifications and become knowledgeable of their content prior to starting any installation.

✓ CHECK YOUR PROGRESS

Can you perform these tasks?

☐ Explain where light fixtures are to be placed and how they are connected.

☐ Determine what branch circuit in a panel supplies a particular light fixture or electrical outlet.

☐ Understand and use the information on a fixture schedule.

☐ Use a schematic diagram to explain how an electrical device is to be wired.

ASSIGNMENT

Refer to the drawings of the School Addition (included in the packet) to complete this assignment.

1. What are the four basic methods of showing electrical circuits?

2. How is a "home run" circuit shown?

3. What is a reflected ceiling plan?

4. In a low voltage lighting system, what does the light switch activate?

5. Where do you find the industry standardized electrical symbols?

6. What are the four basic types of electrical drawings?

7. What is the name of the method used to show the path of wiring or raceway from one level of a building to another?

8. What is a schematic wiring diagram?

9. What is used to systematically list equipment, loads, devices, and information?

10. Where do you find the quality of material intended to be used on a project?

11. How many F17T8/SP30 lamps are required on the first floor of the addition?

12. Explain why one of the switches in room 109 is listed as S_{3M} and the other is simply S_3.

13. How are the lights turned on and off in the first floor corridor?

14. What circuit carries the lights for stair 2?

15. What is the approximate total wattage of the lamps in the boy's toilet room on the first floor?

16. What is the total load for the circuit that serves the lights in the boy's toilet room on the first floor?

17. What is indicated by the D in a square near the doors from the existing building into the addition?

18. What is on circuit L10, 24?

19. Where are the devices on circuit L10, 15?

20. Explain what each of the colored terminals on a classroom lighting control are to be connected to:

    Green   _____

    Orange   _____

    Black   _____

    White   _____

    Blue (inner terminal)   _____

    Blue (outer terminal)   _____

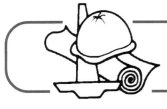

PART IV   Test

A. Which of the symbols shown in Column II is used on the School Addition drawings to represent the objects or materials listed in Column I?

	I			II

1. Room number

a.

2. Building section

b.   (hexagon with D)

3. Rigid insulation

c.   (circle with 4)

4. Window type

d.  (J-6 / A201)

5. Door number

e. 209

6. Interior elevation

f.   (A-8 / A502)

7. Wood blocking

g.

8. Detail key

h.
G14/A610

B. Refer to the door schedule on School Addition Sheet A001 to write the door numbers from Column II that are associated with the material in Column I.

	I		II

9.   Wood   216
10. Steel hollow metal   104
11. Plastic   101

C.   Name the material for each of the numbered items in the drawing of a bond beam.

12. _____
13. _____
14. _____
15. _____
16. _____

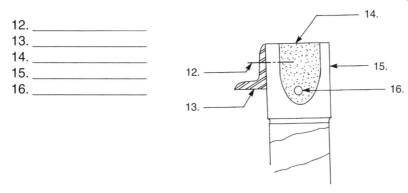

D.   Describe the material at each of the following locations in the School Addition.

17.   Door frame 103
18.   South wall of corridor 106
19.   Structural portion of west wall of classroom 104
20.   Exterior surface of west wall at corners
21.   Exterior surface of west wall below windows
22.   Stair treads
23.   Vertical reinforcement in elevator pit walls
24.   Beam at the top of the west wall of classroom 204
25.   Lintel over the windows in the west wall
26.   Door frame at entrance to stair #1

E.   Refer to the School Addition drawings and write the elevation at the following locations:

27.   Second finish floor
28.   Top of roof
29.   Top of bricks beneath window in classroom 103
30.   Top of concrete blocks in first floor corridor walls
31.   Top of tackboard on north wall of classroom 103
32.   Finish floor in elevator pit
33.   Top of foundation wall at typical locations
34.   Top of footing for column F3
35.   Top of the beams in the second floor framing between the columns in row C

F.   Refer to the School Addition drawings to answer the following questions:

36.   What is the height of the rough opening for the grille that allows outside air to enter classroom 103?
37.   What supports the outer few inches of the roofing membrane in the area where the roof-edge coping meets it?
38.   What size are the concrete masonry units in the north wall?
39.   What are the three materials that make up the roof decking on the canopy over door 101?
40.   What are the drawing and sheet numbers for information about typical column base plates?
41.   Give as much information as possible about the welds that fasten the roof framing bar joists to the beams at the tops of the north and south walls.
42.   On what branch circuit are the light fixtures in stair 1?
43.   How are the light fixtures in the second floor corridor turned on and off?
44.   How many lamps are required for all of the fixtures in the second floor corridor?
45.   What branch circuit supplies the convenience outlets in the second floor corridor?
46.   Which terminal on the corridor lighting control unit is connected to the neutral leg of the supply?
47.   What is the diameter of the sanitary drain where it leaves the building?
48.   What are the pipe sizes that storm water passes through as it flows from the roof drain on the north canopy to the point where it exits the building?
49.   Where is the nearest cleanout to the roof drain at the west end of the addition?
50.   What type of unit provides heat in storage room 109?

Master Keynotes

2.11 Demolition, Removals & Relocation

2.20 Site Preparation & Earthwork
  2.20A Select Fill - Bank Run Gravel
  2.20B Select Granular Material
  2.20D Topsoil
  2.20F Crushed Gravel
  2.20G 4" Perforated P.V.C.
  2.20I Compacted Subgrade

2.22 Structural Excavation, Backfill and Compaction (Building Area)

2.60 Pavement and Walks
  2.60A Vehicular Area Sub-Base Course Granular
  2.60B Asphaltic Concrete - Binder Course
  2.60C Asphaltic Concrete Top - Wearing Course
  2.60D Concrete Walk/Paving
  2.60E Precast Concrete Curb
  2.60G Reinforcing Mesh
  2.60H Expansion Joint Filler
  2.60I Control Joint - 4'-0" O.C. Maximum Saw Cut or Tooled
  2.60J Stabilization Fabric
  2.60P Asphaltic Concrete-Base Course
  2.60Q Detectable Warning Pavers

2.61 Pavement Markings
  2.61B Painted ANSI Handicap Symbol
  2.61C Painted Traffic Control Lines

2.80 Landscaping
  2.80A Seeding

2.83 Chain Link Fence
  2.83D 2.875" O.C. Corner Post
  2.83E 2.875" O.C. Line Post
  2.83F 1-1/2" Top Rail
  2.83G 2" Mesh Fabric
  2.83H 7 Gauge Tension Wire Continuous
  2.83I Post Cap
  2.83J 11 Gauge Rail Clamps
  2.83K Sleeve
  2.83L 11 Gauge Bands
  2.83M 1/4" X 3/4" Turnbuckle
  2.83N Stretcher Rod
  2.83O 3/8" Diameter Truss Rod

2.83P   Pass Thru Cap
2.83Q   Gate Frame
2.83R   Top Hinge
2.83S   Hinge
2.83T   Fork Latch

3.30 Concrete
See Structural Drawings*
  3.30A   Cast-In-Place-Concrete
  3.30B   Reinforcing Bar
  3.30D   Welded Wire Mesh, Size as Noted

3.32 Concrete Slab On Grade

3.325 Concrete Slab on Metal Deck

3.40 Precast Concrete
  3.40D   Precast Dry Pit
  3.40E   Precast Concrete Wall Unit
  3.40F   Precast Concrete Cap Unit
  3.40G   Geogrid

4.10 Mortar and Masonry Grout
  4.10A   Grout Solid
  4.10B   Mortar Fill

4.20 Unit Masonry
  4.20A   Face Brick - Standard Modular
  4.20C   Concrete Masonry Units
  4.20D   Horizontal Joint Reinforcing
  4.20E   Flexible Masonry Anchors
  4.20G   Premolded Control Joint Strips
  4.20I   Reinforcing Bars
  4.20J   Flexible Masonry Column Tie Anchors
  4.20K   Weep Holes
  4.20M   Control Joint
  4.20P   Slate Sill - 1/4" Chamfer at Exposed Edges
  4.20S   Precast Concrete Sill
  4.20Y   Precast Concrete Band
  4.20AA Expansion Joint Closure

5.10 Structural Steel
See Structural Drawings*
  5.10A   Steel Beam
  5.10B   Steel Angle
  5.10D   Anchor Bolt
  5.10E   Steel Tube Column
  5.10F   Miscellaneous Steel - Shape and Size as Noted

5.10G   Steel Tee

5.20 Steel Joists
See Structural Drawings*

5.30 Metal Decking
See Structural Drawings*
5.30A   Metal Roof Decking

5.50 Miscellaneous Metals
5.50A   Miscellaneous Steel - Shape and Size as Noted
5.50B   Steel Angle
5.50C   1-1/4" Nominal Steel Pipe (1.66" O.D.) Post
         4'-0" O.C. Maximum
5.50D   10 Gauge Metal Riser, Tread and Landing Pan
5.50E   MC 12 X 10.6 Stringer
5.50F   3-1/2" X 2-1/2" X 1/4" - LLV - Pan Support at 3"-0" O.C.
         Maximum
5.50G   1-1/2" X 1-1/2" X 1/4" Steel Angle Carriers
5.50M   Galvanized Steel Bent Plate
5.50N   Galvanized Steel Flat Bar Stock
5.50O   3/4" Diameter Galvanized Steel Rungs
5.50S   Steel Bollards - Fill With Concrete - Bevel Concrete at Top
5.50U   1-1/4" Nominal Steel Pipe (1.66" O.D.)
5.50Y   Countersunk Machine Head Screw/Expansion
5.50AF Steel Plate
5.50AM 1" Nominal Steel Pipe Handrail (1.315" O.D.)
5.50AN 1" Nominal Steel Pipe Bracket
5.50AP Steel Wall Bracket, Julius Blum #622, or Equal
5.50AX Aluminum Sleeve (.050")
5.50AY Steel Plate, Size as Noted

5.80 Expansion Joints
5.80A Expansion Joint Cover

6.10 Rough Carpentry
6.10A   Wood Blocking
6.10B   Plywood
6.10C   Wood Furring
6.10I   Wood Framing 16" O.C.
6.10S   Joist Hanger
6.10T   Pressure Treated Bail; Size as Noted
6.10U   Pressure Treated Post; Size as Noted

6.20 Finish Carpentry
6.20F   Softwood Trim

7.15 Dampproofing

7.20 Insulation
7.20B   Batt Insulation
7.20C   Cavity Wall Insulation
7.20H   Metal Z Furring

7.41 Preformed Wall Panels
7.41C   Vertical Siding
7.41F   Trim to Match Siding

7.50 Roofing System (EPDM)
7.50A   Roofing Membrane
7.50B   Membrane Flashing
7.50C   Insulation
7.50D   Thermal Barrier
7.50E   Tapered Insulation
7.50F   Vapor Barrier/Air Seal
7.50G   Ballast and Protective Mat
7.50H   Fascia System
7.50K   Fascia Extender
7.50M   Aluminum Counterflashing
7.50N   Sealant
7.50P   Termination Bar
7.50Q   Expansion Joint Support
7.50R   Closed Cell Backer Rod
7.50T   Insert Drain
7.50V   Recovery Board
7.50W   Aluminum Trim - .032" Thickness

7.51 Roofing System (Silicone)
7.51A   Silicone Coating
7.51B   Urethane Insulation
7.51D   Sealant
7.51E   Aluminum Foam Stop - 0.050" Thickness
7.51J   Aluminum Trim - 0.32" Thickness
7.51K   Insert Drain

7.53 Roofing System (EPDM-Foam Adhesive)
7.53A   Roofing Membrane
7.53B   Foam Adhesive
7.53C   Membrane Flashing
7.53D   Insulation
7.53H   Fascia System
7.53K   Fascia Extender
7.53O   Sealant
7.53Q   Termination Bar
7.53R   Closed Cell Backer Rod
7.53T   Insert Drain

7.60 Flashing and Sheetmetal
  7.60A   Thru-wall Flashing - Turn-up at Ends at Wall Openings to Form Dam

7.72 Roof Accessories

7.90 Caulking and Sealants
  7.90A   Sealant and Backing Material (Size as Required to Fill Void)
  7.90B   Foam Joint Filler - Width Shown X Size Required to Fill Void)
  7.90C   Sealant Bead
  7.90G   Expansion Joint Filler

8.10 Metal Doors and Frames
  8.10A   Steel Hollow Metal Door Frame
  8.10B   Steel Hollow Metal Door
  8.10E   Metal Access Door
  8.10F   16 Gauge Hollow Metal Pipe Enclosure

8.15 Plastic Doors
See Door Schedule*
  8.15A   Plastic Door
  8.15B   Aluminum Door Frame System
  8.15C   .090" Break Metal to Match Door Frame System
  8.15G   Aluminum Angle (Size as Noted)

8.20 Wood Doors
See Door Schedule*
  8.20A   Wood Door

8.33 Rolling Counter Fire Door
  8.33B   Vertical Sliding Pass Window
  8.33C   Stainless Steel Sill

8.50 Metal Windows
  8.50A   Aluminum Windows
  8.50B   1/8" Break Metal to Match Windows
  8.50C   Extruded Aluminum Sill and Clip
  8.50D   Extruded Aluminum Head/Jamb Panning
  8.50E   Extruded Aluminum Trim
  8.50K   Fixed Aluminum Window
  8.50L   Horizontal Rolling Aluminum Window
  8.50R   Foam Tape
  8.50S   Foam Insulation
  8.50T   Double Hung Window

8.70 Hardware and Specialties
  See Door Schedule*
  8.70D   Aluminum Threshold Set in Mastic

8.80 Glazing
  8.80A   Tempered Insulating Glass
  8.80B   Insulating Glass
  8.80C   Safety Wire Glass
  8.80D   Tempered Glass
  8.80I   Muntins Inside Insulated Glass
  8.80K   Obscure Insulating Glass

09805   Encapsulation of Asbestos Containing Material

9.25 Gypsum Wallboard
  9.25A   5/8" Type 'X' Gypsum Wallboard
  9.25B   "J" Casing Bead
  9.25C   Metal Stud System
  9.25D   Metal Furring
  9.25G   Metal Stud Runner - 2" Deep to Provide 1" Expansion for Studs
  9.25J   Gypsum Sheathing
  9.25k   Carrying Channel
  9.25L   5/8" Exterior Gypsum Ceiling Board
  9.25N   Metal Angle Runner

9.30 Tile
See Room Finish Schedule*
  9.30A   Glazed Wall Tile
  9.30B   Unglazed Ceramic Mosaic Tile
  9.30D   Glazed Ceramic Tile Wall Base
  9.30E   Marble Thresholds - See Door Schedule*
  9.30F   Accent Title - Continuous Around Room
  9.30G   Unglazed Ceramic Mosaic Tile Base

9.40 Terrazzo
  9.40A   Thin Set Terrazzo
  9.40C   Terrazzo Cove Base

9.50 Acoustical Treatment
See Room Finish Schedule*
  9.50A   Suspended Ceiling System

9.65 Resilient Flooring
See Room Finish Schedule*
  9.65A   Vinyl Composition Tile
  9.65B   Vinyl Cove Base
  9.65D   Rubber Stair Treads and Risers
  9.65E   Molded Rubber Tile
  9.65J   Rubber Base

9.80 Special Coating System
  9.80A   Special Coating System

9.80B   Finish Coat
9.80C   Base Coat
9.80D   Reinforcing Mesh
9.80E   Insulation
9.80F   Gypsum Sheathing
9.80G   Metal Studs
9.80H   Sealant and Backer Rod
9.80I   Waterproof Base Coat
9.80J   Routed Joint
9.80K   Insulation Board Below Grade
9.80L   Below Grade Waterproofing
9.80M   Slide Clip
9.80N   Bent Galvanized Metal (Size and Gauge as Noted)
9.80O   Expansion Joint

9.90 Painting
See Room Finish Schedule*
9.90E-1 Paint E-1
9.90E-2 Paint E-2
9.90P-1 Paint P-1
9.90P-2 Paint P-2
9.90P-3 Paint P-3
9.90P-4 Paint P-4
9.90P-6 Paint P-6
9.90P-8 Paint P-8

10.10 Chalkboards and Tackboards
10.10B Liquid Marker Board
10.10C Tackboard
10.10F Projection Screen

10.25 Firefighting Devices
10.25A Fire Extinguisher Cabinet. Paint P-2. Model #2409-R2 By Larson
        Manufacturing Co., or Approved Equal

10.42 Signage and Graphics
10.42A 12" X 18" Aluminum Handicap Sign

10.50 Lockers
10.50A Single Tier Lockers
10.50G Sloping Top
10.50H Metal Base
10.50I Recessed Trim

10.62 Folding Partitions
10.62.A Folding Partitions
10.62B Tracks, Support Brackets, Hangar Rods and Finish Trim
10.62D Liquid Marker Board

10.62E Tackboard

10.80 Toilet Accessories
10.80C Mirror
10.80D Grab Bars

11.46 Unit Kitchens
11.46A Compact Kitchen Unit

12.17 Entrance Mats
12.17A Recessed Entrance Mat

12.30 Casework
See Drawings for Casework
12.30A Plastic Laminate Counter with 4" Backsplash
12.30D Filler Strip - Size as Required
12.30G Base Cabinet
12.30I Lab Countertop
12.30K Plastic Laminate Top
12.30L Wall Cabinet

14.20 Elevators
14.20A Elevator

*See reference noted for information elaborating on or work in addition to materials noted here. Provide all materials as required for a complete and proper installation.

Glossary

Addendum A change or modification to the bid documents, plans, and specifications that is made prior to the contractor's bid date

Aggregate Hard materials such as sand and crushed stone used to make concrete

Ampere (AMP) Unit of measure of electric current

Anchor Bolt A bolt placed in the surface of concrete for attaching wood framing members

Apron Concrete slab at the approach to a garage door—also the wood trim below a window stool

Architect's Scale A flat or triangular scale used to measure scale drawings

Ash Dump A small metal door in the bottom of a fireplace

Awning Window A window that is hinged near the top so the bottom opens outward

Backfill Earth placed against a building wall after the foundation is in place

Backsplash The raised lip on the back edge of a countertop to prevent water from running down the backs of the cabinets

Balloon Framing Type of construction in which the studs are continuous from the sill to the top of the wall—upper floor joists are supported by a let-in ribbon

Balusters Vertical pieces which support a handrail

Batt Insulation Flexible, blanket like pieces, usually of fiberglass used for thermal or sound insulation

Batten Narrow strip of wood used to cover joints between boards of sheet materials

Batten Boards An arrangement of stakes and horizontal pieces used to attach lines for laying out a building

Beam Any major horizontal structural member

Beam Pocket A recessed area to hold the end of a beam in a concrete or masonry wall

Board Foot One hundred forty-four cubic inches of wood or the amount contained in a piece measuring 12" × 12" × 1"

Bottom Chord The bottom horizontal member in a truss

Box Sill The header joist nailed across the ends of floor joists at the sill

Branch Circuit The electrical circuit that carries current from the distribution panel to the various parts of the building

British Thermal Unit (BTU) The amount of heat required to raise the temperature of one pound of water one degree Fahrenheit

Building Lines The outside edge of the exterior walls of a building

Casement Window A window that is hinged at one side so the opposite side opens outward

Casing The trim around a door or window

Centerline An actual or imaginary line through the exact center of any object

Change Order A change or modification to the contract documents, plans, and specifications that is made after the contract has been awarded to the selected trade contractor

Cleanout A pipe fitting with a removable plug that allows for cleaning the run of piping in which it is installed or an access door at the bottom of a chimney

Collar Beam Horizontal members that tie opposing rafters together, usually installed about half way up the rafters

Column A metal post to support an object above

Common Rafter A rafter extending from the top of the wall to the ridge

Concrete Building material consisting of find and coarse aggregates bonded together by portland cement

Conductor Electrical wire—a cable may contain several conductors

Contour Lines Lines on a topographic map or site plan to describe the contour of the land

Contract Any agreement in writing for one party to perform certain work and the other party to pay for the work

Convenience Outlet Electrical outlet provided for convenient use of lamps, appliances, and other electrical equipment

Cornice The construction which encloses the ends of the rafters at the top of the wall

Cornice Return The construction where the level cornice meets the sloping rake cornice

Course A single row of building units such as concrete blocks or shingles

Cove Mold Concave molding used to trim an inside corner

Damper A door installed in the throat of a fireplace to regulate the draft

Dampproofing Vapor barrier or coating on foundation walls or under concrete slabs to prevent moisture from entering the house

Datum A reference point from which elevations are measured

Dormer A raised section in a roof to provide extra headroom below

Detail A drawing showing special information about a particular part of the construction—details are usually drawn to a larger scale than other drawings and are sometimes section views

Double-hung Window A window consisting of two sash which slide up and down past one another

Drip Cap A wood ledge over wall openings to prevent water from running back under the frame or trim around the opening

Drip Edge   Metal trim installed at the edge of a roof to stop water from running back under the edge of the roof deck

Drywall   Interior wall construction using gypsum wallboard

Elevation   A drawing that shows vertical dimensions—it may also be the height of a point, usually in feet above sea level

Fascia   The part of a cornice that covers the ends of the rafters

Firestop   Blocking or noncombustible material between wall studs to prevent vertical draft and flamespread

Flashing   Sheet metal used to cover openings and joints in walls and roofs

Float   To level concrete before it begins to cure—floating is done with a tool called a float

Floor Plan   A drawing showing the arrangement of rooms, the locations of windows and doors, and complete dimensions—a floor plan is actually a horizontal section through the entire building

Flue   The opening inside a chimney—the flue is usually formed by a terra cotta flue liner

Flush Door   A door having flat surfaces

Footing   The concrete base upon which the foundation walls are built

Footing Drain (See *Perimeter Drain*)

Frieze   A horizontal board beneath the cornice and against the wall above the siding

Frost Line   The maximum depth to which frost penetrates the earth

Furring   Narrow strips of wood attached to a surface for the purpose of creating a plumb or level surface for attaching the wall, ceiling, or floor surface

Gable   The triangular area between the roof and the top plate walls at the ends of a gable roof

Gable Studs   The studs placed between the end rafters and the top plates of the end walls

Gauge   A standard unit of measurement for the diameter of wire or the thickness of sheet metal

Girder   A beam which supports floor joists

Grout   A thin mixture of high-strength concrete or mortar

Gypsum Wallboard   Drywall materials made of gypsum encased in paper to form boards

Header   A joist fastened across the ends of regular joists in an opening, or the framing member above a window or door opening

Hearth   Concrete or masonry apron in front of a fireplace

Hip   Outside corner formed by intersecting roofs

Hip Rafter   The rafter extending from the corner of a building to the ridge at a hip

Hose Bibb   An outside faucet to which a hose can be attached

Insulated Glazing   Two or more pieces of glass in a single sash with air space between them for the purpose of insulation

Isometric   A kind of drawing in which horizontal lines are 30 degrees from true horizontal and vertical lines are vertical

Jack Rafter   Rafter between the outside wall and a hip rafter or the ridge and a valley rafter

Jamb   Side members of a door or window frame

Joists   Horizontal framing members that support a floor or ceiling

Lintel   Steel or concrete member that spans a clear opening—usually found over doors, windows, and fireplace openings

Masonry Cement   Cement which is specially prepared for making mortar

Mil   A unit of measure for the thickness of very thin sheets—one mil equals .001"

Miter   A 45-degree cut so that two pieces will form a 90-degree corner

Mortar   Cement and aggregate mixture for bonding masonry units together

Mullion   The vertical piece between two windows that are installed side by side—window units that include a mullion are called mullion windows

Muntin   Small vertical and horizontal strips that separate the individual panes of glass in a window sash

Nailer   A piece of wood used in any of several places to provide a nailing surface for other framing members

Nominal Size   The size by which a material is specified—the actual size is often slightly smaller

Nosing   The portion of a stair tread that projects beyond the riser

Orthographic Projection   A method of drawing that shows separate views of an object

Panel Door   A door made up of panels held in place by rails and stiles

Parging   A thin coat of portland-cement plaster used to smooth masonry walls

Penny Size   The length of nails

Perimeter Drain (Also *Footing Drain*)   An underground drain pipe around the footings to carry ground water away from the building

Pilaster   A masonry or concrete pier built as an integral part of a wall

Pitch   Refers to the steepness of a roof—the pitch is written as a fraction with the rise over the span

Plate   The horizontal framing members at the top and bottom of the wall studs

Platform Framing (Also called *Western Framing*)   A method of framing in which each level is framed separately—the subfloor is laid for each floor before the walls above it are formed

Plenum   A chamber within a forced-air heating system that is pressurized with warm air

Plumb   Truly vertical or true according to a plumb bob

Portland Cement   Finely powdered limestone material used to bond the aggregates together in concrete and mortar

R-value   The ability of a material to resist the flow of heat

Rafter   The framing members in a roof

Rail   The horizontal members in a door, sash, or other panel construction

Rake   The sloping cornice at the end of a gable roof

Resilient Flooring   Vinyl, vinyl-asbestos, and other man-made floor coverings that are flexible yet produce a smooth surface

Ridge Board   The framing member between the tops of rafters which runs the length of the ridge of a roof

Rise   The vertical dimension of a roof or stair

Riser   The vertical dimension of one step in a stair—the board enclosing the space between two treads is called a riser

Rowlock   Position of bricks in which the bricks are laid on edge

Run   The horizontal distance covered by an inclined surface such as a rafter or stair

Sash   The frame holding the glass in a window

Saturated Felt Paperlike felt which has been treated with asphalt to make it water resistant

Screed A straight board used to level concrete immediately after it is placed

Section View A drawing showing what would be seen by cutting through a building or part

Setback The distance from a street or front property line to the front of a building

Sheathing The rough exterior covering over the framing members of a building

Shim The framing member in contact with a masonry or concrete foundation

Sill The framing member in contact with a masonry or concrete foundation

Sill Sealer Compressible material used under the sill to seal any gaps

Site Constructed Built on the job

Site Plan The drawing that shows the boundaries of the building, its location, site utilities

Sliding Window A window with two or more sash that slide horizontally past one another

Soffit The bottom surface of any part of a building, such as the underside of a cornice or lowered portion of a ceiling over wall cabinets

Soldier Brick position in which the bricks are stood on end

Span The horizontal dimension between vertical supports—the span of a beam is the distance between the posts that support it

Specifications Written description of materials or construction

Square The amount of siding or roofing materials required to cover 100 square feet

Stack The main vertical pipe into which plumbing fixtures drain

Stair Carriage The supporting framework under a stair

Stile The vertical members in a sash, door, or other panel construction

Stool Trim piece that forms the finished window sill

Stop Molding that stops a door from swinging through the opening as it is closed—also used to hold the sash in place in a window frame

Stud Vertical framing members in a wall

Subfloor The first layer of rough flooring applied to the floor joists

Sweat Method of soldering used in plumbing

Termite Shield Sheet-metal shield installed at the top of a foundation to prevent termites from entering the wood superstructure

Thermal-break Window Window with a metal frame that has the interior and exterior separated by a material with a higher R-value

Thermostat An electrical switch that is activated by changes in temperature

Top Chord The top horizontal member of a truss

Trap A plumbing fitting that holds enough water to prevent sewer gas from entering the building

Tread The surface of a step in stair construction

Trimmers The double framing members at the sides of an opening

Truss A manufactured assembly used to support a load over a long span

Underlayment Any material installed over the subfloor to provide a smooth surface over which floor covering will be installed

Valley The inside corner formed by intersecting roofs

Valley Rafter The rafter extending from an inside corner in the walls to the ridge at a valley

Vapor Barrier Sheet material used to prevent water vapor from passing through a building surface

Veneer A thin covering—in masonry, a single wythe of finished masonry over a wall—in a woodwork, a thin layer of wood

Vent Pipe A pipe, usually through the roof, that allows atmospheric pressure into the drainage system

Vertical Contour Interval The difference in elevation between adjacent contour lines on a topographic map or site plan

Volt The unit of measurement for electrical force

Water Closet A plumbing fixture commonly called *toilet*

Watt The unit of measurement of electrical power—one watt is the amount of power from one ampere of current with one volt of force

Weep Hole A small hole through a masonry wall to allow water to pass

Western Framing (See *Platform Framing*)

Wythe A single thickness of masonry construction

Math Reviews

MATH REVIEW 1
FRACTIONS AND MIXED NUMBERS
—MEANINGS AND DEFINITIONS

- A *fraction* is a value which shows the number of equal parts taken from a whole quantity. A fraction consists of a numerator and a denominator.

$\dfrac{7}{16}$ ←Numerator
      ←Denominator

- *Equivalent fractions* are fractions which have the same value. The value of a fraction is not changed by multiplying the numerator and denominator by the same number.

  Example   Express $\dfrac{5}{8}$ as thirty-seconds.

  Determine what number the denominator is multiplied by to get the desired denominator. (32 ÷ 8 = 4)

  $\dfrac{5}{8} = \dfrac{?}{32}$

  Multiply the numerator and denominator by 4.

  $\dfrac{5}{8} \times \dfrac{4}{4} = \dfrac{20}{32}$

- The *lowest common denominator* of two or more fractions is the smallest denominator which is evenly divisible by each of the denominators of the fractions.

  Example 1   The lowest common denominator of $\dfrac{3}{4}$, $\dfrac{5}{8}$, and

  $\dfrac{13}{32}$ is 32, because is the smallest number evenly divisible by 4, 8, and 32.

  $32 \div 4 = 8$
  $32 \div 8 = 4$
  $32 \div 32 = 1$

  Example 2   The lowest common denominator of $\dfrac{2}{3}$, $\dfrac{1}{5}$, $\dfrac{7}{10}$ and is

  30, because is 30 is the smallest number evenly divisible by 3, 5, and 10.

  $30 \div 3 = 10$
  $30 \div 5 = 6$
  $30 \div 10 = 3$

- *Factors* are numbers used in multiplying. For example, 3 and 5 are factors of 15.

  $3 \times 5 = 15$

- A fraction is in its *lowest terms* when the numerator and the denominator **do not** contain a common factor.

  Example   Express $\dfrac{12}{16}$ in lowest terms.

  Determine the largest common factor in the numerator and denominator. The numerator and the denominator can be evenly divided by 4.

  $\dfrac{12 \div 4}{16 \div 4} = \dfrac{3}{4}$

- *A mixed number* is a whole number plus a fraction.

  6 $\dfrac{15}{16}$

  Whole Number ⟶   ⟵ Fraction

  $6 + \dfrac{15}{16} = 6\dfrac{15}{16}$

- *Expressing fractions as mixed numbers.* In certain fractions, the numerator is larger than the denominator. To express the fraction as a mixed number, divide the numeratorby the denominator. Express the fractional part in lowest terms.

    Example   Express $\dfrac{38}{16}$ as a mixed number.

    Divide the numerator 38 by the denominator 16.

    $$\dfrac{38}{16} = 2\,\dfrac{6}{16}$$

    Express the fractional part $\dfrac{6}{16}$ in lowest terms.

    $$\dfrac{6 \div 2}{16 \div 2} = \dfrac{3}{8}$$

    Combine the whole number and fraction.

    $$\dfrac{38}{16} = 2\,\dfrac{3}{8}$$

- *Expressing mixed numbers as fractions.* To express a mixed number as a fraction, multiply the whole number by the denominator of the fractional part. Add the numerator of the fractional part. The sum is the numerator of the fraction. The denominator is the same as the denominator of the original fractional part.

    Example   Express $7\,\dfrac{3}{4}$ as a fraction.

    $$\dfrac{7 \times 4 + 3}{4} = \dfrac{31}{4}$$

    or

    Multiply the whole number 7 by the denominator 4 of the fractional part ($7 \times 4 = 28$). Add the numerator 3 of the fractional part to 28. The sum 31 is the numerator of the fraction. The denominator 4 is the same as the denominator of the original fractional part.

    $$\dfrac{7}{1} \times \dfrac{4}{4} = \dfrac{28}{4}$$

    $$\dfrac{28}{4} + \dfrac{3}{4} = \dfrac{31}{4}$$

MATH REVIEW 2
ADDING FRACTIONS

- Fractions must have a common denominator in order to be added.

- To add fractions, express the fractions as equivalent fractions having the lowest common denominator. Add the numerators and write their sum over the lowest common denominator. Express the fraction in lowest terms.

    Example   Add: $\dfrac{3}{8} + \dfrac{3}{4} + \dfrac{3}{16} + \dfrac{1}{32}$

    Express the fractions as equivalent fractions with 32 as the denominator.

    Add the numerators.

    $$\dfrac{3}{8} = \dfrac{3}{8} \times \dfrac{4}{4} = \dfrac{12}{32}$$
    $$\dfrac{1}{4} = \dfrac{1}{4} \times \dfrac{8}{8} = \dfrac{8}{32}$$
    $$\dfrac{3}{16} = \dfrac{3}{16} \times \dfrac{2}{2} = \dfrac{6}{32}$$
    $$+\dfrac{1}{32} = \qquad \dfrac{1}{32}$$
    $$\qquad\qquad\qquad \dfrac{27}{32}$$

- After fractions are added, if the numerator is greater than the denominator, the fraction should be expressed as a mixed number.

    Example   Add: $\dfrac{1}{2} + \dfrac{3}{4} + \dfrac{15}{16} + \dfrac{11}{16}$

    Express the fractions as equivalent fractions with 16 as the denominator.

    Add the numerators.

    $$\dfrac{1}{2} = \dfrac{1}{2} \times \dfrac{8}{8} = \dfrac{8}{16}$$
    $$\dfrac{3}{4} = \dfrac{3}{4} \times \dfrac{4}{4} = \dfrac{12}{16}$$
    $$\dfrac{15}{16} = \qquad \dfrac{15}{16}$$
    $$+\dfrac{11}{16} = \qquad \dfrac{11}{16}$$
    $$\qquad\qquad\qquad \dfrac{46}{16}$$

    Express $\dfrac{46}{16}$ as a mixed number in lowest terms.

    $$\dfrac{46}{16} = 2\,\dfrac{14}{16} = 2\,\dfrac{7}{8}$$

MATH REVIEW 3
ADDING COMBINATIONS OF FRACTIONS, MIXED NUMBERS, AND WHOLE NUMBERS

* To add mixed numbers or combinations of fractions, mixed numbers, and whole numbers, express the fractional parts of the numbers as equivalent fractions having the lowest common denominator. Add the whole numbers. Add the fractions. Combine the whole number and the fraction and express in lowest terms.

Example 1   Add: $3\dfrac{7}{8} + 5\dfrac{1}{2} + 9\dfrac{3}{16}$

Express the fractional parts as equivalent fractions with 16 as the common denominator. Add the whole numbers. Add the fractions. Combine the whole number and the fraction. Express the answer in lowest terms.

Example 2   Add: $6\dfrac{3}{4} + \dfrac{9}{16} + 7\dfrac{21}{32} + 15$

Express the fractional parts as equivalent fractions with 32 as the common denominator. Add the whole numbers. Add the fractions. Combine the whole number and the fraction. Express the answer in lowest terms.

$$3\dfrac{7}{8} = 3\dfrac{14}{16}$$
$$5\dfrac{1}{2} = 5\dfrac{8}{16}$$
$$+9\dfrac{3}{16} = 9\dfrac{3}{16}$$
$$17\dfrac{25}{16} = 17 + 1\dfrac{9}{16} = 18\dfrac{9}{16}$$

$$6\dfrac{3}{4} = 6\dfrac{24}{32}$$
$$\dfrac{9}{16} = \dfrac{18}{32}$$
$$7\dfrac{21}{32} = 7\dfrac{21}{32}$$
$$+15 = 15$$
$$28\dfrac{63}{32} = 28 + 1\dfrac{31}{32} = 29\dfrac{31}{32}$$

MATH REVIEW 4
SUBTRACTING FRACTIONS FROM FRACTIONS

* Fractions must have a common denominator in order to be subtracted.

* To subtract a fraction from a fraction, express the fractions as equivalent fractions having the lowest common denominator. Subtract the numerators. Write their difference over the common denominator.

Example   Subtract $\dfrac{3}{4}$ from $\dfrac{15}{16}$

Express the fractions as equivalent fractions with 16 as the common denominator. Subtract the numerator 12 from the numerator 15. Write the difference 3 over the common denominator 16.

$$\dfrac{15}{16} = \dfrac{15}{16}$$
$$-\dfrac{3}{4} = -\dfrac{12}{16}$$
$$\dfrac{3}{16}$$

MATH REVIEW 5
SUBTRACTING FRACTIONS AND MIXED NUMBERS FROM WHOLE NUMBERS

* To subtract a fraction or a mixed number from a whole number, express the whole number as an equivalent mixed number. The fraction of the mixed number has the same denominator as the denominator of the fraction which is subtracted. Subtract the numerators of the fractions and write their difference over the common denominator. Subtract the whole numbers. Comnine the whole number and fractiopn. Express the answer in lowest terms.

Example 1   Subtract $\frac{3}{8}$ from 7.

Express the whole number as an equivalent mixed number with the same denominator as the denominator of the fraction which is subtracted ($7 = 6\frac{8}{8}$).

$$
\begin{aligned}
7 &= 6\,\frac{8}{8} \\
-\frac{3}{8} &= -\frac{3}{8} \\
\hline
&\quad\; 6\,\frac{5}{8}
\end{aligned}
$$

Subtract $\frac{3}{8}$ from $\frac{8}{8}$

Combine whole number and fraction.

Example 2   Subtract $5\frac{15}{32}$ from 12

Express the whole number as an equivalent mixed number with the same denominator as the denominator of fraction which is sunbtracted ($12 = 11\frac{32}{32}$).

$$
\begin{aligned}
12 &= 11\,\frac{32}{32} \\
-5\,\frac{15}{32} &= -5\,\frac{15}{32} \\
\hline
&\quad\; 6\,\frac{17}{32}
\end{aligned}
$$

Subtract fractions.

Subtract whole numbers.

Combine whole number and fraction.

MATH REVIEW 6
SUBTRACTING FRACTIONS AND
MIXED NUMBERS FROM MIXED NUMBERS

- To subtract a fraction or a mixed number from a mixed number, the fractional part of each number must have the same denominator. Express fractions as equivalent fractions having a common denominator. When the fraction subtracted is larger than the fraction from which it is subtracted, one unit of a whole number is expressed as a fraction with the common denominator. Combine the whole number and fractions. Subtract fractions and subtract whole numbers.

Example 1   Subtract $\frac{7}{8}$ from $4\frac{3}{16}$

Express the fractions as equivalent fractions with the common denominator 16. Since 14 is larger than 3, express one unit of $4\frac{3}{16}$ as a fraction and comnine whole number and fractions.

$(4\,\frac{3}{16} = 3 + \frac{16}{16} + \frac{3}{16} = 3\,\frac{19}{16})$.

$$
\begin{aligned}
4\,\frac{3}{16} &= 4\,\frac{3}{16} = 3\,\frac{19}{16} \\
-\frac{7}{8} &= -\frac{14}{16} = -\frac{14}{16} \\
\hline
&\qquad\qquad\quad 3\,\frac{5}{16}
\end{aligned}
$$

Subtract.

Example 2   Subtract $13\frac{1}{4}$ from $20\frac{15}{32}$

Express the fractions as equivalent fractions with the common denominator 32.

$$
\begin{aligned}
20\,\frac{15}{32} &= 20\,\frac{15}{32} \\
-13\,\frac{1}{4} &= -13\,\frac{8}{32} \\
\hline
&\quad\; 7\,\frac{7}{32}
\end{aligned}
$$

Subtract fractions.

Subtract whole numbers.

MATH REVIEW 7
MULTIPLYING FRACTIONS

- To multiply two or more fractions, multiply the numerators. Multiply the denominators. Write as a fraction with the product of the numerators over the product of the denominators. Express the answer in lowest terms.

Example 1 Multiply $\dfrac{3}{4} \times \dfrac{5}{8}$

Multiply the numerators.

Multiply the denominators.

Write as a fraction.

$$\frac{3}{4} \times \frac{5}{8} = \frac{15}{32}$$

Example 2 Multiply $\dfrac{1}{2} \times \dfrac{2}{3} \times \dfrac{4}{5}$

Multiply the numerators.

Multiply the denominators.

Write as a fraction and express answer in lowest terms.

$$\frac{1}{2} \times \frac{2}{3} \times \frac{4}{5} = \frac{8}{30} = \frac{4}{15}$$

MATH REVIEW 8
MULTIPLYING ANY COMBINATION OF
FRACTIONS, MIXED NUMBERS, AND WHOLE NUMBERS

- To multiply any combination of fractions, mixed numbers, and whole numbers, write the mixed numbers as fractions. Write whole numbers over the denominator 1. Multiply numerators. Multiply denomionators. Express the answer in lowes terms.

Example 1 Multiply $3\dfrac{1}{4} \times \dfrac{3}{8}$

Write the mixed number $3\dfrac{1}{4}$ as the fraction $\dfrac{13}{4}$.

Multiply the numerators.

Multiply the denominators.

Express as a mixed number.

$$3\frac{1}{4} \times \frac{3}{8} = \frac{13}{4} \times \frac{3}{8} = \frac{39}{32} = 1\frac{7}{32}$$

Example 1 Multiply $2\dfrac{1}{3} \times 4 \times \dfrac{4}{5}$

Write the mixed number $2\dfrac{1}{3}$ as the fraction $\dfrac{7}{3}$.

Write the whole number 4 over 1.

Multiply the numerators.

Multiply the denominators.

Express as a mixed number.

$$2\frac{1}{3} \times 4 \times \frac{4}{5} = \frac{7}{3} \times \frac{4}{1} \times \frac{4}{5} = \frac{112}{15}$$
$$\frac{112}{15} = 7\frac{7}{15}$$

MATH REVIEW 9
DIVIDING FRACTIONS

- Division is the inverse of multiplication. Dividing by 4 is the same as multiplying by $\frac{1}{4}$. Four is the inverse of $\frac{1}{4}$ and $\frac{1}{4}$ is the inverse of 4. The inverse of $\frac{5}{16}$ is $\frac{16}{5}$.

- To divide fractions, invert the divisor, change to the inverse operation and multiply. Express the answer in lowest terms.

  Example Divide: $\frac{7}{8} \div \frac{2}{3}$

  Invert the divisor $\frac{2}{3}$

  $\frac{2}{3}$ inverted is $\frac{3}{2}$.

  Change to the inverse operation and multiply.

  Express as a mixed number.

$$\frac{7}{8} \div \frac{2}{3} = \frac{7}{8} \times \frac{3}{2} = \frac{21}{16} = 1\frac{5}{16}$$

MATH REVIEW 10
DIVIDING ANY COMBINATION OF
FRACTION, MIXED NUMBERS, AND WHOLE NUMBERS

- To divide any combination of fractions, mixed numbers, and whole numbers, write the mixed number as fractions. Write whole numbers over the denominator 1. Invert the divisor. Change to the inverse operation and multiply. Express the answer in lowest terms.

  Example 1 Divide: $6 \div \frac{7}{10}$

  Write the whole number 6 over the denominator 1.

  Invert the divisor $\frac{7}{10}$; $\frac{7}{10}$ inverted is $\frac{10}{7}$.

  Change to the inverse operation and multiply.

  Express as a mixed number.

$$\frac{6}{1} \div \frac{7}{10} =$$
$$\frac{6}{1} \times \frac{10}{7} = \frac{60}{7} = 8\frac{4}{7}$$

  Example 2 Divide: $\frac{3}{4} \div 2\frac{1}{5}$

  Write the mixed number divisor $2\frac{1}{5}$ as the fraction $\frac{11}{5}$.

  Invert the divisor $\frac{11}{5}$; $\frac{11}{5}$ inverted is $\frac{5}{11}$.

  Change to the inverse operation and multiply.

$$\frac{3}{4} \div \frac{11}{5} =$$
$$\frac{3}{4} \times \frac{5}{11} = \frac{15}{44}$$

  Example 3 Divide: $4\frac{5}{8} \div 7$

  Write the mixed number $4\frac{5}{8}$ as the fraction $\frac{37}{8}$.

  Write the whole number divisor over the denominator 1.

  Invert the divisor $\frac{7}{1}$; $\frac{7}{1}$ inverted is $\frac{1}{7}$.

  Change to the inverse operation and multiply.

$$\frac{37}{8} \div \frac{7}{1} =$$
$$\frac{37}{8} \times \frac{1}{7} = \frac{37}{56}$$

MATH REVIEW 11
ROUNDING DECIMAL FRACTIONS

- To round a decimal fraction, locate the digit in the number that gives the desired number of decimal places. Increase that digit by 1 if the digit which directly follows is 5 or more. Do not change the value of the digit if the digit which dollows is less than 5. Drop all digits which follow.

Example 1   Round 0.63861 to 3 decimal places.

Locate the digit in the third place (8). The fourth decimal-place digit, 6, is greater than 5 and increases the third decimal-place digit 8, to 9. Drop all digits which follow.

$0.63\underline{8}61 \approx 0.639$

Example 2   Round 3.0746 to 2 decimal places.

Locate the digit in the second decimal place (7). The third decimal-place digit 4 is less than 5 and does not change the value of the second decimal-place digit 7. Drop all digits which follow.

$3.0\underline{7}46 \approx 3.07$

MATH REVIEW 12
ADDING DECIMAL FRACTIONS

- To add decimal fractions, arrange the numbers so that the decimal points are directly under each other. The decimal point of a whole number is directly to the right of the last digit. Add each column as with whole numbers. Place the decimal point in the sum directly under the other decimal points.

Example   Add: 7.65 + 208.062 + 0.009 + 36 + 5.1037

Arrange the numbers so that the decimal points are directly under each other.

Add zeros so that all numbers have the same number of places to the right of the decimal point.

Add each column of numbers.

Place the decimal point in the sum directly under the other decimal points.

$$
\begin{array}{r}
7.6500 \\
208.0620 \\
0.0090 \\
36.0000 \\
+ \quad 5.1037 \\
\hline
256.8247
\end{array}
$$

MATH REVIEW 13
SUBTRACTING DECIMAL FRACTIONS

• To subtract decimal fractions, arrange the numbers so that the decimal points are directly under each other. Subtract each column as with whole numbers. Place the decimal point in the difference directly under the other decimal points.

Example   Subtract: 87.4 − 42.125

Arrange the numbers so that the decimal points are directly under each other. Add zeros so that the numbers have the same number of places to the right of the decimal point.

Subtract each column of numbers.

Place the decimal point in the difference directly under the other decimal points.

$$\begin{array}{r} 87.400 \\ -42.125 \\ \hline 45.275 \end{array}$$

MATH REVIEW 14
MULTIPLYING DECIMAL FRACTIONS

• To multiply decimal fractions, multiply using the same procedure as with whole numbers. Count the number of decimal places in both the multiplier and the multiplicand. Begin counting from the last digit on the right of the product and place the decimal point the same number of places as there are in both the multiplicand and the multiplier.

Example   Multiply: 50.216 × 1.73

Multiply as with whole numbers.

Count the number of decimal places in the multiplier (2 places) and the multiplicand (3 places).

Beginning at the right of the product, place the decimal point the same number of places as there are in both the multiplicand and the multiplier (5 places).

50.216 ← Multiplicand (3 places)
× 1.73 ← Multiplier (2 places)

$$\begin{array}{r} 150648 \\ 351512 \\ 50216 \\ \hline 86.87368 \end{array}$$ (5 places)

• When multiplying certain decimal fractions, the product has a smaller number of digits than the number of decimal places required. For these products, add as many zeros to the left of the product as are necessary to give the required number of decimal places.

Example   Multiply: 0.27 × 0.18

Multiply as with whole numbers.

The product must have 4 decimal places.

Add one zero to the left of the product.

$$\begin{array}{r} 0.27 \text{ (2 places)} \\ \times \ 0.18 \text{ (2 places)} \\ \hline 216 \\ 27 \\ \hline 0.0486 \text{ (4 places)} \end{array}$$

MATH REVIEW 15
DIVIDING DECIMAL FRACTIONS

• To divide decimal fractions, use the same procedure as with whole numbers. Move the decimal point of the divisor as many places to the right as necessary to make the divisor a whole number. Move the decimal point of the dividend the same number of places to the right. Add zeros to the dividend if necessary. Place the decimal point in the answer directly above the decimal point in the dividend. Divide as with whole numbers. Zeros may be added to the dividend to five the number of decimal places required in the answer.

Example 1   Divide: 0.6150 ÷ 0.75

Move the decimal point 2 places to the right in the divisor.

Move the decimal point 2 places in the dividend.

Place the decimal point in the answer directly above the decimal point in the dividend.

Divide as with whole numbers.

$$
\begin{array}{r}
0.82 \\
\text{Divisor} \rightarrow 0\,75.\,)0\,61.50 \leftarrow \text{Dividend} \\
\underline{60\ 0} \\
1\ 50 \\
\underline{1\ 50}
\end{array}
$$

Example 2   Divide: 10.7 ÷ 4.375. Round the answer to 3 decimal places.

Move the decimal point 3 places to the right in the divisor.

Move the decimal point 3 places in the dividend, adding 2 zeros.

Place the decimal point in the answer directly above the decimal point in the dividend.

Add 4 zeros to the dividend. One more zero is added than the number of decimal places required in the answer.

Divide as with whole numbers.

$$
\begin{array}{r}
2.4457 \approx 2.446 \\
4\,375.\,)10\,700.0000 \\
\underline{8\ 750} \\
1\ 950\ 0 \\
\underline{1\ 750\ 0} \\
200\ 00 \\
\underline{175\ 00} \\
25\ 00 \\
\underline{21\ 875} \\
3\ 1250 \\
\underline{3\ 0625} \\
625
\end{array}
$$

MATH REVIEW 16
EXPRESSING COMON FRACTIONS AS DECIMAL FRACTIONS

• A common fraction is an indicated division. A common fraction is expressed as a decimal fraction by dividing the numerator by the denominator.

Example   Express $\frac{5}{8}$ as a decimal fraction.

Write $\frac{5}{8}$ as an indicated division.

Place a decimal point after the 5 and add zeros to the right of the decimal point.

Place the decimal point for the answer directly above the decimal point in the dividend.

Divide.

$$
\begin{array}{r}
8\overline{)5} \\
\ \\
8\overline{)5.000} \\
\ \\
8\overline{)5.000} \\
\ \\
0.625 \\
8\overline{)5.000}
\end{array}
$$

- A common fraction which will not divide evenly is expressed as a repeating decimal.

  Example   Express $\frac{1}{3}$ as a decimal fraction.

  Write $\frac{1}{3}$ as an indicated division.

  $3\overline{)1}$

  Place a decimal point after the 1 and add zeros to the right of the decimal point.

  $3\overline{)1.0000}$

  Place the decimal point for the answer directly above the decimal point in the dividend.

  $3\overline{)1.0000}^{\,\cdot}$

  Divide.

  $\begin{array}{r} 0.3333 \\ 3\overline{)1.0000} \end{array}$

MATH REVIEW 17
EXPRESSING DECIMAL FRACTIONS AS COMMON FRACTIONS

- To express a decimal fraction as a common fraction, write the number after the decimal point as the numerator of a common fraction. Write the denominator as 1 followed by as many zeros as there are digits to the right of the decimal point. Express the common fraction in lowest terms.

  Example 1   Express 0.9 as a common fraction.

  Write 9 as the numerator.

  Write the denominator as 1 followed by 1 zero. The denominator is 10.

  $\frac{9}{10}$

  Example 2   Express 0.125 as a common fraction.

  Write 125 as the numerator.

  Write the denominator as 1 followed by 3 zeros. The denominator is 1000.

  $\frac{125}{1000}$

  Express the fraction in lowest terms.

  $\frac{125}{1000} = \frac{1}{8}$

MATH REVIEW 18
EXPRESSING INCHES AS FEET AND INCHES

- There are 12 inches in 1 foot.

- To express inches as feet and inches, divide the given length in inches by 12 to obtain the number of whole feet. The remainder is the number of inches in addition to the number of whole feet. The answer is the number of whole feet plus the remainder in inches.

  Example 1   Express $176\frac{7}{16}$ inches as feet and inches.

  Divide $176\frac{7}{16}$ inches by 12.

$$
\begin{array}{r}
14 \quad \text{(feet)} \\
12\overline{)176\frac{7}{16}} \\
12 \\
\overline{56} \\
48 \\
\overline{8\frac{7}{16}} \leftarrow \text{Remainder (feet)}
\end{array}
$$

There are 14 feet plus a remainder of $8\frac{7}{16}$ inches.

$14' - 8\frac{7}{16}''$

Example 2   Express 54.2 inches as feet and inches.

Divide 54.2 inches by 12.

There are 4 feet plus a remainder of 6.2 inches.

$$
\begin{array}{r}
4 \quad \text{(feet)} \\
12 \overline{)\,54.2} \\
\underline{48} \\
6.2
\end{array}
$$
← Remainder (inches)

4 feet 6.2 inches

MATH REVIEW 19
EXPRESSING FEET AND INCHES AS INCHES

- There are 12 inches in one foot.

- To express feet and inches as inches, multiply the number of feet in the given length by 12. To this product, add the number of inches in the given length.

    Example   Express 7 feet 9 $\frac{3}{4}$ inches as inches.

    Multiply 7 feet by 12. There are 84 inches in 7 feet.

    Add 9 $\frac{3}{4}$ inches to 84 inches.

$7 \times 12 = 84$
7 feet = 84 inches

84 inches + 9 $\frac{3}{4}$ inches =

93 $\frac{3}{4}$ inches

MATH REVIEW 20
EXPRESSING INCHES AS DECIMAL FRACTIONS OF A FOOT

- An inch is of $\frac{1}{12}$ a foot. To express whole inches as a decimal part of a foot, divide the number of inches by 12.

    Example   Express 7 inches as a decimal fraction of a foot.

    Divide 7 by 12.

$7 \div 12 = 0.58$
0.58 feet

- To express a common fraction of an inch as a decimal fraction of a foot, express the common fraction as a decimal, then divide the decimal by 12.

    Example 1   Express $\frac{3}{4}$ inch as a decimal fraction of a foot.

    Express $\frac{3}{4}$ as a decimal.

    Divide the decimal by 12.

$3 \div 4 = 0.75$
$0.75 \div 12 = 0.06$
0.06 feet

    Example 2   Express 4 $\frac{3}{4}$ inches as a decimal fraction of a foot.

    Express 4 $\frac{3}{4}$ as a decimal.

    Divide the decimal inches by 12.

$4 + \frac{3}{4} = 4 + 0.75 = 4.75$

$4.75 \div 12 = 0.39$

0.39 feet

MATH REVIEW 21
EXPRESSING DECIMAL FRACTIONS OF A FOOT AS INCHES

• To express a decimal part of a foot as decimal inches multiply by 12.

    Example Express 0.62 feet as inches.

    Multiply 0.62 by 12.

$0.62 \times 12 = 7.44$
7.44 inches

• To express a decimal fraction of an inch as a common fraction, see Math Review 17.

MATH REVIEW 22
AREA MEASURE

• A surface is measured by determining the number of surface units contained in it. A surface is two dimensional. It has length and width, but no thickness. Both length and width, but no thickness. Both length and width must be expressed in the same unit of measure. Area is expressed in square units. For example, 5 feet $\times$ 8 feet equals 40 square feet.

• *Equivalent Units of Area Measure:*

    1 square foot (sq ft) =

        12 inches $\times$ 12 inches = 144 square inches (sq in)

    1 square yard (sq yd) =

        3 feet $\times$ 3 feet = 9 square feet (sq ft)

• To express a given unit of area as a larger unit of area, divide the given area by the number of square units contained in one of the larger units.

    Example 1 Express 648 square inches as square feet.

    Since 144 sq in = 1 sq ft, divide 648 by 144.

$648 \div 144$ 4.5
648 square inches = 4.5 square feet

    Example 2 Express 28.8 square feet as square yards.

    Since 9 sq ft = 1 sq yd, divide 28.8 by 9.

$28.8 \div 9 = 3.2$
28.8 sqaure feet = 3.2 square yards

• To express a given unit of area as a smaller unit of area, multiply the given area by the number of square units contained in one of the larger units.

    Example 1 Express 7.5 square feet as square inches.

    Since 144 sq in = 1 sq ft, divide 7.5 by 144.

$7.5 \times 144 = 1080$
23 square yards = 207 square feet

    Example 2 Express 23 square yards as square feet.

    Since 9 sq ft = 1 sq yd, multiply 23 by 9.

$23 \times 9 = 207$
23 square yards = 207 square feet

- *Computing Areas of Common Geometric Figures:*

    1. **Rectangle**   A rectangle is a four-sided plane figure with 4 right (90°) angles.

        The area of a rectangle is equal to the product of its length and its width.

        Area = length × width (A = 1 × w)

        Example   Find the area of a rectangle 24 feet long and 13 feet wide.

        A = l × w

        A = 24 ft × 13 ft

        A = 312 square feet

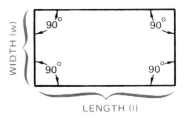

    2. **Triangle**   A triangle is a plane figure with 3 sides and 3 angles.

        The area of a trinagle is equal to one-half the product of its base and altitude.

        $A = \dfrac{1}{2}$ base × altitude (A = b × a)

        Example   Find the area of a trinagle with a base of 16 feet and an altitude of 12 feet.

        $A = \dfrac{1}{2}$ b × a

        $A = \dfrac{1}{2}$ × 16 ft × 12 ft

        A = 96 square feet

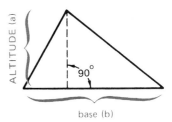

    3. **Circle**   The area of a circle is equal to π times the square of its radius.

        Area = π × radius² (A = π × r²)

        Note: π (pronounced "pi") is approximately equal to 3.14. Radius squared (r²) means r × r.

        Example   Find the area of a circle with a 15-inch radius.

        A = π × r²

        A = 3.14 × (15 in)²

        A = 3.14 × 225 sq in

        A = 706.5 square inches

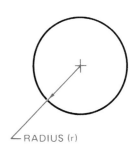

MATH REVIEW 23
VOLUME MEASURE

- A solid is measured by determining the number of cubic units contained in it. A solid is three dimensional; it has length, width, and thickness or height. Length, width, and thickness must be expressed in the same unit of measure. Volume is expressed in cubic units. For example, 3 feet $\times$ 5 feet $\times$ 10 feet = 150 cubic feet.

- *Equivalent Units of Volume Measure:*

    1 cubic foot (cu ft) =

    12 in $\times$ 12 in $\times$ 12 in = 1728 cubic inches (cu in)

    1 cubic yard (cu yd) =

    3 ft $\times$ 3 ft $\times$ 3 ft = 27 cubic feet (cu ft)

- To express a given unit of volume as a larger unit of volume, divide the given volume by the number of cubic units contained in one of the larger units.

    Example 1   Express 6048 cubic inches as cubic feet.

    Since 1728 cu in = 1 cu ft, divide 6048 by 1728.

    $6048 \div 1728 = 3.5$
    6048 cubic inches = 3.5 cubic feet

    Example 2   Express 167.4 cubic feet as cubic yards.

    Since 17 cu ft = 1 cu yd, divide 167.4 by 27.

    $167.4 \div 4\ 27 = 6.2$
    167.4 cubic feet = 6.2 yards

- To express a given unit of volume as a smaller unit of volume, multiply the given volume by the number of cubic units contained in one of the larger units.

    Example 1   Express 1.6 cubic feet as cubic inches.

    Since 1728 cu in = 1 cu ft, divide 1.6 by 1728.

    $1.6 \times 1728 = 2764.8$
    1.6 cubic feet = 2764.8 cubic inches

    Example 2   Express 8.1 cubic yards as cubic feet.

    Since 27 cu ft = 1 cu yd, divide 8.1 by 27.

    $8.1 \times 27 = 218.7$
    8.1 cubic yards = 218.7 cubic feet

- *Computing Volumes of Common Solids*

- A prism is a solid which has teo identical faces called bases and parallel lateral edges. In a right prism, the lateral edges are perpendicular (at 90°) to the bases. The altitude or height (h) of a prism is the perpendicular distance between its two bases. Prisms are named according to the shapes of their bases.

- The volume of any prism is equal to the product of the area of its base and altitude or height.

    Volume = area of base $\times$ altitude ($V = A_B \times h$)

- *Right Rectangular Prism:*

    A right rectangular prism has rectangular bases.

    Volume = area of base × altitude

    $V = A_B \times h$

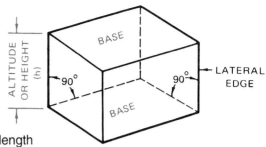

Example   Find the volume of a rectangular prism with a base length of 20 feet, a base width of 14 feet and a height (altitude) of 8 feet.

   $V = A_B \times h$

   Compute the area of the base (A_B):

   Area of base = length × width

   $A_B = 20 \text{ ft} \times 14 \text{ ft}$

   $A_B = 280 \text{ sq ft}$

   Compute the volume of the prism:

   $V = A_B \times h$

   $V = 280 \text{ sq ft} \times 8 \text{ ft}$

   $V = 2240 \text{ cu ft}$

- *Right Triangular Prism:*

    A right triangular prism has triangular bases.

    Volume = area of base × altitude

    $V = A_B \times h$

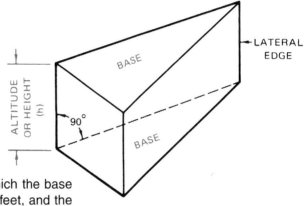

Example   Find the volume of a triangular prism in which the base of the triangle is 5 feet, the altitude of the triangle is 3 feet, and the altitude (height) of the prism is 4 feet. Refer to the accompanying figure.

   Volume = area of base × altitude

   $V = A_B \times h$

   Compute the area of the base:

   Area of base = $\frac{1}{2}$ base of triangle × altitude of triangle

   $A_B = \frac{1}{2} b \times a$

   $A_B = \frac{1}{2} \times 5 \text{ ft } 3 \times \text{ ft}$

   $A_B = 7.5 \text{ sq ft}$

   Compute the volume of the prism:

   $V = A_B \times h$

   $V = 7.5 \text{ sq ft} \times 4 \text{ ft}$

   $V = 30 \text{ cubic feet}$

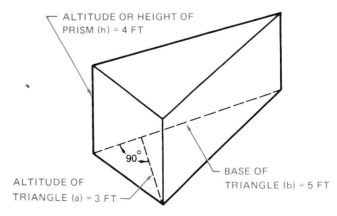

- *Right Circular Cylinder:*

    A right circular cylinder has circular bases.

    Volume = area of base × altitude

    $$V = A_B \times h$$

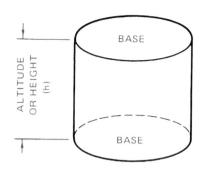

Example   Find the volume of a circular cylinder 1 foot in diameter and 10 feet high.

Note: Radius = $\frac{1}{2}$ Diameter; Radius = $\frac{1}{2}$ × 1 ft = 0.5 ft.

$$V = A_B \times h$$

Compute the area of the base:

Area of base = π × radius squared

$$A_B = 3.14 \times (0.5)^2$$

$$A_B = 3.14 \times 0.5 \text{ ft } 3 \text{ } 0.5 \text{ ft}$$

$$A_B = 3.14 \times 0.25 \text{ sq ft}$$

$$A_B = 0.785 \text{ sq ft}$$

Compute the volume of the cylinder:

$$V = A_B \times h$$

$$V = 0.785 \text{ sq ft } 3 \text{ } 10 \text{ ft}$$

$$V = 7.85 \text{ cubic feet}$$

MATH REVIEW 24
FINDING AN UNKNOWN SIDE OF A RIGHT TRIANGLE, GIVEN TWO SIDES

- If one of the angles of a triangle is a right (90°) angle, the figure is called a right triangle. The side opposite the right angle is called the hypotenuse. In the figure shown, c is opposite the right angle; c is the hypotenuse.

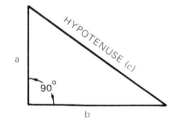

- In a right triangle, the square of the hypotenuse is equal to the sum of the squares of the other two sides:

    $$C^2 = a^2 + b^2$$

If any two sides of a right triangle are known, the length of the third side can be determined by one of the following formulas:

$$c = \sqrt{a^2 + b^2}$$

$$a = \sqrt{c^2 - b^2}$$

$$b = \sqrt{c^2 - a^2}$$

Example 1   In the right triangle shown, a = 6 ft, b = 8 ft, find c.

$$c = \sqrt{a^2 + b^2}$$

$$c = \sqrt{6^2 + 8^2}$$

$$c = \sqrt{36 + 64}$$

$$c = \sqrt{100}$$

$$c = 10 \ \text{feet}$$

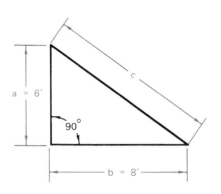

Example 2   In the right triangle shown, c = 30 ft, b = 20 ft, find a.

$$a = \sqrt{c^2 - b^2}$$

$$a = \sqrt{30^2 - 20^2}$$

$$a = \sqrt{900 - 400}$$

$$a = \sqrt{500}$$

$$a = 22.36 \ \text{feet (to 2 decimal places)}$$

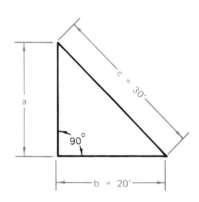

Example 3   In the right triangle shown, c = 18 ft, a = 6 ft, find b.

$$b = \sqrt{c^2 - a^2}$$

$$b = \sqrt{18^2 - 6^2}$$

$$b = \sqrt{324 - 36}$$

$$b = \sqrt{288}$$

$$b = 16.97 \ \text{feet (to 2 decimal places)}$$

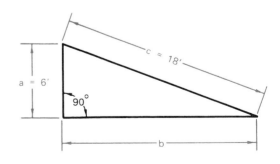

Material Symbols in Sections

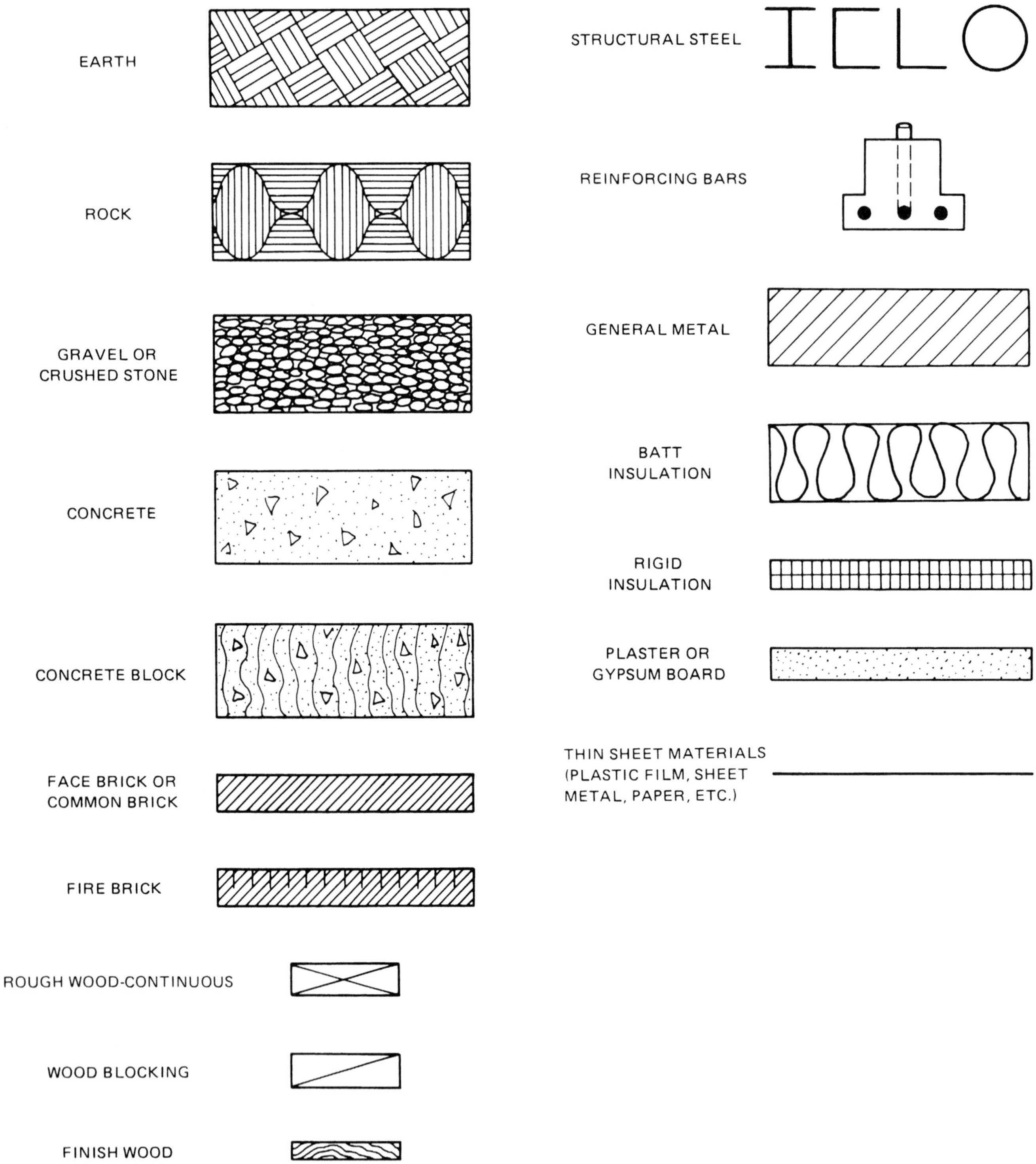

EARTH

ROCK

GRAVEL OR
CRUSHED STONE

CONCRETE

CONCRETE BLOCK

FACE BRICK OR
COMMON BRICK

FIRE BRICK

ROUGH WOOD-CONTINUOUS

WOOD BLOCKING

FINISH WOOD

STRUCTURAL STEEL

REINFORCING BARS

GENERAL METAL

BATT
INSULATION

RIGID
INSULATION

PLASTER OR
GYPSUM BOARD

THIN SHEET MATERIALS
(PLASTIC FILM, SHEET
METAL, PAPER, ETC.)

Plumbing Symbols

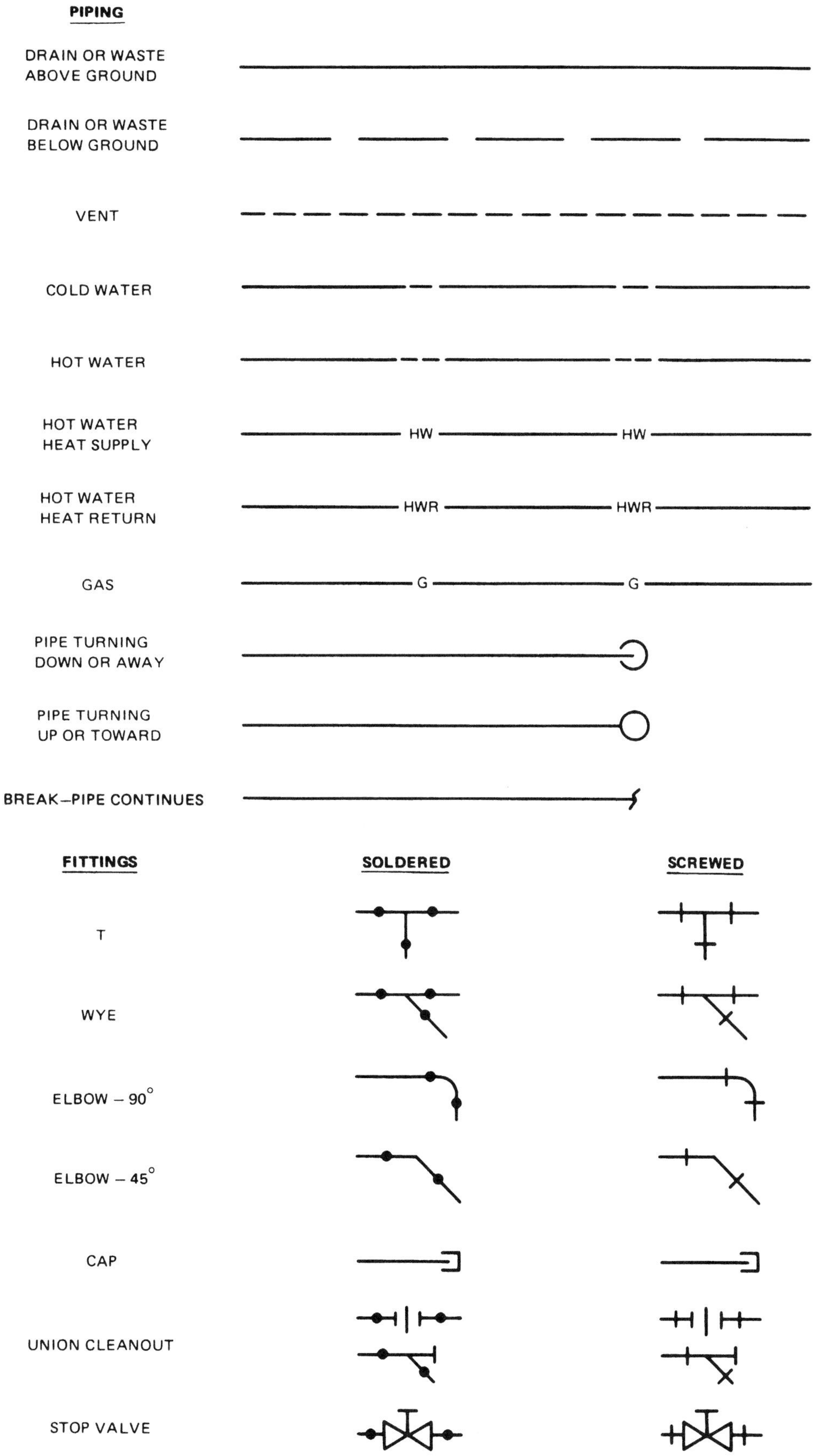

PIPING

DRAIN OR WASTE
ABOVE GROUND

_____

DRAIN OR WASTE
BELOW GROUND

— — — — — — — —

VENT

- - - - - - - - - - -

COLD WATER

—— — —— — —— —

HOT WATER

—— — — —— — — ——

HOT WATER
HEAT SUPPLY

——— HW ——— HW ———

HOT WATER
HEAT RETURN

——— HWR ——— HWR ———

GAS

——— G ——— G ———

PIPE TURNING
DOWN OR AWAY

PIPE TURNING
UP OR TOWARD

BREAK—PIPE CONTINUES

FITTINGS	**SOLDERED**	**SCREWED**
T		
WYE		
ELBOW – 90°		
ELBOW – 45°		
CAP		
UNION CLEANOUT		
STOP VALVE		

CEILING FIXTURE			TELEPHONE	
CEILING FIXTURE WITH PULL SWITCH	PS		INTERCOM	
WALL MOUNTED FIXTURE			TELEVISION ANTENNA	TV
RECESSED CEILING FIXTURE — OUTLINE SHOWS SHAPE			SMOKE DETECTOR	
FLOURESCENT FIXTURE			DISTRIBUTION PANEL	
FAN OUTLET	F or		JUNCTION BOX	J
CONVENIENCE DUPLEX OUTLET			SINGLE-POLE SWITCH	S
SPLIT WIRED DUPLEX OUTLET			THREE-WAY SWITCH	S_3
WEATHERPROOF OUTLET	WP		SWITCH WITH PILOT LIGHT	S_p
OUTLET WITH GROUND FAULT INTERRUPTER	GFI		WEATHERPROOF SWITCH	S_{wp}
SPECIAL-EQUIPMENT OUTLET			SWITCH WIRING	or
RANGE OUTLET	R			
PUSH BUTTON				
CHIME	CH			
TRANSFORMER	T			

Abbreviations

A.B.—anchor bolt
A.C.—air conditioning
AL. or ALUM.—aluminum
BA—bathroom
BLDG.—building
BLK.—block
BLKG.—blocking
BM.—beam
BOTT.—bottom
B.PL.—base plate
BR—bedroom
BRM.—broom closet
BSMT.—basement
CAB.—cabinet
℄—centerline
CLNG. or CLG.—ceiling
C.M.U.—concrete masonry unit
    (concrete block)
CNTR.—center or counter
COL.—column
COMP.—composition
CONC.—concrete
CONST.—construction
CONT.—continuous
CORRUG.—corrugated
CRNRS.—corners
CU—copper
d—penny (nail size)
DBL.—double
DET.—detail
DIA or ⊘ —diameter
DIM.—dimension
DN.—down
DO—ditto
DP.—deep or depth
DR.—door
D.W.—dishwasher
ELEC.—electric
ELEV.—elevation
EQ.—equal

EXP.—exposed or expansion
EXT.—exterior
F.G.—fuel gas
FIN.—finish
FL. or FLR.—floor
FOUND. or FDN.—foundation
F.P.—fireplace
FT.—foot or feet
FTG.—footing
GAR.—garage
G.F.I.—ground fault interrupter
G.I.—galvanized iron
GL.—glass
GRD.—grade
GYP.BD.—gypsum board
H.C.—hollow core door
H.C.W.—hollow core wood
HDR.—header
H.M.—hollow metal
HORIZ.—horizontal
HT. or HGT.—height
H.W.—hot water
H.W.M.—high water mark
IN.—inch or inches
INSUL.—insulation
INT.—interior
JSTS.—joists
JT.—joint
LAV.—lavatory
L.H.—left hand
LIN.—linen closet
LT.—light
MANUF.—manufacturer
MAS.—masonry
MATL.—material
MAX.—maximum
MIN.—minimum
MTL.—metal
NAT.—natural
N/F—now or formerly

N.I.C.—not in contract
o/—overhead or over
O.C.—on centers
O.H. DOOR—overhead door
PERF.—perforated
℞L—plate
PLYWD.—plywood
P.T.—pressure-treated lumber
R—risers
REF.—refrigerator
REINF.—reinforcement
REQ.—requirement
R.H.—right hand
RM—room
R.O.B.—run of bank (gravel)
R.O.W.—right of way
SCRND.—screened
SHT.—sheet
SHTG.—sheathing
SHWR.—shower
SIM.—similar
SL.—sliding
S&P—shelf and pole
SQ. or ▱ —square
STD.—standard
STL.—steel
STY.—story
T&G—tongue and groove
THK.—thick
T'HOLD.—threshold
TYP.—typical
V.B.—vapor barrier
w/—with
WARD.—wardrobe
W.C.—water closet
WD.—wood
WDW.—window
W.H.—water heater
W.I.—wrought iron

Index